W9-DEP-120

THE UPPER PALAEOLITHIC OF BRITAIN

THE UPPER PALAEOLITHIC OF BRITAIN

A Study of Man and Nature in the Late Ice Age

JOHN B. CAMPBELL

VOLUME I

CLARENDON PRESS · OXFORD
1977

Oxford University Press, Walton Street, Oxford OX2 6DP

OXFORD LONDON GLASGOW NEW YORK
TORONTO MELBOURNE WELLINGTON CAPE TOWN
IBADAN NAIROBI DAR ES SALAAM LUSAKA ADDIS ABABA
KUALA LUMPUR SINGAPORE JAKARTA HONG KONG TOKYO
DELHI BOMBAY CALCUTTA MADRAS KARACHI

British Library Cataloguing in Publication Data
Campbell, John B
The Upper Palaeolithic of Britain.
1. Palaeolithic period – Great Britain 2. Great Britain – Antiquities
I. Title
936.1'01 GN805

ISBN 0-19-813188-7

Printed in Great Britain by
Billing & Sons Ltd., Guildford and London

PREFACE

This book is a co-ordinated study of the archaeology and palaeoecology of the Upper Palaeolithic peoples of Britain. It is a completely new work based on a thorough re-analysis of all extant old evidence and on the results of my own excavations. The time range concerned is from almost 40,000 years ago to about 10,000 years ago, during which span there are substantial changes in the natural environment and consequently in the human responses to those variations. As is to be shown, I see what I call the Earlier Upper Palaeolithic (or E.U.P.) peoples as having ultimately disappeared from Britain owing to the maximum ice advances of about 18,000 to 16,000 years ago, whilst the Later Upper Palaeolithic (or L.U.P.) peoples went on to become the ancestors of the native Mesolithic population and in a sense the first 'British' population who managed to stay.

This book is in fact the result of 10 years of extensive research. As a potential doctoral project it was first conceived in late 1965. The research was then begun officially in October 1966 at Oxford. After comprehensive library, museum, field and laboratory studies, a D.Phil. thesis was written up in late 1971 and submitted in early 1972. Subsequently, my examiners Dr. Charles B.M. McBurney and Professor Christopher F.C. Hawkes, as well as my supervisor Dr. Derek A. Roe, recommended that I seek to have it published in full as a book. Thus a contract was eventually signed in February 1973 with the Delegates of Oxford University Press. The basic manuscript was in fact then expanded rather than reduced, as certain additional studies were by that point nearly finished. The final main manuscript was submitted in June 1975, being reasonably up to date as of then, whilst the outstanding appendices were submitted in 1976. Various comparatively brief preliminary accounts on certain aspects had already been published (see Campbell, 1969 and 1970; Campbell and Sampson, 1971; Tratman, Donovan and Campbell, 1971).

Compared with the massive amount of work which has been and is continuing to be done on the continental European Upper Palaeolithic, especially that of France, with but few exceptions the British Upper Palaeolithic has been virtually ignored. The main exceptions in this century, aside from the efforts of many amateurs of varying interest, include those of the late Professor Dorothy A.E. Garrod in the 1920s and Dr. Charles McBurney in 1958–60, as well as my former Edinburgh student Mr. Simon Collcutt's current research at Oxford. Before me, Professor Garrod was the only person who had actually carried out and published a fully comprehensive study on the subject (see Garrod, 1926). Her work was also published by Oxford.

Since the completion of the manuscript for this book, a number of related archaeological researches have been finished and in some cases published at least in part (e.g., Roger M. Jacobi, 1976, 'Britain inside and outside Mesolithic Europe', *Proceedings of the Prehistoric Society*, 42, pp. 67-84). My Belgian colleagues and friends Michel Dewez and Marcel Otte have completed their doctoral theses, entitled respectively 'Prodrome du Paléolithique final dans les grottes de la Belgique' (1975, Liège) and 'Le Paléolithique supérieur ancien en Belgique'. (1975–76, Liège). The Upper Palaeolithic of Belgium now appears in many ways almost identical to that of Britain (see also Appendices 3 and 4 in this present book). Many other relevant continental works are summarized and/or cited in the forthcoming *Actes* of the IXth Congress of the 'Union Internationale des Sciences Préhistoriques et Protohistoriques', which was held at Nice in September 1976. The *Actes* also include a brief paper by me in French on 'Le Paléolithique supérieur britannique et européen: socioéconomies et leurs paléoenvironnements'.

A number of related studies in physical anthropology have also been completed recently. Perhaps most relevant is that by Dr. David W. Frayer on 'Metric Dental Change in the European Upper Paleolithic and Mesolithic' (1977, *American Journal of Physical Anthropology*, 46, pp. 109-20). Dr. Frayer has done a detailed metrical and statistical analysis of British and European samples of Upper Palaeolithic and Mesolithic human teeth. For the Upper Palaeolithic he conveniently, though somewhat differently, employs my labels 'EUP' and 'LUP', although he does not actually cite my work. It may be of great significance that he has detected what appears to be a substantially greater difference between 'EUP' and 'LUP' samples than between 'LUP' and Mesolithic samples. Furthermore, he has recognized a Middle Palaeolithic (Neanderthal) to 'EUP' transition on the one hand, and an 'LUP' to Mesolithic transition on the other. Quite independently, on the basis of a combination of cultural, environmental and stratigraphic/radiocarbon evidence, rather than human biological evidence, I had already recognized broadly comparable differences and transitions, particularly in the case of a break between Earlier and Later Upper Palaeolithic and then a Later Upper Palaeolithic/Mesolithic transition, which I have normally referred stratigraphically to as Meso./L.U.P. (see especially pp. 36-80, 134-39, 156-57, 171-75 and 186-201 in this Volume). Frayer's work has been written up in full in his Ph.D. thesis, 'Evolutionary Dental Changes in Upper Paleolithic and Mesolithic Human Populations' (1976, Michigan). Taken together our separate lines of evidence might suggest a much more complex and drawn-out interplay between human biology and culture during the Last Glacial and Early Post Glacial than has previously been seen.

Still further papers of related interest, principally in palaeoecology, palaeogeography etc., are to be presented at the Xth Congress of the International Union for Quaternary Research (INQUA), which is to be held at Birmingham in August 1977.

Finally, where there are discrepancies between my thesis and my book, the book should be taken as essentially correct. For example, further osteological studies have slightly increased the numbers of fauna actually identified from my Robin Hood's Cave 1969 excavations. Also my identification of small mammals has been altered slightly (e.g., *Microtus gregalis* has in some cases now been attributed to *M. ratticeps*) as a result of careful checking by Dr. John C. Pernetta, a small mammal specialist. But generally speaking, the process of bringing the book up to date has been one of addition rather than correction. Nevertheless, if there are still any errors or misinterpretations, I certainly accept full responsibility for them.

Department of Behavioural Sciences
James Cook University
Queensland, Australia
April 1977

John B. Campbell

ACKNOWLEDGEMENTS

It is unfortunately not possible in the space available to single out every individual and every organization whom I should like to thank for the help which they provided at various stages in the course of this research and ultimately in the preparation of this book. All together literally hundreds of people became involved in one way or another. To begin with, sincere thanks are certainly extended to all museums and private collectors who provided their time and facilities, and to all excavation assistants, without whom the field-work could not have been accomplished. Grateful acknowledgement is also given for a Pre-doctoral Fellowship which was awarded by the Wenner-Gren Foundation for Anthropological Research (New York) in 1967, a Meyerstein Research Award which was granted by the Committee for Archaeology at Oxford University in 1969, and Excavation Grants which were provided by the Ministry of Public Building and Works (now the Department of the Environment, London) for work at Creswell Crags and Hengistbury Head in 1969 and by the Munro Foundation of the University of Edinburgh for related work at La grotte du Coléoptère in Belgium in 1974. At the start of this research in 1966–67 the President and Fellows of St. John's College (Oxford) very kindly awarded me Special Travel Grants so that I might gain a first-hand knowledge of French Palaeolithic excavation and laboratory techniques, which were then and which definitely still are amongst the most advanced in the world.

Amongst those whom I am able to single out by name, certainly foremost are my Oxford doctoral supervisor, Dr. Derek A. Roe, who gave many hours of careful guidance and friendship, and my American grandfather, Mr. Harold C. Seward, who financed most of my many years at Oxford and who gave guidance and friendship of another sort, to both of whom I now record my heart-felt thanks. Sincere thanks are also extended to the following colleagues and friends for many helpful suggestions and much useful information: the late Mr. Donald F.W. Baden-Powell, Professor François Bordes, Mr. Donald Bramwell, Mr. Simon Collcutt, Dr. G. Russell Coope, Dr. Michel C. Dewez, Professor Geoffrey W. Dimbleby, Dr. John G. Evans, Professor Burkhard Frenzel, the late Professor Dorothy A.E. Garrod, Dr. Joachim Hahn, Professor Christopher F.C. Hawkes, Madame Arlette Leroi-Gourhan, Professor Henry de Lumley-Woodyear, Dr. Charles B.M. McBurney, Dr. Paul A. Mellars, Miss Theya Molleson, Professor Hansjürgen Müller-Beck, Dr. Kenneth P. Oakley, Dr. Marcel Otte, Professor Stuart Piggott, Dr. Andrée Rosenfeld, Dr. C. Garth Sampson, Dr. Derek A. Sturdy, Professor E.K. Tratman, Dr. Richard G. West and Mr. John J. Wymer.

I should also like to acknowledge gratefully permission by the following to excavate at sites on their properties: the Council of the County Borough of Bournemouth (for Hengistbury Head), the South Wales Conservancy of the Forestry Commission (for Cathole), Messrs. W.S. Hodgkinson and Co. Ltd. (for Badger Hole), Mr. J.M. Jones (for Long Hole), the National Trust for Places of Historic Interest or Natural Beauty (for Sun Hole) and His Grace the Duke of Portland (for Mother Grundy's Parlour and Robin Hood's Cave).

Radiocarbon age estimates were determined by Mr. Richard Burleigh at the British Museum Research Laboratory and by Dr. W.G. Mook at the 'Natuurkundig Laboratorium der Rijks-Universiteit te Groningen', and thermoluminescent studies were conducted by Dr. Martin J. Aitken at the Oxford University Research Laboratory for Archaeology and the History of Art, to all of whom I am greatly indebted. Identification of charred wood fragments was done by Dr. J.D. Brazier, that of land snail shells by Dr. John G. Evans, that of bird bones by Mr. Donald Bramwell and that of human bones and teeth by Miss Rosemary Powers, whose reports are included in the Appendices and to all of whom I am equally indebted. Brief reports on the interrelationships between the Belgian and British Upper Palaeolithic were done by Dr. Marcel Otte and Dr. Michel Dewez (see Appendices 3 and 4), to both of whom I am most grateful.

All of my own original laboratory studies were conducted in the Department of Ethnology and Prehistory at the Pitt Rivers Museum, Oxford University, to whose academic and technical staff and recent Head, Mr. Bernard E.B. Fagg, I shall be forever indebted. The writing up of my thesis was carried out in that Department, whilst the subsequent writing up of the present book was done in many different places: the Department of Archaeology at the University of Edinburgh in 1973-74, the Department of Anthropology at the University of Pennsylvania in 1974-75, the Pitt Rivers at Oxford again in 1975 and finally the Department of Behavioural Sciences at James Cook University in 1975-77, to all of whom I extend sincere thanks.

Lastly, I should like to record my thanks to my Oxford typist, Mrs. Peggy Broadbent, who worked on my entire thesis and also did a substantial portion of this book, to the staff of the Oxford University Press, to my wife, Dott. Camilla Bettoni Campbell, and again to Dr. Roe and Dr. McBurney, all of whom really pushed to see this book completed and yet showed incredible patience with my meddlesome idiosyncrasies.

CONTENTS

VOLUME I

CONTENTS

VOLUME II

CHAPTER I

INTRODUCTION

A. GENERAL NATURE OF THE UPPER PALAEOLITHIC IN BRITAIN

In the traditional archaeological sense, the Upper Palaeolithic is the last and shortest major division of the European Old Stone Age. It is characterized usually by a predominance of stone tools made on blades (or long flakes) as against the usual core and flake tools of the earlier divisions of the Palaeolithic, and it often has much more developed bone and antler tools. It is a time when man is physically fully evolved into his modern form, *Homo sapiens sapiens*, and if one may interpret the relevant anthropological/archaeological and ecological evidence thus, it is a time when man attains his greatest degree of sophistication as a Palaeolithic hunter and gatherer.

In terms of the Upper Palaeolithic of Britain, there are perhaps four particularly relevant ecological and chronological factors about the Last Glacial:

1. A severe climatic/ecological cycle with deterioration from steppe-tundra conditions to high tundra and in places even to "Arctic desert", and then amelioration from the latter through tundra and steppe-tundra stages to steppe-forest;
2. The wide exposure of parts of the present southern North Sea and English Channel beds as part of the North European Plain, perhaps allowing at times a greater terrestrial "biomass" than would otherwise have been possible;
3. The British last maximum cold glacial episode from about 20,000 to about 15,000 years Before Present in relation to the known time ranges of the various continental European Upper Palaeolithic sequences from about 38,000 to about 10,000 years Before Present;
4. The apparent and probably sporadic availability of mainly England and Wales for suitable hunting, gathering and camping.

The research programme described in this book suggests that in possible response to these factors the British Upper Palaeolithic seems to fall into two main groups of industries, an Earlier and a Later one, which an increasing body of evidence divides to before and after the maximum Last Glacial (Devensian/Weichselian/Würmian) ice advance, respectively.

The complex French Upper Palaeolithic sequence of industries including Lower Perigordian, Aurignacian, Upper Perigordian, "Proto-Magdalenian", Solutrean and Magdalenian and the German Upper Palaeolithic sequence of industries including late "Blattspitzen-Gruppen", Aurignacian, "Eastern Gravettian", Hamburgian, "Rissen/Tjongerian" ("Federmesser-Gruppen") and Ahrensburgian are not repeated in Britain. Instead the Earlier and Later British Upper Palaeolithic industries, which have little resemblance to each other, apparently consist of what in continental terms would be a mixture of tool-types. The Earlier Upper Palaeolithic (or British "Aurignacian/Proto-Solutrean") stone tools include notably unifacial and bifacial leaf-shaped points, which sometimes approach "Blattspitzen" and Solutrean forms and sometimes retain a distinctively British appearance. These leaf-points are sometimes found associated with large, but still somewhat Perigordian Va-like, tanged flakes and blades, simple lateral and oblique burins, Aurignacian II-like "burins busqués", Mousterian-like side and side-and-end scrapers, more Upper Palaeolithic-like end-of-blade scrapers, Mousterian, Aurignacian and Solutrean-like denticulated, serrated and/or notched flakes and blades, Aurignacian-like laterally retouched flakes and blades, and crudely chipped pebbles. They are also occasionally associated with simple bone awls/points, as well as possibly some more elaborate bone and ivory work. The Later Upper Palaeolithic (or Creswellian/"Cheddarian") stone tools include, notably, backed forms: obliquely truncated blades, Tjongerian-like convex backed blades, typically British (and possibly Tjongerian) sub-triangular and trapeziform backed blades (respectively, so-called Creswell and Cheddar points), Hamburgian-like and Magdalenian V-like shouldered points, Rissen and Ahrensburgian-like tanged points, and Upper Perigordian and Tjongerian-like "penknife" points (convex backed blades with an opposed basal oblique truncation), the angular forms being the most common. These backed tools are sometimes found associated with simple to multiple lateral, oblique and transverse burins, end-of-flake and end-of-blade scrapers, Hamburgian-like ?"Zinken", various borers and awls, denticulated, serrated and/or notched flakes and blades, laterally retouched flakes and blades, and a wide variety of multiple-type tools. They are also occasionally associated with elaborate bone and antler work including Magdalenian V-like and Maglemosian-like uniserial "harpoons" (or barbed points), Magdalenian-like bevelled points, Magdalenian VI-like biserial "harpoons", Magdalenian-like "bâtons", decorated rib (or other bone) segments and eyed needles. Convergence may, of course, account for some of the similarities between British and continental tools, but some may well be due to direct or indirect relationships, the best parallels for Britain apparently being found in the Belgian Upper Palaeolithic sequence, lying as it does at the crossroads of the British, French and German traditions.

In general the British Earlier Upper Palaeolithic may date from as early as 40,000 years B.P. to 20,000 years B.P., and certainly to no later than 18,000 years B.P., whilst the British Later Upper Palaeolithic may date from about 15,000 years B.P. to no later than 9,000 years B.P. The Earlier group (or E.U.P.) is associated with a steppe-tundra to high tundra flora and a large mammal fauna that includes as herbivores: woolly mammoth, woolly rhinoceros, wild horse, red deer, giant Irish deer, reindeer and possibly bison; and as carnivores: hyena, lion, wolf, fox and bear. The Later group (or L.U.P.) is associated with a steppe-tundra/steppe-forest to low tundra flora and a large mammal fauna that includes as herbivores: woolly rhinoceros (infrequently), wild horse (very abundantly), red deer, giant Irish deer, reindeer (sometimes abundantly) and possibly bison; and as carnivores: hyena (very infrequently), northern lynx, wolf, common and arctic foxes and brown bear. Mammoth and lion are absent from the Later group. Typological, metrical and statistical differences in the artifacts of both groups are present. These may represent stylistic and/or functional changes both in time and space, perhaps reflecting various human re-adaptations to the constantly changing conditions of the Last Glacial.

B. BRIEF HISTORY OF WORK ON THE BRITISH UPPER PALAEOLITHIC

In 1823 Dean Buckland uncovered a human skeleton associated with bone and ivory artifacts at Paviland Cave on the Gower Peninsula in South Wales (Buckland, 1823). The skeleton, which had been stained with red iron ochre, became known as the "Red Lady of Paviland" and was interpreted by Buckland as Roman. As a result of radiocarbon measurements on the skeleton itself (Oakley, 1968; Barker, Burleigh and Meeks, 1969) and further excavations at Paviland (Sollas, 1913), it is now seen to be at least Upper Palaeolithic, if not in fact Earlier Upper Palaeolithic. Although Buckland was averse, on theological grounds, to the idea that man may at one time have lived in contemporaneity with now extinct mammals such as mammoth, the Revd. John MacEnery could not but believe the evidence he found when he excavated at Kent's Cavern intermittently between 1825 and 1829 (Alexander, 1964). MacEnery, however, refrained from publishing his observations as the orthodox views of his day were set firmly against him. A severely edited version of his manuscripts was published by Edward Vivian in 1859, eighteen years after his death, and a much fuller version was published by William Pengelly in 1869.

From 1859 to 1863, William Boyd Dawkins conducted the excavation of Wookey Hole Hyena Den in the Mendip Hills of Somerset (Dawkins, 1862, 1863a and b and 1874). He found and published Palaeolithic stone and bone tools in association with a Late Pleistocene fauna. From 1865 to 1880, William Pengelly directed excavations at Kent's Cavern for the British Association for the Advancement of Science (Pengelly, 1865–80, 1884). His work was remarkably systematic for its day; he laid out an elaborate grid and spit system by means of which he carefully recorded the three-dimensional position of every bone and artifact and marked appropriate find-numbers on the individual specimens. But he regrettably failed to publish his meticulous field observations in the detail worthy of them. He did, however, publish monthly and annual reports which clearly demonstrated what he set out initially to find: more irrefutable evidence for the contemporaneity of ancient man with extinct mammals; an hypothesis which he had already tested and proved to some of the scientific community by his 1858–59 excavations at Brixham Cave. And happily soon after their discovery

Sir John Evans figured some of the more notable Middle and Upper Palaeolithic tools from Kent's Cavern (Evans, 1872).

From 1874 to 1879, the Revd. J. Magens Mello, guided by William Boyd Dawkins, excavated at the four main cave-sites at Creswell Crags on the Derbyshire and Nottinghamshire border (Mello, 1875, 1876 and 1877; Dawkins, 1876 and 1877). The field-work was not as carefully done as that by Pengelly at Kent's Cavern, but the published reports on Creswell were fairly good for their time. Middle and Upper Palaeolithic artifacts were found and partly figured, and differences in selection of raw materials were noted.

In 1903 a nearly complete human skeleton was found by a workman in Gough's Cave in Cheddar Gorge, Somerset (Davies, 1904; Seligman and Parsons, 1914). This skeleton became known as "Cheddar Man" and was supposedly found in association with Upper Palaeolithic artifacts including a "bâton". Radiocarbon measurements have recently been done on the skeleton itself (Campbell, 1970; Barker, Burleigh and Meeks, 1971) and it is now thought to be either Later Upper Palaeolithic or Early Mesolithic. From 1919 to 1922, the University of Bristol Spelaeological Society excavated Later Upper Palaeolithic/Early Mesolithic human bones, artifacts and faunal remains from Aveline's Hole in Burrington Coombe, Somerset (Davies, 1921, 1922, 1923 and 1925). A noteworthy antler tool discovered during this work was a biserial "harpoon", but it, as well as practically all of the other material from Aveline's Hole, was destroyed by bombing during the Second World War.

In 1924 A.L. Armstrong excavated Later Upper Palaeolithic and Mesolithic evidence from the entrance platform of Mother Grundy's Parlour, Creswell Crags, Derbyshire (Armstrong, 1925). His field-work was apparently not as careful as Pengelly's had been half a century earlier, but his published account was fairly comprehensive.

Miss D.A.E. Garrod then published in 1926 a site by site account of mainly first-hand observations on all of the British Upper Palaeolithic material known to her. Her book, *The Upper Palaeolithic Age in Britain*, has remained the basic reference work on the subject ever since, even though it has become very much out of date.

During the nineteen-twenties and thirties, A.L. Armstrong conducted extensive excavations inside Pin Hole at Creswell Crags, finding and publishing a stratified sequence of Middle Palaeolithic, Earlier Upper Palaeolithic and Later Upper Palaeolithic material (Armstrong, 1925, 1928, 1931a and 1956). At about the same time, J. Reid Moir found and published a number of supposedly Earlier Upper Palaeolithic open-sites in East Anglia (Moir, 1923, 1927, 1931 and 1938). His judgements are evaluated later in this book (see Chapter IV. H): the only really convincing site seems to be Bramford Road in Ipswich, Suffolk. Miss N.F. Layard also found a site at Pit I near White Colne, Essex (Layard, 1927).

From 1927 to 1931, R.F. Parry conducted systematic excavations at Gough's

Cave in most of what was left of the deposits (Parry, 1928, 1929 and 1931). He recovered over 7,000 artifacts, by far the largest extant Upper Palaeolithic collection yet obtained in Britain. He also excavated the neighbouring site of Soldier's Hole where he found a Later Upper Palaeolithic backed blade assemblage clearly stratified above an Earlier Upper Palaeolithic leaf-point assemblage (Parry, 1931).

From 1925 to 1927 and in 1955, excavations were carried out by the University of Bristol Spelaeological Society at King Arthur's Cave on the Herefordshire side of the Wye Valley (Hewer, 1926; Taylor, 1928; ApSimon, 1955). Later Upper Palaeolithic evidence was found stratified above Earlier Upper Palaeolithic. From 1926 to 1928 and from 1951 to 1953, the same society excavated at Sun Hole in Cheddar Gorge (Tratman and Henderson, 1928; Tratman, 1955). Later Upper Palaeolithic artifacts were found here, but all the material from both Sun Hole and King Arthur's Cave that had been obtained before the Second World War was almost completely destroyed by bombing (E.K. Tratman, personal communication).

Since about 1950, D. Bramwell and other members of the Peakland Archaeological Society have been excavating various Later Upper Palaeolithic cave-sites in the Peak District of Derbyshire and North Staffordshire (Jackson, 1962; Bramwell, 1973 and personal communications). A number of other excavations at caves and rock-shelters here and elsewhere in England and Wales have taken place since 1950, but have remained either unpublished or incompletely published. Further comments on them are made in Chapters IV and V.

In 1957 Mrs. Angela Mace excavated an apparently Later Upper Palaeolithic assemblage of flint artifacts from an open-site at Hengistbury Head near Bournemouth in Hampshire (Mace, 1959). She also began a study of British Upper Palaeolithic material in general which, due to illness, she was unable to complete (A. Mace and C.B.M. McBurney, personal communications). From 1958 to 1960, Dr. C.B.M. McBurney excavated at a number of cave-sites in England and Wales, including notably Cathole on the Gower Peninsula, Glamorganshire, Little Hoyle near Tenby in Pembrokeshire and Mother Grundy's Parlour at Creswell Crags, Derbyshire (McBurney, 1959 and personal communications).

The present study of the British Upper Palaeolithic was begun in 1966 on collections at the University of Oxford, and then extended in 1967 to 1970 with the aid of a Pre-doctoral Fellowship from the Wenner-Gren Foundation for Anthropological Research, to include a survey of museum collections throughout England and Wales, laboratory studies at Oxford and excavations (in 1968 and 1969) at cave-sites in England and Wales and an open-site in Hampshire. In 1971 this entire study was written up in thesis form and subsequently submitted at Oxford in 1972 for the Degree of Doctor of Philosophy. From 1970 to 1975 a scatter of additional Later Upper Palaeolithic finds came to light, including notably those from High Furlong, Poulton-le-Fylde, Lancashire (Barnes,

Edwards, Hallam and Stuart, 1971; Hallam, Edwards, Barnes and Stuart, 1973) and Sproughton near Ipswich, Suffolk (J.J. Wymer and P.R.J. Saunders, personal communications).

C. PREVIOUSLY TO PRESENTLY KNOWN SITE DISTRIBUTION

Sir Charles Lyell considered the "Post-pliocene" evidence for man and the environment (Lyell, 1863) only four years after Charles Darwin published his *The Origin of Species, by Means of Natural Selection*. But the number of British sites available to him was quite limited. By 1865 Sir John Lubbock was able to include Kent's Cavern and Wookey Hole Hyena Den in his popular work on *Pre-historic Times* (Lubbock, 1865). Lubbock coined the terms Palaeolithic and Neolithic for the evidence for successive stages of the Stone Age as he saw it, and Sir John Evans sensibly employed them in his much more exhaustive account of *The Ancient Stone Implements, Weapons and Ornaments of Great Britain* seven years later (Evans, 1872). Evans was able to review not only Kent's Cavern and the Hyena Den, but Long Hole and 'Oyle Cave or Hoyle's Mouth of South Wales, and King Arthur's Cave of Herefordshire. For comparison he also had available the south-western French sequence which had recently been recognized by Lartet and Christy in their *Reliquiae Aquitanicae*: 1. "Age of Le Moustier", 2. "Age of Laugerie Haute", 3. "Age of Cro-Magnon" and 4. "Age of La Madelaine". Evans noted both the similarities and differences between the series at Kent's Cavern and those at the French sites. His work was revised in 1897. In 1874 Sir William Boyd Dawkins published his *Cave Hunting* which gave fairly detailed accounts of the above-mentioned British sites.

However, the first attempt at actually mapping the distribution of Upper Palaeolithic sites in Europe, including Britain, was not done until 1911, when Prof. W.J. Sollas mapped by "culture" the Aurignacian, Solutrean and Magdalenian sites of Europe in his *Ancient Hunters and their Modern Representatives* (Sollas, 1911). In 1915 he published a revised edition of this book in which he reported upon his efforts at Paviland. The second, and hitherto final distribution map of all of the known British Upper Palaeolithic sites, was that presented by Miss Dorothy A.E. Garrod in 1926 in her book, *The Upper Palaeolithic Age in Britain*. The sites known to her have been plotted in the present account on Map 1; Mesolithic, or what she termed "Epipalaeolithic", sites have also been shown as these were considered by her either on her map or in her text. It is worth noting that all of the definitely Upper Palaeolithic sites were confined to England and Wales. Even with the addition of many "new" sites, the distribution pattern in general has not changed very much since her time. For comparison, Map 2 has been included to show all Upper Palaeolithic sites known to the present author in 1975. Mesolithic sites have not been included here as such, but it should be noted that they now number in the thousands (R. Jacobi, P.A. Mellars and J.J. Wymer, personal communications). Earlier U.P. sites have been distinguished from Later U.P. ones in Map 2.

Lesser accounts of the local distribution of sites have appeared since 1926 in the "County Archaeologies" and more recently in the "Regional Archaeologies". H. Schwabedissen relied mainly upon Garrod in his review of the British "Creswellian" sites as part of the general North-west European pattern within his "Federmesser-Gruppen" (Schwabedissen, 1954), whilst A. Bohmers detected what he thought were regional differences within the "Creswellian" and suggested the term "Cheddarian" for his southern variant (Bohmers, 1956 and 1963). Angela Mace considered the general pattern of Later Upper Palaeolithic sites in relation to her work at Hengistbury Head (Mace, 1959), and Dr. C.B.M. McBurney divided the British Upper Palaeolithic into two main groups in time relating them to some of the continental evidence and pointing out their rather thin geographical distribution in Britain (McBurney, 1959, 1961 and 1965). T.G. Manby listed all of the sites in northern England that he thought were "Creswellian" in relation to his analysis of the supposedly "Creswellian" flint assemblage from Brigham, East Yorkshire (Manby, 1966), and Philip E.L. Smith briefly reviewed the British "Proto-Solutrean" in relation to the French Solutrean (Smith, 1966). The British evidence has also been very briefly considered by H. Müller-Karpe (1966), François Bordes (1968), John M. Coles and Eric S. Higgs (1969), Derek A. Roe (1970) and Desmond Collins (1973) in their various general accounts of the Old Stone Age. Finally, a still general but more up-to-date account has just recently been published by Paul A. Mellars (1974), in which he includes a distribution map showing at least the main sites.

D. EXTANT COLLECTIONS

A large proportion of the artifact assemblages and faunal materials from the initial excavations at the most productive British cave-sites of Paviland, Kent's Cavern, Gough's Cave and the Creswell caves has been apparently lost, as well as in practically every case a portion of that from the less productive ones. Buckland, normally a careful observer, reported only one worked flint from Paviland (Buckland, 1823), whereas Sollas reported over 3,600 stone artifacts from his excavations ninety years later (Sollas, 1913). Buckland's undetected artifacts probably passed out of the cave with his tip and may now lie scattered on the bed of the Bristol Channel awaiting possible misinterpretation by some uninformed future observer. Sollas' sample was divided mainly between the National Museum of Wales and the Oxford University Museum, and it is still for the most part available. A few of the flint artifacts found by MacEnery at Kent's Cavern found their way into the Torquay, British and Oxford museums, but for any detailed analysis of the site one must refer to Pengelly's finds and Diary (Pengelly, 1865–80), both of which are still available at the Torquay Natural

History Society Museum, with the exception of some finds which went to the British Museum (Natural History), as well as a number of other museums, and some finds which appear to have been lost. Nonetheless, there are even discrepancies in Pengelly's accounts (Rogers, 1955; but see Campbell and Sampson, 1971, as well). Parry excavated over 7,000 artifacts from Gough's Cave, of which 4,505 are still available, mainly at the Cheddar Caves Museum (Parry, 1931). There is reason to suppose that Gough found at least as many artifacts which have since been lost (Davies, 1904; Donovan, 1955; E.K. Tratman, personal communications). The Creswell caves have suffered from ransacking by people such as Dr. Laing of Newcastle from time to time (Armstrong, 1925), and a good deal of the material found by Mello and Dawkins has been so widely dispersed, including beyond Britain, that it has become impossible to trace all of it.

However, at least some of the artifacts from nearly all of the recorded sites (both open and cave) are still available either in museum or private collections in England and Wales. In fact, complete assemblages are available from some of the published and unpublished sites that have been excavated within the past thirty years, for example, Badger Hole (Somerset), Hengistbury Head (Hampshire), Dead Man's Cave or "Anston Stones Cave" (Yorkshire) and Sproughton (Suffolk).

A list of extant collections by site with their present whereabouts is given in the Gazetteers at the end of this book. All or at least most of the known extant artifacts have been examined as part of this research, and in many cases, despite the known gaps, they have been included in metrical and statistical analyses, as they are all that one has to go on. Extant faunal collections from earlier excavations have not been examined in quite such detail, as there are undoubtedly even greater gaps connected with them, and a precise analysis of them would usually be extremely difficult or impossible to correlate with the artifacts as they so often lack sufficient stratigraphic records. Small numbers of unstratified artifacts, on the other hand, may be compared rather more usefully, at least in the case of the British Upper Palaeolithic, with known stratified series. Unstratified cold-loving mammals could perfectly well belong to the Early Last Glacial (Early Devensian), if not indeed in a few cases to the Penultimate Glacial (Wolstonian), both periods which lie beyond the range of this enquiry. As regards floral collections, apart from the analysis of fossil pollen samples collected by the author during his excavations in 1968–69, floral records have not been studied in full detail, owing to limitations in funds and time. But future work ought to include a check of the Cambridge data bank (see Deacon, 1972).

CHAPTER II

METHODS AND CONCEPTS OF STUDY

A. CHRONOLOGY

The basic framework for the chronology of the British Upper Palaeolithic must be obtained by combining the evidence of what few clearly stratified sites there are. It may then be extended by comparisons of known associated environmental/ecological evidence with that from other Upper Palaeolithic sites as well as from Last Glacial (Devensian) environmental/ecological sites in Britain, and by comparisons of artifact typology and morphology.

One additional line of evidence may then be obtained by radiocarbon age estimates of organic material from both Upper Palaeolithic sites and environmental Last Glacial sites. However, even radiocarbon age estimates are often inconsistent and unreliable due to fluctuations in rate of 14-carbon formation in the upper atmosphere, sampling errors in the field, differences in the amount of residual 14-carbon in different kinds of samples and contamination by 14-carbon that is either older or younger than the true age of the sample concerned. The range of error in one standard deviation also increases either with an increase in true age or a decrease in actual sample size or both.

Another method of "absolute dating" for the Upper Palaeolithic might eventually be offered by an extension of thermoluminescent measurements to teeth, bones, stalagmite and burnt stone artifacts. If reasonable glow curves are obtained from thermoluminescent samples, then they might well be used as a completely independent check on radiocarbon samples from the same sites and layers, as well as a means of arriving at age estimates for artifact assemblages for which radiocarbon has been unsuccessful. But thermoluminescence also often has fairly wide margins of error, and these again increase with age. A quite separate line of chronological evidence may eventually be offered by palaeomagnetism (e.g. see Mörner, Lanser and Hospers, 1971 on a Late Last Glacial "reversal").

Yet another chronological framework may be obtained by comparing British Upper Palaeolithic artifact assemblages with roughly similar continental ones that have already been "dated" by radiocarbon or environmental context, or both. This approach assumes an approximate contemporaneity which may, of course, not really exist: it should never be used on its own, or uncritically, but might certainly be of interest from time to time in a supporting role.

B. ECOLOGY

The ecology of the British Upper Palaeolithic may be studied by combining geomorphological, floral, faunal and related lines of evidence.

1. Geomorphology

The geomorphology of Britain at various times during the Last Glacial (Devensian) may be reconstructed on a regional and general basis by (a) considering glacial and periglacial features that have been observed by various Quaternary specialists and sometimes radiocarbon dated by various laboratories (e.g. Embleton and King, 1968; West, 1968; Mitchell, Penny, Shotton and West, 1973), and by (b) granulometric analyses of stratified deposit samples from the entrance platforms of selected British Upper Palaeolithic cave-sites. Continental workers have been granulometrically analysing cave deposits for sometime (e.g., Vértes, 1959; Miskovsky, 1969), whilst in Britain work has been carried out, for example, by Cornwall (in Lacaille and Grimes, 1955), the present author (in Tratman, Donovan and Campbell, 1971) and Shackley (1972), amongst others.

The granulometry of British Upper Palaeolithic cave deposits may be done simply by carefully dry-sieving the samples through a sealed column of graded sieves, as the main components seem generally to be scree and sand rather than silt and clay. The method is basically quantitative as percentages are calculated for each sample on the basis of the total of the weights read for the amount of material caught in each sieve, and changes illustrated by plotting on linear diagrams for each site. The meshes of the sieves employed are as follows: 1.4 (15.4 mm.), 6 (3.1 mm.), 10 (1.73 mm.), 20 (0.775 mm.), 30 (0.500 mm.), 40 (0.376 mm.), 60 (0.251 mm.), 80 (0.187 mm.), 100 (0.150 mm.), 120 (0.124 mm.) and 200 (0.074 mm.). The British standard grades of gravel, sand, silt and clay are as follows: stones or cobbles >60 mm., coarse gravel 60 to 20 mm., medium gravel 20 to 6 mm., fine gravel 6 to 2 mm., coarse sand 2.0 to 0.6 mm., medium sand 0.6 to 0.2 mm., fine sand 0.2 to 0.06 mm., silt 0.06 to 0.002 mm. and clay <0.002 mm. (Cornwall, 1958). As these grades are meant mainly for alluvial and fine wind-blown deposits, the grades used in this study for scree, sand, silt and clay are as follows: coarse scree 100 to 15.4 mm., medium scree 15.4 to 3.1 mm., fine scree 3.1 to 1.73 mm., coarse sand 1.73 to 0.500 mm., medium sand 0.500 to 0.187 mm., fine sand 0.187 to 0.124 mm., silt 0.124 to 0.074 mm. and clay <0.074 mm. Screes are further sub-divided into thermoclastic scree (sharp freeze/thaw-formed) and weathered scree (rounded, chemically eroded), both of which may easily be recognized macroscopically. In sieving, all organic materials (shells, bones and roots), artifacts and possibly derived materials (e.g. stalagmitic fragments) are removed before weights are read. Rises in thermoclastic scree are interpreted as indicative of cold conditions, whilst rises in weathered scree are interpreted as indicative of damp, milder conditions. Substantial rises in medium and fine sand, silt and clay are thought indicative of dry, very cold conditions, as they are sometimes associated with a complete or nearly complete absence of scree at hillside sites and might be considered British equivalents of loess. Rises in coarse sand associated with weathered scree are interpreted as indicative of somewhat damp, milder conditions. Qualitative field and laboratory observations on colour and structure of deposits are also discussed (see Chapter III.B.1.). In future work much more rigorous analyses might be conducted (e.g. see Shackley, 1972).

2. Floras

The British Last Glacial (Devensian) floras may be reconstructed by reviewing the radiocarbon dated plant fragments and pollen that have been identified by various specialists (e.g. Godwin, 1956; West, 1968; Pennington, 1969; Walker and West, 1970; Mitchell, Penny, Shotton and West, 1973), and by actual pollen analyses, carried out by the present writer, of stratified deposit samples from entrance platforms of selected Upper Palaeolithic cave-sites. Although pollen has been obtained from calcareous continental cave deposits for a quarter of a century (Welten, 1944; Derville and Firtion, 1951; Schütrumpf, 1951; Leroi-Gourhan, 1956, 1959, 1964, 1965, 1966 and 1968; Donner and Kurtén, 1958; Beaulieu, 1969; Paquereau, 1969; Renault-Miskovsky, 1972; Damblon, 1974; to cite merely a selection of references), prior to this study it had not been extracted from the mostly inorganic, calcareous British cave deposits. A preliminary publication of part of this study appeared in Tratman, Donovan and Campbell (1971). At about the same time pollen was successfully extracted from Bronze Age/Iron Age deposits at Fox Hole (Derbyshire) by D.W. Shimwell (in Bramwell, 1971), but he in fact studied only the more humic or organic levels.

The method of pollen extraction from inorganic calcareous samples employed by the author is as follows:

1. Collect about 500 gm. of freshly exposed deposit in a polythene bag and seal with a metal label on which the site reference, date, metre square, layer, exact position and number of the sample are recorded.
2. Transfer sample to a 1,000 ml. beaker and dry overnight in an oven at 80° to 90 °C.
3. Crush sample lightly with a pestle through a 1 mm. (mesh no.16) sieve into a large mortar.
4. Weigh out 50 gm. of $<$1 mm. component and transfer remainder of entire sample back to polythene bag for storage.
5. Transfer fine component to 50 ml. "nalgene" centrifuge tube and place tube at an angle in a 600 ml. low form beaker.
6. Add 50% hydrochloric (HC1) acid adjusting pouring and amount to the rate of reaction between the HC1 and the calcium carbonate ($CaCO_2$). This reaction gives off hydrogen gas which causes a certain amount of effervescence which in turn causes the "microflotation" of any pollen present and some of the silt and clay up out of the tube and into the beaker.
7. Once all of the sample in the tube has reacted with the HC1, remove and clean the tube.
8. Transfer the microfloted portion from the beaker to a clean 50 ml. "nalgene" centrifuge tube, centrifuge at 4,000 r.p.m. for 10 minutes and then decant the liquid.
9. Add about 20 ml. of distilled water to the tube, mix with the residue, centrifuge as above and decant.

10. Add about 20 ml. of 10% sodium hydroxide (NaOH), mix and bring to boil for about 10 minutes in a water bath. Centrifuge and decant.
11. Transfer residue to a 15 ml. copper crucible with about 5 to 10 ml. of 40% hydrofluoric (HF) acid and bring to boil for about 10 minutes, stirring frequently with a copper rod. Allow to cool, then centrifuge and decant. 15 ml. "nalgene" centrifuge tubes should be used.
12. If much silt and clay seems to remain, repeat step 11.
13. Add about 20 ml. of 10% HC1, warm in water bath for about 5 minutes, centrifuge and decant.
14. Dehydrate residue with about 10 ml. of glacial acetic (CH_3COOH) acid, centrifuge and decant.
15. Add about 10 ml. of freshly made acetolysis mixture (1 part concentrated sulphuric (H_2SO_4) added drop by drop from a burette to 9 parts acetic anhydride ($C_4H_6O_3$) in a constantly agitated 150 ml. conical flask; for four samples a convenient ratio is 5 ml. sulphuric to 45 ml. acetic anhydride). Bring to boil in water bath for about 15 minutes, centrifuge and decant.
16. Add about 5 ml. of glacial acetic acid, centrifuge and decant.
17. Add about 10 ml. of 10% NaOH to residue, bring to boil in water bath for 20 minutes, centrifuge and decant.
18. Wash, centrifuge and decant twice with water.
19. Add 3 ml. of 1:1 glycerol and water with 1:50 safranin stain, mix, transfer to 5 ml. glass specimen tube, close with plastic cap and label.
20. Place an 0.05 ml. drop with a calibrated syringe on a microscope slide, cover with a coverglass and scan at 100 and 400 magnifications with a binocular microscope.

This method allows about four to eight samples to be prepared in more or less one day. It is based, aside from my idea of microflotation by the initial application of HC1 (steps 5 to 8 above), partly on the methods employed by Prof. G.W. Dimbleby (under whom the author studied in 1966), P. Dumait, L. Marceau, C. Devin and M. van Campo (1963), B. Frenzel (1964 and personal communications) and K. Faegri and J. Iversen (1964). For British cave-sites, it has thus far produced counts of pollen and spores per sample of about 50 to 200, and rarely to about 450. Such counts are usually insufficient for fully reliable detailed analyses of pollen spectra but at least permit percentages of total trees and shrubs to be plotted against percentages of total herbs and ferns, which provide rough indications of how open or closed the vegetation may have been at various times during the Last Glacial (Devensian). Presence or absence of certain genera might also be considered important. For the moment it is hard to imagine how higher counts could be obtained from these cave-platform samples, but future work should improve the method. One should also certainly keep in mind how many sources of error there can be even in pollen analyses of peats and lake sediments (e.g. Faegri and Iversen, 1964; Davis, 1968; Tyldesley, 1973) which generally produce counts of 500 to 1,000 plus.

Although one may fairly quickly learn to recognize some of the main tree types of pollen, identification of pollen from many shrub and herb genera requires patiently gained experience and constant comparison with reference microscope slides of modern pollen. Such slides were kindly provided, together with reference photographs, by Professor G.W. Dimbleby of the Department of Human Environment at the London University Institute of Archaeology. To these the author has added certain types which were absent, and which he has personally collected, such as *Dryas octopetala* from the garden of the late Mr. D.F.W. Baden-Powell in Oxford.

3. Faunas

The British Last Glacial (Devensian) faunas may be reconstructed by carefully reviewing the lists of material identified by various specialists (e.g. see Jackson, 1962; Kurtén, 1968), and by analyses of the samples of bones, antlers and teeth actually collected by the author during his own field-work at selected Upper Palaeolithic cave-sites. For the large mammal analyses, all teeth, antlers and identifiable bones are included, whilst for the small mammals only teeth and jaws are normally included. Identifications of mammal bones and teeth, especially those of bones, by palaeontologists, and more especially by amateurs, before 1950, if indeed not 1960, should be accepted only with great caution as random checks by the author and others have recently revealed a number of blatant errors. As with the identification of pollen and spores, comprehensive reference collections of bones, antlers and teeth are required before one may suggest with any confidence the possible genus and species of a particular specimen. It is not sufficient to attempt final identification by comparison with figures or written descriptions, at least in most cases. The mammal material excavated by the author has been carefully examined and measured next to reference specimens in the University Museum and Department of Zoology at Oxford.

Minimum numbers of individuals represented per species per layer per site excavated by the author have been calculated approximately, but the reader is advised to accept these estimates with due caution. There are both advantages and limits to this method (e.g., see Ryder, 1968; Grayson, 1973; Casteel, 1974), but for the moment it remains one of the most powerful alternatives to listing simply the number of bones found, or worse still, stating simply qualifications such as "rare", "common" or "abundant". However, as there are nonetheless some advantages in reporting numbers of bones, teeth and antlers, these have been recorded in this book as well. Perkins for example (in Matolcsi, 1973, pp. 367–69) has even suggested that each bone identified should be assumed to represent a different individual, mainly because bone survival is often so poor at archaeological sites. Thus, following Perkins to some extent, maximum number of bones represented per species per layer might be taken as indicative of "maximum" number of individuals. Hopefully, future osteological/zoological research might arrive at a method for estimating "optimum" numbers of individuals, i.e. somewhere between "minimum" and "maximum", but whether that would be a valid statistic would seem to me uncertain.

"Dead weights" and "meat weights" have also been calculated for certain species at selected sites, at least where the evidence available from the author's own excavations allows such estimates to be made. "Dead weight" refers to an estimate of total body weight at the time of death for a given animal, usually in this present study a land mammal that was either definitely or possibly hunted by Upper Palaeolithic man. Approximate total body weights of the relevant modern mammals have been used, the information having been extracted principally from Hall and Kelson (1959), Martin and Guilday (1967), Mohr (1971) and Clark (1972), allowances having been made for any apparent major differences in size between Last Glacial (Devensian) mammals in Britain and their modern representatives in either Britain or abroad.

"Meat weights", on the other hand, are obviously less than "dead weights". I have chosen the somewhat arbitrary figure of $\frac{1}{2}$ or 50% of estimated "dead weight" as estimated "meat weight". This, if anything, would probably be an underestimate of the amount of meat actually obtained and consumed by British Upper Palaeolithic hunter/gatherers. I say this whilst thinking of Grahame Clark's re-analysis of British Mesolithic activities and cultural ecology at Star Carr in Yorkshire (Clark, 1972), as well as Joe Ben Wheat's splendid ethnographic reconstruction of Palaeo-indian activities at the Olsen-Chubbuck bison-kill site in Colorado, where daily meat consumption per person for a few days may have been as high as c. 3 kilogrammes or more (Wheat, 1972). The exact amounts consumed at Star Carr are less certain, but according to Clark's reconstruction they are much lower than in the North American example. Whether British Upper Palaeolithic and Mesolithic peoples actually consumed, at least at times, as much meat per person per day as prehistoric and historic Plains Indians appear to have often done unfortunately remains quite uncertain, largely owing to a lack of similar, direct ethnographic data. Until more is known, conservative estimates both of "meat weights" and meat actually consumed would seem to be the best approach to the British evidence. But then an average estimate of c. 1 kilogramme per person per day for people who presumably subsisted largely on meat may not be all that unreasonable for the time and place under study, and so it is the arbitrary but feasible 1 kilogramme figure which I have employed. Even this amount might well seem large to the "average" European, North American or Australian today who might, for example, eat about $\frac{1}{20}$ to $\frac{1}{5}$ of a kilogramme of meat (i.e., c. 50 to 200 gm.) in a mixed diet, and yet still be considered a heavy meat eater by much of the rest of the present-day world!

In addition to land mammals birds and snails have been identified from two of the sites excavated by the present author. The identification and interpretation of bird bones has been carried out by Mr. Donald Bramwell of Bakewell, Derbyshire, whilst that of the land snail shells has been done by Dr. John G. Evans of the Department of Archaeology at University College, Cardiff. Their reports have been included amongst the Appendices to this book. As they are each amongst the foremost authorities in Britain on their respective subjects, I am indeed most grateful to them for having undertaken this work. In addition to their own comments, their results are further discussed in the appropriate section of Chapter III (namely, III.B.3).

In the identification and interpretation of large land mammals, I was aided at times by the late Mr. Donald F.W. Baden-Powell, formerly of the Departments of Geology and of Ethnology and Prehistory at the University of Oxford, as well as by Mr. Philip Powell of the Oxford University Museum. In that of the smaller land mammals (i.e., so-called "microfauna"), I was assisted by Dr. John C. Pernetta at the Department of Zoology at Oxford. Although I certainly wish to record my thanks for their help, I must also stress that any errors or inconsistencies in this work are largely, if not entirely, my own. Of course, I have endeavoured to keep errors to a minimum, but much of the evidence is very fragmentary and thus difficult to identify. In assigning scientific names I have followed Kurtén (1968) in my choice of generic and specific labels for the mammals.

4. Ecosystems and Climatic Cycles

Ecology may be defined as the "biology of ecosystems" (Margalef, 1968, p. 4), that is as the study of all or most aspects of the interrelationships and/or interactions of life-forms, whether man, other animal or plant, and whether past, present or future, with their natural and physical environments, and in the case of man with his cultural environment as well, as part of an "ecosystemic" whole in a given region or in the world at large. Whatever we label them, "ecosystems" clearly exist wherever there is life (e.g. see Clapham, 1973). Such systems may often tend towards a state of balance or equilibrium for the exchange of energy within and between their sub-systems and super-systems, even though the "ideal" state of completely balanced exchange (i.e., a "steady state" or "negative feedback relationship") may never be attained. In contrast to a "steady state", the concept of a "constant flux state" (with "positive feedback relationships" and thus self-sustaining change) with incomplete but continual re-balancing of the ecosystem(s) may better approximate reality, or rather reality is presumably more often betwixt the two. There is much jargon in the study of ecology, as in most any field these days, but for useful, basic accounts of the discipline the reader is referred to Boughey (1971), Odum (1971) and Clapham (1973).

The study of ecology is both simple and complex in that it is mostly "holistic" in its approaches. The many complexities involved in attempts to analyse, to explain and ultimately to predict the nature and multi-facetted trajectories of ecosystems in time (past–present–future) are often so great that the aid of a computer is generally necessary, particularly for simulation of possibilities and probabilities. "Cybernetics" and "systems theory" come into play (e.g., see Margalef, 1968), and so one important approach now in use is "general systems analysis" (e.g., see Dale, 1970; Patten, 1971). This latter method might someday be applied to the British Upper Palaeolithic.

In short, the "holistic" aspects of ecological theory and method provide a much more rounded, multi-dimensional picture when applied to problems in time and space than that normally obtained by traditional anthropology (including archaeology), biology, geology, etc. But then the accuracy of such studies is only as good as that of the information with which they have to work. Ideally, had the evidence with which this present book is concerned been more reliable and had there been more time and funds, I should have liked to have employed some sort of computer-aided systems analysis, attempting various computer simulations of British Last Glacial/Upper Palaeolithic ecosystems and their changes in time and space. For the moment, I simply present more or less intuitive graphic models of Last Glacial/Upper Palaeolithic ecosystems in "steady state" and in "self-sustaining change". These are figured and discussed in Chapter III (section III.B.4; Figs. 84 and 85). It is of course to be hoped that future research will obtain much more reliable data and conduct much more rigorous analyses.

A climatic interpretation of the environmental/ecological evidence presented in this book is also included in section III.B.4, where it is compared with a July mean thermal curve based on beetle remains (i.e., palaeotemperatures for the Coleoptera) from British Last Glacial (Devensian) deposits, as well as with a world-wide estimated sea-level curve. Severe climatic cycles obviously affect human, animal and plant behaviour, and so it is interesting if not crucial to try to determine at least in broad terms the interplay between ecosystems and climatic cycles. For example, a major advance of glaciers over much of Britain would obviously have a drastic effect on the behaviour, not to mention survival, of most if not all local life-forms, be they human or otherwise. Whether man in particular would successfully re-adapt to such a situation would depend on many factors ranging from technological to social and ecological.

C. ARTIFACT ASSEMBLAGES

1. Previous Work

Sir John Evans figured some of the stone, bone and antler tools from Kent's Cavern very soon after they were uncovered by Pengelly's excavations (Evans, 1872). He also drew attention to some extent to their typology and the nature of the raw materials from which they had been made. At the beginning of this century Prof. Sollas, with the guidance of the Abbé H. Breuil, figured a representative selection of the stone artifacts which they had excavated from Paviland Cave (Sollas, 1913). Sollas considered their typology in relation to known French material and pointed out their general Aurignacian aspect with certain particularities. Despite a lack of any clear stratigraphy, he further sub-divided them on typological grounds into "Mousterian", "Pseudo-Mousterian", "Oldest Aurignacian", "Middle Aurignacian", "Upper Aurignacian" and "Proto-Solutrean", but then concluded rather wisely that the entire assemblage may have been "Aurignacian". He also considered their raw materials and noted a preponderance of chert.

Leslie Armstrong, in his interpretation of the stone artifacts which he had excavated from the entrance platform of Mother Grundy's Parlour, again relied heavily upon known French material for his typology of his stratified series (Armstrong, 1925). He did, however, denote the British aspect of the assemblages and was even able to suggest on the basis of some of the evidence he found at this site and at the neighbouring one of Pin Hole, that the earlier backed tool assemblages were "Late Magdalenian" in age but "Developed Aurignacian" in character. The later backed tool assemblages were thought more "Azilio-Tardenoisian" in character and were assigned to the "Early Transition Period". He supported these allocations partly by the associated faunal series, and this division into Later Upper Palaeolithic and Mesolithic with a possible transition phase still seems not unreasonable on the grounds of my recent excavations at Mother Grundy's Parlour (see Campbell, 1969; as well as the relevant Chapters, III and V, in the present book).

In 1926 when Miss Dorothy Garrod brought out her book on the British Upper Palaeolithic, it was the first, and for some considerable time last, attempt to bring together all, or nearly all, of the known evidence about the subject. She illustrated relatively fair selections of stone, bone and antler tools from all of the sites known to her, and she cautiously strove to detect whatever patterns they might fall into. She compared the material with the known continental evidence and then went on to point out some of its particularly British characteristics, and, qualitatively, which tool-types seemed most common in her various sub-divisions. The groups of implements and sites which she recognized were as follows:

"Aurignacian, Lower: (?) Paviland; Middle: Paviland, Kent's Cavern, Ffynnon Beuno, Cae Gwyn; Upper: Paviland, Creswell Crags, Langwith, King Arthur's Cave, (?) Hoyle's Mouth, (?) Cat's Hole, (?) North Cray, (?) Halling;

"Proto-Solutrean: Kent's Cavern, Bench Cavern, Paviland, Uphill, Wookey Hole, King Arthur's Cave, Ffynnon Beuno, Creswell Crags;

"Solutrean: Creswell Crags, (?) Constantine Road, Ipswich, surface finds in East Anglia;

"Magdalenian with important Aurignacian survivals ('Creswellian'): Creswell Crags (M. 3–6), Victoria Cave (M. 4?), Kent's Cavern (M. 5–6a), Aveline's Hole (M. 6b), Cheddar, (?) Hoyle's Mouth, (?) Wangford Warren, (?) Halling" (Garrod, 1926; N.B., parentheses are hers).

Fifty years after her work, I would still agree with the basic idea of two main groups, regardless of the labels one may attach to them, i.e. " 'Middle' Aurignacian/Proto-Solutrean" and "Magdalenian/Creswellian" or simply "Earlier Upper Palaeolithic" and "Later Upper Palaeolithic". However, I would remove the more misleading terms "Lower Aurignacian/Upper Aurignacian" (in France now equal to Lower Perigordian or Chatelperronian and Upper Perigordian or Gravettian, respectively) and "Solutrean" as there appears to be little or no acceptable evidence for either in Britain. Also all or nearly all of the open-sites in Garrod's list now seem either Mesolithic or Neolithic/Beaker.

The term "Creswellian" was suggested by Garrod in 1926 to differentiate the British backed blade industries from the Magdalenian and Upper Perigordian of France. It is regrettable in some ways that she did not suggest a second term for the "Aurignacian/Proto-Solutrean" of Britain, but I have decided from my observations that simply "Earlier" and "Later" would be most convenient and least subjective for the apparent differences and separation of the two main groups. The continued use of the term "Creswellian" would to a certain extent be quite acceptable, except that it has suffered much misapplication in recent years (e.g. Manby, 1966; but also see Radley, 1969). But it is at least pleasing to state that Garrod's work remains very useful.

As a result of his excavations from 1927 to 1930 at Gough's Cave and Soldier's Hole, R.F. Parry obtained a large sample of Later Upper Palaeolithic material from the former site which enabled him, with the initial aid of J.A. Davies, to point out its Magdalenian and "Upper Aurignacian" (i.e., Upper Perigordian) aspects, and a small stratified series from the latter site which provided him with what he regarded as "Magdalenian" backed tools overlying "Solutrean" bifacial leaf-points (Parry, 1931). He (and Davies in the case of the former site) figured a fair selection of tool-types from both sites and considered the relative abundance of the various tool-types and their differences and similarities in comparison with other British and French material. Davies also noted possible sub-divisions at Gough's Cave within the Later Upper Palaeolithic, the "Upper Zone" being somewhat like Aveline's Hole with "pigmy knives" and the "Lower Zone" somewhat like the lower series from Mother Grundy's Parlour (Parry, 1929). Parry excavated in six inch (c. 15 cm.) "working layers" (or spits), the numbers of which have been marked on the actual specimens, and his recovery of Later Upper Palaeolithic at Soldier's Hole overlying Earlier Upper Palaeolithic remains of tremendous value.

After the Second World War Schwabedissen in his *Die Federmesser-Gruppen des nordwesteuropäischen Flachlandes* relied heavily upon Armstrong, Garrod and Parry's work for his basic knowledge of "Creswellian" tool-types (Schwabedissen, 1954). He further elaborated their typologies and suggested that the backed tools may have been the equivalents of modern penknives. He also noted that many of the British backed forms also appeared at Dutch and Belgian sites, whilst some other tool forms which seemed more Magdalenian-like appeared at the British sites. Hence he concluded that the "Magdalenian" helped give rise to the "Creswellian", and the latter became a source of the "Tjongerian". A. Bohmers disagreed with this scheme of origins to some extent and on the basis of his own statistical analyses of the artifacts, recommended that the British sites be split into separate "cultures". He suggested the continued use of "Creswellian" for the northern sites which supposedly had more frequent "Creswell points", and the adoption of his term "Cheddarian" for the southern sites which supposedly possessed more frequent "Cheddar points" as well as showing differences in other tool-types (Bohmers, 1956). He also grouped certain Dutch sites with either the "Creswellian" or the "Cheddarian" on similar evidence, removing them completely from the "Tjongerian". Bohmers was the first archaeologist to apply statistical and metrical analyses to north-west European Later Upper Palaeolithic artifact assemblages, although it should be stressed that some of his samples seem to have been unrepresentative of the total tools available from the British sites which he dealt with. For example, the total tools from Gough's Cave, Cheddar, number 799, whereas Bohmers only employed 275 of these. On the side of terminology for Upper Palaeolithic backed tools, Bohmers preferred "point" to Schwabedissen's "Federmesser", as his colleague, Aq. Wouters, found a "Tjonger" or "Gravette" point deeply embedded in a mandible of *Megaloceros* at a site near Roermond (Bohmers, 1956, p.25).

Following his excavations at Cathole, Gower Peninsula, in 1958, Dr. C.B.M. McBurney figured by type a selection of "Creswellian" stone tools and plotted the horizontal distribution of individual artifacts on and within his Layer B: thermoclastic scree (McBurney, 1959). Mrs. Angela Mace also figured a selection of artifacts from her 1957 excavations at Hengistbury Head, Bournemouth, and went on to develop an elaborate morphological typology for them in which she listed actual counts and percentages (Mace, 1959). She interpreted her sample of 2,263 flint artifacts, including 251 tools but excluding any pieces smaller than 15 mm., as Late Upper Palaeolithic-like and thought her 22 typed backed forms most comparable to the backed tools from Rissen near Hamburg. She did not, however, report in any detail on their vertical distribution at Hengistbury, and only gave their horizontal occurrence by yard squares. Nor did she give any more than a "typical section" of the site, drawn by someone else, which she seems to have regarded as sufficient stratigraphic detail; McBurney, on the other hand, recorded and published detailed section drawings for all of his trenches at Cathole 1958. As a result of excavations by the author at Cathole in 1968 and at Hengistbury in 1968 and 1969, there is some indication that both McBurney and Mace may have unwittingly mixed Later Upper Palaeolithic with Mesolithic material. This is indicated by stratigraphic and environmental evidence, as well as metrical and statistical analyses of the artifacts, and the point is considered in detail in later sections (see Chapters III.A.1, III.B.1–3, V.D, V.L and V.M).

T.G. Manby published counts and some percentages of tools and waste material from an assemblage of 4,461 flint artifacts which had been rescued by C. and E. Grantham from the topsoil of a gravel pit at Brigham Hill, East Yorkshire, during 1962 and 1963 (Manby, 1966). He also figured a reasonably fair selection of these and diagnosed them as "Creswellian". The present author has concluded, on the other hand, from the results of his own metrical and statistical analysis, that their typological affinities lie rather with known Early Mesolithic assemblages such as those from the not too distant sites in the Vale of Pickering (Clark, 1954; Moore,

1950), than with the definite Later Upper Palaeolithic assemblages from Creswell Crags and elsewhere further to the south-west (a similar conclusion was reached by Radley, 1969 in his brief review of Brigham). A reasonably close comparison might perhaps be made between the Brigham Hill artifacts and some of the material from Mother Grundy's Parlour, but only with that which is now seen as Early Mesolithic or "transitional to Mesolithic" anyway (Campbell, 1969). But to be fair to Manby, he did at least point out some of the similarities and dissimilarities between Brigham, Star Carr and Flixton Carr, and he really only aligned Brigham with Mother Grundy's Parlour and the Dutch site of Neer II which Bohmers had already classified as "Creswellian" (Bohmers, 1956).

In 1971 B. Barnes, B.J.N. Edwards, J.S. Hallam and A.J. Stuart published a preliminary account of two uniserially barbed bone points which had been found in 1970 in association with the skeleton of a male elk (*Alces alces*) at High Furlong, Poulton-le-Fylde, Lancashire in a Late Last Glacial lake or pond mud. In their subsequent full report (Hallam, Edwards, Barnes and Stuart, 1973) they demonstrate quite clearly that this deposit and its finds belong to Zone II ("Allerød"). Further, they present an intriguing list of possible artifacts which they infer from lesions on the elk bones (e.g., stone-tipped arrows and stone axes!). They interpret the barbed points as "Maglemosian" or Earlier Mesolithic on the basis of similarities with those in "group D" at the younger, Zone IV (or "Pre-boreal") site of Star Carr (also see Clark, 1954 and 1972). But they seem to have ignored what little evidence there is for the more closely contemporaneous barbed points (or "harpoons") of the Later Upper Palaeolithic (e.g., as in Garrod, 1926: Kent's Cavern and Aveline's Hole). Of course, the labels "Mesolithic" and "Upper Palaeolithic", convenient as they may often seem, may well mask many of the realities of the relevant human adaptations and re-adaptations. In other words, the High Furlong finds could reflect a "transitional to Mesolithic" adaptation which was useful for a time during the slightly wooded, milder conditions of the "Allerød Interstadial", and which became useful if not vital again during the "Pre-boreal" as at Star Carr.

Also in 1971 I wrote up my thesis on the British Upper Palaeolithic which I submitted at Oxford early in 1972. The present book is based largely on that work (Campbell, 1971). Finally, in 1974 Dr. Paul A. Mellars considered some of the British Upper Palaeolithic artifact assemblages as part of a general survey of the British Palaeolithic and Mesolithic (Mellars, 1974). Having read my doctoral thesis, his account broadly follows mine.

2. Classification and Typology

The method of classification and typology used by the author in this study of British Upper Palaeolithic cultural material is primarily morphological. Terms such as "point", "burin", "scraper", etc. for stone "tools", and "awl", "needle", "bâton", "harpoon" and/or "barbed point", etc. for bone "tools", are used merely for simplicity and to achieve continuity with the studies of previous and contemporaneous writers; they are not meant to imply a definite knowledge of the artifacts' original functions, study of which unfortunately lies outside the scope of this current work, though it would make an interesting and important research topic in its own right. Terms such as "tool" and "waste" are also used for simplicity and are not meant to imply a knowledge of the intentions of their makers.

The system of grouping and splitting the artifact forms that seem to exist or might exist in Britain is arranged by an alphabetical and numerical code that is based in part on the work of Mace (1959). It is also based to some extent on Schwabedissen (1954), Bohmers (1956 and 1963), Sonneville-Bordes (1960), Smith (1966) and Bosinski (1967). This code is given alongside the descriptions of artifact forms in Table 1. As it is arranged at the different levels of artifact material (Roman numeral), artifact class or type-group (single letter), artifact type (double letters) and artifact sub-type (double letters and Arabic numeral, etc.), it remains flexible and may be added to or subtracted from during future research without the necessity of altering the entire system. The code is also designed for brevity when referring either to actual specimens, to the descriptive list of all artifacts, or to the ideal types and sub-types of tools in Figures 1 and 2.

Brief names such as "shouldered point" and "leaf-point" are also used in the text and are fully described in the list, but no culturally value-laden names such as "pointe de la Gravette" are employed, with the exception of "Creswell point" and "Cheddar point", which are British names created by Bohmers (1956) for certain specific backed forms in the Later Upper Palaeolithic. "Unfinished tools" and/or "roughouts" are not generally allowed for as such in this study. An "obliquely truncated blade", for example, could hypothetically be either an unfinished "Creswell point" or other obliquely truncated backed blade or an unfinished "burin on oblique truncation"; as its oblique truncation is normally formed by very steep retouch (or so-called "backing"), such an artifact is classified, instead, by the author under the class of "backed tools" at its own type level. If, on the other hand, larger samples of Earlier Upper Palaeolithic "leaf-points" than those at present known are ever found in Britain, then possible "roughouts" might be distinguished from "finished" ones; such a distinction is apparently already being made in France for the Solutrean (e.g. Smith, 1966), and in Germany for the "Altmühlgruppe" (e.g. Bohmers, 1951).

Briefly, the artifact material and class levels recognized and/or assumed by the author are as follows:

I. *Stone Artifacts*: (A–H "Tools")

A. *Backed Tools.*
B. *Burins.*
C. *Scrapers.*
D. *"Zinken".*
E. *Borers and Awls.*
F. *Saws and Notches or "Spokeshaves".*
G. *Multiple Class or "Composite" Tools.*
H. *Retouched Flakes and Blades (HA–HD); Leaf-points (HE–HF).*

(I–R "Waste")

I. *Cores (but IF = Cores Adapted as "Tools").*
J. *Core Trimming Flakes.*
K. *Unbroken Unretouched Flakes and Blades.*
L. *Broken Unretouched Flakes and Blades.*
M. *Worn End Unretouched Flakes and Blades.*
N. *"By-products"*: Burin Spalls, "Microburins" and Axe-sharpening Flakes.
O. *"Hammerstones" and "Anvils".*
P. *"Pebble Tools".*
Q. *Unstruck Pebbles/Cobbles*: "Manuports".
R. *Engraved or Decorated Stones and Stone Surfaces.*

II. *Bone, Ivory, Tooth or Antler Artifacts*:

(A–I "Tools")

A. *Awls.*
B. *Eyed Needles.*
C. *Carved-base Points*: e.g. Split-base Points and Bevelled-base Points.
D. *Decorated Segments*: e.g. Incised Ribs.
E. *"Bâtons-de-Commandements".*
F. *Barbed "Harpoons" or Points.*
G. *Spatulas.*
H. *Ornaments*: e.g. Perforated Teeth and Bracelets.
I. *Musical Instruments*: e.g. Whistles and Drums.

(J "Waste")

J. *"Waste"*: e.g. Flakes, Splinters, "Polishers" and "Percuteurs".

III. *Shell Artifacts*:
A. *Perforated Shells*: e.g. Shell Beads.
B. *"Scoops" and "Scrapers".*
C. *"Waste"*: e.g. Flakes, Splinters, etc.

IV. *Snow or Ice Artifacts*: Only Presumed Thus Far; May Be Inferred in Future.
V. *Sand or Silt Artifacts*: Possibly Present.
VI. *Clay Artifacts*: Possibly Present.
VII. *Gut, Skin or Fur Artifacts*: Only Presumed and/or Inferred Thus Far.
VIII. *Moss, Herb, Bark or Wood Artifacts*: Only Presumed and/or Inferred Thus Far.

The above list is intended as a guide to the detailed descriptions of British Upper Palaeolithic artifact forms, both recognized and presumed, which may be found in the Descriptive Typological List of Ideal Artifact Forms (Table 1). Lists of actual artifacts found thus far may be found in the Gazetteers, but to interpret these the reader is again referred back to the Descriptive Typological List.

The Descriptive Typological List is considered reasonably comprehensive, but not necessarily exhaustive (particularly for organic artifacts), for surviving British Upper Palaeolithic artifacts as well as most British Later Middle Palaeolithic and Earlier Mesolithic ones. Despite however unwieldy it may in some ways at first seem, the author finds it easy to work with and easy to remember. Again, it may be added to or subtracted from as future research deems necessary, as its structure is more flexible than a simple, numbered list of artifact types, etc. (e.g. Sonneville-Bordes, 1960; Bohmers, 1963). Of course, many of the artifact type and sub-type attributes included by the author in his list may well be of little general importance to the British Upper Palaeolithic. Also some, if not most, of this classification system is doubtless still intuitive, and therefore not quite as objective as one would ideally hope for. But its basic morphological approach is probably sufficiently objective for the immediate aim of this study: an ordering of British Upper Palaeolithic cultural material in its ecological and chronological setting.

3. Metrical and Statistical Approach

Before any description of the methods of artifact analysis employed in this study, it should be stressed that nearly all British Upper Palaeolithic artifact assemblages are extremely small for obtaining reasonably "sound" statistics. Many samples must also be considered inaccurately excavated and/or incompletely collected. Some have even been partly or completely lost since excavation. Keeping these sources of error in mind, the author is nevertheless convinced that it is well worthwhile pursuing certain measurements and computations in an attempt to compare quantitatively what is available and endeavour to establish what may be the nature of any apparent clustering of assemblages.

Typological statistics may be obtained by assigning individual artifacts (in the Gazetteers) to their appropriate positions on the typological list as given in Table 1 and then adding up the totals for each sub-type, type and class in each assemblage. Percentages are sometimes computed from tool-type and class counts and used to compare any two or more reasonably large (in British terms) assemblages, as well as simply to help in the description of a single "large" assemblage. Counts, and where appropriate percentages, for each assemblage are given by site (and where possible by layer) in the Gazetteers and referred to in part or in whole in Chapters IV and V. Because there is clearly a wide range of tool-types with much variation, examples of most tool-types in most assemblages are also figured and discussed in Chapters IV and V. Frequencies of tool-types and classes in the larger undisturbed assemblages are demonstrated by frequency graph diagrams (and also discussed in Chapters IV and V) for greater clarity and ease of comparison.

In addition to studying simple morphological variation, it is sometimes informative to take and compare certain absolute size measurements. The size range of various tool and waste forms may thus be determined, and trends of size change may be observed in stratigraphically and/or chronologically ordered assemblages. The basic measurements taken from most complete stone artifacts include maximum length, breadth and thickness. These are taken with a metric caliper rule and recorded to the nearest millimetre. On stone artifacts with secondary work the following are measured in particular (see Fig.3): maximum breadth of backed tools (breadth is less easily affected by breakage than length), breadth of burin facets, angle of scraper retouch, and angle of invasive retouch and maximum breadth and thickness of leaf-points. Angles are measured with an adjustable protractor and recorded to the nearest five degrees. The breadth of burin facets as near as possible to their striking point is recorded to the nearest half millimetre. No measurements are normally taken on core trimming flakes, broken unretouched flakes and blades, and burin spalls, nor "manuports" (Leakey, 1971) such as unworked pebbles, although the weight of "manuports" is recorded in grammes, as is that of cores, hammerstones and anvils.

Metrical statistics are obtained by calculating the means and standard deviations, in respect of each of these measurements, for each sample. Certain ratios such as thickness over breadth are also calculated, and their means and standard deviations similarly computed. Size and/or ratio ranges are taken to one standard deviation only, i.e. a 66.7% probability. The statistical exposition and comparison of the metrical study of selected assemblages are illustrated in graph-form in Chapters IV and V, and the data is given by site (and layer if possible) for all measured assemblages in Tables 48 to 53.

Elaborate quantitative analyses such as "attribute cluster analysis" (e.g. Sackett's analysis of Aurignacian end scrapers, 1966) and "factor analysis" (e.g. the Binfords' interpretation of selected Mousterian assemblages, 1966b) are not included in this study as it is thought that most of the material concerned is too unreliable to warrant the use of a computer at this stage. Much of the necessary data has, however, been collected and could be included in future studies when and if larger, more reliable samples are added to the extant ones. Less sophisticated analyses such as "linear regression" are included in a few cases and have been conducted on a Casio AL−1000 programmable desk calculator. These are described where employed (Chapters IV and V).

4. Functions and Technology

The possible original functions of recognized and/or presumed artifact forms are sometimes suggested in this study. However, this is not to say that originally intended functions have been determined with any firm certainty. Although it would have been a worthwhile study in its own right, microscopic analysis of edge-damage, wear-traces and/or use-marks has been beyond the scope of this present report. Of course, it is to be hoped that future research on the British material will include not only microscopic analyses of the classic type as begun in Russia by Semenov (1964), but also new methods of the type now being developed at Oxford by Mr. Lawrence Keeley in his work on British Lower and Middle Palaeolithic stone artifacts (Keeley, 1974 and personal communications). In the case of the Upper Palaeolithic, although one could and perhaps should examine all artifacts, burins and scrapers might prove of special interest (e.g., see Pradel, 1973 on burins; Rosenfeld, 1971 on scrapers). In addition, Earlier Upper

Palaeolithic leaf-points and Later Upper Palaeolithic backed tools might well prove of analytic interest; I for one should have thought that these tool-forms served many functions, sometimes points, sometimes knives, sometimes "scrapers", etc., but only proper examination and experimentation would determine their definite or probable functions. Even unretouched flakes and blades may have served many functions as actual tools, despite the fact that we might not normally recognize them as such, or rather classify them as such.

Recent ethnographic work on stone artifact typology, technology, morphology and function is of immense interest. Gould's field-work, for example, amongst the "Yiwara" of the Western Desert of Australia shows quite clearly that a hunter/gatherer's own typology and an archaeologist's typology might at times overlap, but their concepts of typology, function and so forth would nearly always be different (Gould, 1969 and 1971). Further, the "Yiwara" typology is constructed mostly according to function at time of use, rather than morphology at time of manufacture, although both sorts of typology are recognized by them.

Work by White and Thomas (1972) in the New Guinea Highlands suggests broadly comparable results, but at a much more refined level of observation, analysis and interpretation (also see Thomas, 1974). With the aid of a computer they were able to detect differences in similar stone artifacts from separate but neighbouring villages, differences which the New Guineans themselves were unaware of, but which nonetheless seemed real. These slight morphological/metrical differences apparently reflect differences in actual craftsmanship between different individuals. They also seem to reflect socio-psychological differences. In other words, the results obtained by White and Thomas suggest that we might eventually be able not only to distinguish individual craftsman in the Stone Age, but to pursue a form of psychological archaeology!

Returning to the Upper Palaeolithic of Europe, Marshack's recent microscopic work on the "cognitive aspects" of engraved bones and stones suggests a surprisingly formal tradition of apparently intentional notation and symbol usage. His recognition of a possible lunar calendrical system is of direct relevance to the British Upper Palaeolithic, as we may have at least one example of this sort of behaviour in the Later Upper Palaeolithic engraved bone "ruler" from Gough's Cave, Somerset (see Fig. 119, no. 9, as well as Hawkes, Tratman and Powers, 1970). Of course, some of these engraved "calendars" may equally well have served as "games", etc. (e.g., see Dewez, 1974). Marshack's own work is published in a number of accounts (e.g., Marshack, 1971, 1972a, 1972b and 1972c). His most recent paper (Marshack, 1975) even suggests that "art and notation" of a "macaroni" form extend well back into the Middle and Lower Palaeolithic!

As for technology, quantitative experiments in the manufacture of artifacts are unfortunately beyond the range of the present study as well. Techniques of manufacture are sometimes suggested as mere guidelines or possibilities, but more rigorous analyses and experimental manufacture of artifacts similar to those found in the British Upper Palaeolithic are needed before we may be more certain of the actual technology involved. Happily, there is a tradition in British Palaeolithic archaeology of experimental knapping, e.g. Baden-Powell (1949) on Clactonian flaking technique, and more recently Newcomer (1971) on handaxe manufacturing technique. Newcomer has also attempted to replicate the sorts of bone tools found in the Upper Palaeolithic at Ksar Akil in the Lebanon (Newcomer, 1974).

In 1972 I participated in a brief flint-knapping session with Dr. Newcomer, Mr. MacRae, Mr. Jones and the late Mr. Baden-Powell, but really a proper, more formal, research team is needed for future work, both on stone and bone artifacts. Abroad, the experimental work of Dr. Crabtree in the United States and Prof. Bordes in France is now quite famous (e.g. see Bordes, 1969), having even been broadcast there and in Britain on a number of television programmes. But unlike some of the French Solutrean leaf-points (see Smith, 1966 as well), as far as I can tell, British Earlier Upper Palaeolithic leaf-points in general were not pretreated with heat prior to flaking. This, of course, definitely needs to be properly checked, perhaps particularly for those leaf-points manufactured in chert and adinole, the flaking properties of which would probably have been improved by careful preheating.

As for the striking of blades, Tixier's experiments are perhaps of some relevance to the British Upper Palaeolithic, particularly in the production of the sorts of large, long blades which are a common feature in certain Later Upper Palaeolithic assemblages (e.g. Sproughton, Suffolk; Crown Acres, Berkshire; Hengistbury Head, Hampshire; and to some extent Gough's Cave, Somerset) and sometimes present in Earlier Upper Palaeolithic assemblages (e.g. Kent's Cavern, Devonshire; possibly Paviland Cave, Glamorganshire; Ffynnon Beuno Cave, Flintshire; and possibly Bramford Road, Suffolk). The Earlier Upper Palaeolithic large, long blades appear generally more robust (i.e. thicker, broader and to some extent "cruder") and may well have been produced by a technique different from that which produced the more refined Later Upper Palaeolithic blades. Here it is of interest that Tixier's experimental blades and cores produced "sous le pied" (Tixier, 1972) appear nearly identical to some of those made by the Later Upper Palaeolithic flint-knappers at Sproughton, Crown Acres and particularly Hengistbury Head. The good quality of the local English flint may have much to do with these apparent similarities, but it seems also quite possible that large Later Upper Palaeolithic cores were in fact held under the left foot (or right if left-handed) in the manner of Tixier and struck by indirect percussion with antler, bone or wood hammers and punches. Future British research ought to test this kind of behaviour; it might even reveal differences between right and left-handed work!

5. Raw Materials

No chemical analyses of raw materials are included in this study. Raw materials are qualitatively identified and described and quantitatively listed according to the basic categories of inorganic (e.g. stone) and organic (e.g. bone). Inorganic materials are further divided into flint, chert, adinole, quartzite, sandstone, limestone, ironstone, pyrites and ochre, and presumably more perishable materials such as snow and ice. Flint may be sub-divisible into Yorkshire-Lincolnshire Wold flint, East Anglian flint and southern English Downs flint, but many assemblages are made on mixtures of these, probably from a derived context such as river gravel or boulder clay. Future studies might get further with the line of research known as "flint trace element analysis" (e.g. Sieveking, Craddock *et al.*'s analysis of Neolithic flint axes, 1970), but the imponderables would probably be very great indeed. "Trace element analysis" has, however, recently been carried out on other "stone" materials such as obsidian in New Zealand with reasonable success, and this method may yet be extended to still other materials (e.g. see Ward, 1974). Chert is apparently more easily sub-divisible for the present by eye, occurring as southern Welsh Carboniferous black chert, Peak District Carboniferous brown chert, Dorsetshire Jurassic Portland chert and Somerset-Devonshire Cretaceous Upper Greensand Blackdown mottled chert. Adinole is a mixture of quartz and albite and occurs in metamorphosed Cambrian shales in South Wales (Green in Leach, 1918, p.4). It seems to flake as well as good chert. Both chert and adinole may be obtained from local gravels and boulder clays, particularly in South Wales. Sources of quartzite include Devonian Old Red Sandstone, Triassic Bunter Pebble Beds, various river gravels and boulder clays, but it is very difficult to trace an exact source. Sandstone is also widely available but apparently little used. Limestone is mainly exploited for its tendency to form caves, particularly Devonian, Carboniferous and Permian limestones. It may, however, also be used for hearths, anvils and possibly some flakes and cores. Both sandstone and limestone could be used for grinding wild seeds. Ironstone is available from the southern English Eocene "Barton, Bracklesham and Bagshot" Beds. It may be used either for anvils or, as with any other stone, for weights for holding things down. Iron pyrites and other minerals such as quartz crystals may be collected simply as curiosities, although pyrites itself might be used in making fire. Red (or yellow) ochre may be obtained from many sources and may be used as a body pigment at least in burials and for highlighting engraved decoration on bones, stones etc.

Although quite a wide variety of inorganic material is exploited, flint is certainly the most commonly used for stone artifacts. Map 3 shows the general distribution or "economic geology" of most sources of inorganic materials which either were or could have been exploited. The "known exploitation limits" are roughly the same as the northern and western limits of the distribution of known

Earlier and Later Upper Palaeolithic sites. (N.B. as in Map 2, the abbreviations E.U.P. for Earlier Upper Palaeolithic and L.U.P. for Later Upper Palaeolithic are again used in this and subsequent maps and figures for convenience, and they are sometimes employed in tables for the same reason.) Ireland and Scotland are apparently beyond these ranges, although materials from them might be obtained from boulder clays in England and Wales, as might materials from Scandinavia.

Organic raw materials used or available include bone, tooth, ivory, antler, shell, amber, and presumably more perishable items such as gut, skin, fur, wood, bark, leaves, grass, etc. Bones are obtained from large herbivores mostly but also from birds, carnivores and lagomorphs. Teeth are obtained from carnivores and herbivores, ivory from mammoth and antlers from various deer. Both marine and land shells are gathered, presumably from original sources. Amber is rare and the only definite example is of ultimately Baltic origin, although it may of course be from a secondary source such as boulder clay (Gough's Cave in Beck, 1965). The originally intended use of this amber is uncertain. Bones, teeth, ivory and antlers are used for tools (e.g. the well-known bone and antler "harpoons" from Kent's Cavern, Garrod, 1926), and ornaments (e.g. the ivory ?bracelet and perforated teeth from Paviland, Sollas, 1913). Shells are used apparently for ornaments only (e.g. beads at Aveline's Hole in Davies, 1921), although some could be used for tools.

The more or less completely perishable organic materials listed above are used today by such hunting peoples as the Nunamiut Eskimos (Gubser, 1965). The Nunamiut hunt primarily "caribou" (i.e. "reindeer", *Rangifer tarandus*) in an Arctic environment, and in these respects they are not entirely unlike the Upper Palaeolithic hunters of Britain. Gut, skin and fur are obtained from wild herbivores, carnivores, rodents and lagomorphs, and they are used for water-carriers, clothing, tents, etc. Bird skins are also used in making clothing. Wood is obtained from various trees and shrubs including notably willow and birch and is used for fuel, tools, travelling equipment and house and tent building. Bark is collected from birch for baskets and alder for reddening skins. Herbs and mosses are important for house construction and hygiene.

The use of gut, skin and fur by British Upper Palaeolithic hunters is at least implied by their manufacture of bone eyed-needles and awls (e.g. Kent's Cavern, Garrod, 1926). Wood is certainly in part a source of fuel as indicated by the occurrence of birch (*Betula*) charcoal (e.g. Mother Grundy's Parlour, Armstrong, 1925), and by extrapolation from certain well-preserved Ahrensburgian assemblages, it is probably a source of tool material (e.g. the Ahrensburgian arrow-shafts from Stellmoor, Rust, 1943 and 1962).

For a further discussion of some of the possible/probable uses of perishable inorganic (e.g. snow and ice) and organic materials see the end of the artifact description list in Table 1, in particular artifact material levels IV to VIII (*viz.*: IV. Snow or Ice Artifacts; V. Sand or Silt Artifacts; VI. Clay Artifacts; VII. Gut,

Skin or Fur Artifacts; VIII. Moss, Herb, Bark or Wood Artifacts). There and here a number of suggestions have been set down merely as guidelines to or examples of what might eventually be found in Britain, or what might possibly have been made/used in Britain, but which would probably not have survived the rigours of time.

6. Individual Artifact Distribution within Sites

Plotting the horizontal and vertical position of individual artifacts is only possible for a few of the known British Upper Palaeolithic sites. Where this is possible, however, the patterns which emerge are often quite informative. Tool and waste classes may be plotted for an assemblage along with other evidence such as faunal material, hearths and areas of disturbance. In undisturbed occupation areas one often finds clusters of certain tools and waste forms and wide scatters of others. Actual settlement patterns may thus begin to emerge, although one should perhaps term them "abandonment patterns" (for ethnographic examples of "abandonment patterns" see Gould, 1971 and Bonnichsen, 1973; they also consider differences between reality and archaeological reasoning). Of course, if one finds a cluster of scraper forms, for example, it would be logical to assume that scraping activity probably took place in more or less that area (e.g. see Leroi-Gourhan and Brézillon, 1972).

The publication of plans showing the horizontal position of individual artifacts is common enough (e.g. the Cathole excavations, Wales, by McBurney, 1959, plate 19; the Orangia 1 excavations, South Africa, by Sampson, 1968, Figs. 13–18), but there are as yet very few examples of closely spaced sections showing the vertical position of individual artifacts (one is the Lazaret excavations, France, by Lumley *et al.*, 1969, Figs. 43–50). Lumley clearly indicates the multi-facetted values of plotting both horizontal and vertical positions. For example, artifacts found in the same natural layer may sometimes occur vertically as separate archaeological lenses, which would best be reproduced by illustrating both planes, rather than simply the horizontal. Careful examination of previously excavated British Upper Palaeolithic artifacts leads me to think that many originally undisturbed and separate assemblages have been mixed by exceedingly inaccurate field-work. The outstanding partial exception for its time is Pengelly's work in 1865–80 at Kent's Cavern, Devonshire (see pp. 37–42).

In this study, therefore, excavations by the author have included measurement horizontally and vertically of the position of nearly every undisturbed artifact to about the nearest centimetre, as well as the careful recording of site plans and closely spaced sections. The measurements were recorded in field catalogues and on field drawings and when plotted revealed much useful information which would otherwise have been lost. Large mammal faunal material was similarly recorded. These plans and sections are reproduced in this book with the plotted artifacts, faunal material and radiocarbon samples projected on them (see Figs. 10 to 55).

D. CULTURAL ECOLOGY: HUNTING AND GATHERING EVIDENCE

1. Archaeological Sites, Anthropology and Ecology

The presently known definite British Upper Palaeolithic sites are confined to England and Wales (see Maps 2 and 4). Their variable density and relation to topography and water resources reveal much which might be interpreted socio-economically, as well as ecologically. Of course, some of their distribution patterns are probably due in part to highly variable methods of discovery, but some must almost certainly be due to the desires and necessities of the Upper Palaeolithic hunters who exploited the various regions.

In Chapters III, IV and V, archaeological sites are plotted according to the known number of Upper Palaeolithic artifacts recorded from them. Earlier Upper Palaeolithic sites are shown on one series of maps and Later Upper Palaeolithic sites on another. Artifact counts are grouped into various less-than and greater-than amounts. These amounts are meant to indicate both where collection has revealed the most evidence and where Upper Palaeolithic activity may have been greatest. Thus far there may be a bias in favour of cave-sites rather than open-air sites, but even so, how many more open-sites may eventually be found might just be predictable (computer simulation of known and possible distributions is a worthwhile line for future research and might aid in such predictions). Cave-sites and open-sites are indicated by different symbols. The size of the symbol reflects the size of the total site sample. The total site sample is the total number of Earlier Upper Palaeolithic artifacts on the one hand, or Later Upper Palaeolithic on the other, found thus far at a site, regardless of clear or possible separation of assemblages. Assemblages are grouped as either simply Earlier or Later where their sub-division is known because they increase the total site sample, and thereby the impression, in this case in time, of greater site activity. In any case, definite sub-divisions of the Earlier Upper Palaeolithic are still uncertain, and there are only a few examples of them thus far in the Later Upper Palaeolithic. In some cases Earlier and Later material has been mixed by disturbance, artifical or natural, and its separation on the basis of typology presents only a very rough approximation of what the original counts may have been.

Modern rivers and coasts are shown on all of the archaeological maps, but towns, roads and footpaths are only shown on the large scale maps of sites selected for excavation. Selected contours are shown on most of the maps. National Grid lines are indicated on the margins of nearly all maps. Regional maps are arranged in a clock-wise order beginning with Devonshire and proceeding northward through the Mendips and Wales and then across to the eastern coast and down to the southern coast.

So-called "site catchment analysis" (Vita-Finzi and Higgs, 1970) is employed in a general way on the regional maps; it is the term coined by them for the study of the relation between technology and those natural resources lying within economic range of individual sites. In this sense the regional maps are maps of specific "exploitation territories" and their immediate environs. A "site exploitation territory" is taken as the territory surrounding a site which is assumed to have been habitually exploited by the inhabitants of that site. A hunter/gatherer, at least in the case of the !Kung Bushman today (Lee, 1969), often finds it uneconomic to exploit the available resources beyond a distance of about 10 kilometres from the site he is using in a given season, month and/or *particularly* day. The area within that approximate 10 kilometre radius (or 2 hours of walking) would therefore be his site exploitation territory. The territory would have at least one main occupation site, or "home base-site", and probably more than one secondary, or "transit" site. A base-site would be the main focus of exploitation activity and would have relatively dense occupation débris, whilst a transit-site would be one of a number of strategically placed "subsidiary" sites generally within the 10 kilometre range of the base-site if related to it and beyond if concerned with migration between base-sites. A transit-site would have relatively little occupation débris and could be simply a find-spot of a single implement. Finally, an "annual territory" would be the total area exploited by a human group throughout a given year; it would have one or more site exploitation territories.

Location analysis of sites, cities, etc. in relation to their resources, manufacturing, food production, trade or exchange, etc. is a fairly well established method, or series of methods, in demographic, economic, geographical and now anthropological and archaeological studies. Lösch's (1954) work has become a classic in economics, but it and many other studies before and after really go back to the concepts of Thünen's (1826) pioneering appraisal and abstraction of relations between a model city and its environs, i.e. a series of concentric rings or "Thünen rings" not totally unlike those constructed much more recently by Lee (1969) after his field-work on !Kung Bushman camps and their resources (see Map 5B). Lee's work on hunter/gatherer location analysis and that of Chisholm (1962) on *Rural Settlement and Land Use* are two of the main building blocks employed by Vita-Finzi and Higgs (1970) and Jarman (1972) in their construction and application of "site catchment analysis".

In this study *possible* site exploitation territories are indicated on the regional maps by circles with 10 kilometres radii. These circles are drawn round sites which have yielded what would locally be abundant evidence. In some regions of Britain the lower limit of abundance is set at over 50 artifacts, but generally it is set at at least over 100 artifacts. By comparison with the analyses conducted by Vita-Finzi and Higgs (1970) and Jarman (1972) on Mediterranean sites, the British sites allocated the status of base-site very often have much smaller artifact counts. Exceptions include a few British Upper Palaeolithic sites with well over a thousand artifacts each (Paviland Cave, Gough's Cave, Sproughton and Hengistbury Head); it may be that these are the only known "true" base-sites. Those suggested base-sites with less than a hundred artifacts might best be considered as guidelines to areas where more productive related sites might eventually be found, assuming the basic hypotheses of site catchment analysis to be correct.

Under the term cave-site are included the following: cavern, cave, "ogof" (Welsh for cave), "hole" (colloquial English name for many cave-sites), fissure and rock-shelter. A British Upper Palaeolithic cave-site may have one or more artifacts, and the artifacts may be stratified, derived or just attributed to a given cave-site. Under the term open-site are included all open-air or non-cave sites. A British Upper Palaeolithic open-site may also have one or more artifacts, stratified, derived or just attributed to it. Due to subsequent erosion and disturbance by man, other animals and plants, open-sites are probably more often partly or completely destroyed than cave-sites. Both cave- and open-sites may be interpreted as base-sites provided they have a sufficient number of artifacts (i.e., usually over 100 artifacts, ideally over 1000), otherwise they are better interpreted as various kinds of transit-sites, e.g. gathering, knapping, observing, killing or butchering sites.

In considering the possible normal limits of the "annual territories" of British Upper Palaeolithic hunter/gatherers it is essential to review, albeit briefly, the apparent and/or real annual or even total ranges of various living hunter/gatherer populations. Space does not permit us to consider all groups so instead Maps 5 to 8 are presented here as a careful selection of examples of "territories", their implied adaptations and land uses. A noteworthy feature which may have some predictive value is that many hunter/gatherer group "territories" seldom exceed 100 to 200 kilometres across. The "annual territories" of sub-groups would then normally fall within such areas, an "annual territory" itself perhaps often having a radius of no more than about 50 kilometres. I have extracted this and other ethnographic data principally from the following sources: Balikci (1968 and 1970); Burch (1972); Campbell (1968); Coon (1972); Forde (1934); Gubser (1965); Helm (1968); Hickerson (1965 and 1970); Lee (1969 and 1972); Lee and DeVore (1968); Leeds (1965); Paine (1972); Peterson (1975); Sahlins (1972); Service (1966); Spencer (1959); Stanner (1965); Steward (1933 and 1955); Williams (1974); and Woodburn (1968 and 1972).

Using this somewhat arbitrary but not unreasonable figure of 50 kilometres for possible "annual territories", I have drawn circles with radii of that distance round the apparent British Upper Palaeolithic "base-camps". Areas with clusters of lesser sites but no "base-camps" yet certain have dashed circles drawn round them merely as a suggestion of where larger sites might eventually be found. These data on possible land use in the Earlier and Later Upper Palaeolithic are presented, respectively, in Maps 33 and 46. Map 46 also includes a suggested socio-economic "buffer zone" between northern and southern Later Upper Palaeolithic sites, though it should be noted this is not meant to imply the sort of warfare which helped to produce the game reservoir between the Chippewa and Sioux tribes (see Map 6B; also Hickerson, 1965 and 1970). Although there might have been some social tension in the Later Upper Palaeolithic, the value of a "food reservoir" might have been realized without warfare.

2. Faunas, Floras and Cultural Ecology

Occurrences of definitely Last Glacial (Devensian) fauna and flora have been recorded more or less throughout the British Isles with apparent concentrations in certain parts of England, Wales and southern Ireland. Some of the faunal and floral sites in England and Wales are also British Upper Palaeolithic sites.

Reasonably accurate faunal frequencies and actual "biomasses" are difficult to estimate as most of the larger assemblages are improperly recorded, and those excavated by the author are all statistically small. Nonetheless, some stratigraphic evidence is available and various patterns of faunal distribution at different levels of the Last Glacial are beginning to emerge. These distributions of fauna are considered in relation to archaeological sites, flora, topography, water resources and climate. Possible migration routes are suggested, where applicable. It is worth noting here that reindeer or caribou are not always as easy to hunt as many might assume (see Burch, 1972). The contrast between Eskimo ranges and caribou ranges may be seen quite clearly in Map 8.

Pollen analysis of selected sites allows the reconstruction of possible vegetation zones which are mapped according to the apparent and relevant stages of the Last Glacial (Devensian): Middle; Full; and Late Zone I, II and III (see pp. 136–39 and Maps 19 to 23). Pollen and spore percentages for each relevant site are indicated on "clock-face" diagrams which are plotted as near the original site locations as possible. Roughly contemporaneous ice fronts, coastlines and large lakes are also shown on the vegetation maps. The distribution of floras is considered in relation to archaeological sites, fauna, topography, water resources, climate and ecology in general.

The distribution of faunas and floras as natural resources for Upper Palaeolithic hunter/gatherers is discussed in detail in Chapters III, IV and V. Earlier and Later Upper Palaeolithic patterns of faunal exploitation are becoming partly clear by virtue of certain direct associations with faunal evidence, but there are as yet no absolutely definite floral exploitation patterns as there are no known direct associations with plant remains, apart from charcoal, of *Quercus, Ulmus* and ?*Rhamnus* at Kent's Cavern (see Appendix 1 and p. 95). and of *Betula* at Mother Grundy's Parlour (Armstrong, 1925). The presence of pollen in all samples studied thus far is apparently of natural rather than artificial origin (the Paviland burial, Buckland, 1823, was unfortunately discovered much too early to reveal the sort of evidence found for example at Shanidar in Iraq, i.e. the Mousterian "flower burial", Solecki, 1971).

Man presumably exploited some plant materials and food directly, but he was probably largely dependent upon other mammals for his resources. Amongst the Nunamiut Eskimos today only about 5% of their diet consists of plants, whilst most of it is from large herbivores, particularly reindeer (Gubser, 1965). In fact, at a general level the importance of meat as a constituent of diet seems to increase as one goes from equator to pole (Lee, 1968; Isaac, 1971); such an effect could be very important in periglacial Britain. Lee (1968) shows further that meat frequently constitutes at least about 20–35% for most hunter/gatherer groups today. In short, I think one may safely assume that British Upper Palaeolithic people had somewhere between 50% and 95% meat in their diet.

However, even if plants contributed a small proportion of the Upper Palaeolithic diet it is worth noting that today well over 50% of the Nunamiut artifacts are made from plant materials (Gubser, 1965). It is certainly therefore hoped that British Upper Palaeolithic sites with conditions allowing for the preservation of wood, bark, leaves, grasses, etc. might be found. Such conditions would be more likely to occur in a water-logged open-site, rather than any sort of cave-site (e.g. as in the British Early Mesolithic open-site of Star Carr in East Yorkshire, Clark, 1954).

E. SITES SELECTED FOR EXCAVATION IN 1968–69

As two of the main concerns of this study are chronology and ecology, it was decided to excavate for new and hopefully more reliable evidence. Rather than ignore the known sites and spend a great deal of time surveying for completely new sites, certain known sites that were thought likely still to have some undisturbed deposits were considered for the excavation programme. Thus mainly cave-sites were selected for re-excavation, as they had already provided the best stratigraphic and faunal evidence as well as the largest number of artifacts. It was not expected that large artifact samples would be obtained during a re-excavation of the six selected cave-sites, merely a few diagnostic tool-forms in undisturbed contexts. Reasonable radiocarbon samples were also expected from these sites, as were deposit samples for pollen analysis. Three previously unexcavated rock-shelters in the vicinity of some of the selected cave-sites were also chosen for sounding. It was not known what sort of evidence they might yield. The only open-site selected for excavation was Hengistbury Head, where it was hoped to find and date a large artifact assemblage.

Map 9 shows the names and locations of the sites selected for excavation. Field-work was carried out in 1968 and 1969. The counties and National Grid references for these sites are given in the Gazetteers of Sites. The full excavation results are given in Chapters III, IV and V (as well as in the Appendices where appropriate) according to the location and nature of the material found, but a summary is given here for convenience. Badger Hole was excavated in August 1968 and yielded apparently Earlier Upper Palaeolithic evidence. Sturdy Shelter was sounded in December 1968 and was sterile. Sun Hole was re-opened in September and December 1968 and yielded apparently Later Upper Palaeolithic evidence. Cathole was excavated in July–August 1968 and yielded Later Upper Palaeolithic and Mesolithic evidence. Long Hole was excavated in August–September 1969 and yielded apparently Middle Palaeolithic and Earlier Upper Palaeolithic evidence. Shallow Shelter was sounded in August 1969 and was sterile. Holly Shelter was sounded in July 1969 and was virtually sterile except for a nineteenth century hearth. Mother Grundy's Parlour was excavated in July 1969 and yielded Later Upper Palaeolithic and Mesolithic evidence. Robin Hood's Cave was excavated in July 1969 as well and yielded Middle Palaeolithic and Later Upper Palaeolithic evidence. Hengistbury Head was excavated in April–May and June–July 1968 and April 1969 and yielded Later Upper Palaeolithic and Mesolithic evidence. The names of the three sterile sites mentioned above were assigned by the author.

Future field-work ought to include a search for new sites in all parts of the British Isles, but perhaps particularly on a well-organized regional basis with an emphasis on areas most likely to yield organic artifacts and food débris *in situ*, e.g. peaty deposits. Future surveys might also do well to include underwater cave excavations of the sort being conducted by Bonifay (1970) in the French Mediterranean.

CHAPTER III

CHRONOLOGY AND ECOLOGY OF THE BRITISH UPPER PALAEOLITHIC

A. CHRONOLOGICAL FRAMEWORK FOR THE UPPER PALAEOLITHIC AND THE LAST GLACIAL IN BRITAIN

1. Stratigraphic Evidence (including results of excavations in 1968–69)

The clearest stratigraphic evidence for the sub-division of the British Upper Palaeolithic is provided by the following four sites: Kent's Cavern (Devonshire), Soldier's Hole (Somerset), King Arthur's Cave (Herefordshire) and Pin Hole (Derbyshire) (Map 4). None of these has been excavated by truly accurate standards, but the records made are certainly sufficient to indicate their basic stratigraphic pattern: two main Upper Palaeolithic groups, an Earlier one characterized by leaf-points and a Later one characterized by backed tools. By correlation the bracketing sequence is also clearly indicated as Later Middle Palaeolithic (or "Handaxe Mousterian") followed by Earlier Upper Palaeolithic and then Later Upper Palaeolithic sealed by a somewhat multifarious Mesolithic. The twin sites of Cae Gwyn and Ffynnon Beuno Caves (Flintshire) (Map 4) provide a different but crucial point of reference for the Upper Palaeolithic: characteristic Earlier Upper Palaeolithic tools are overlain (or presumably overlain, in the case of Ffynnon Beuno) by glacial deposits from the last major advance of northern Welsh and Irish Sea ice attributable to the Full Last Glacial (Full Devensian). In other words, the Earlier Upper Palaeolithic belongs before the Last Glacial (Devensian) maximum ice advance, whilst the Later Upper Palaeolithic, which has no sites later sealed by glacial deposits, belongs after the maximum advance. The main aspects of the above evidence are summarized in Table 2, and references are given in the more detailed descriptions of these and other sites which follow in this and subsequent Chapters, as well as in the Gazetteers.

Stratigraphic evidence for the British Last Glacial (Devensian) is not completely clear, but a general pattern is beginning to emerge from a rapidly growing body of information. This basic pattern appears to be Last Interglacial (Ipswichian) marine or other earlier deposits overlain by Last Glacial (Devensian) periglacial and/or till deposits, which are in turn overlain by Post Glacial (Flandrian) marine or other later deposits (Oakley and Baden-Powell, 1963; West, 1968; Sparks and West, 1972; Mitchell, Penny, Shotton and West, 1973). The sequence within the Last Glacial (Devensian) appears to be at least tripartite: periglacial-glacial-periglacial deposits, although there are also many minor associated complications. The first periglacial series can be quite long and varied, and it is therefore generally divided into the Early and Middle Last Glacial (West, 1968). The last periglacial series is much shorter but nonetheless quite varied; it is now usually assigned to the Late Last Glacial (or Late Devensian) and sub-divided into Zones I, II and III (Godwin, 1956; West, 1968; Mitchell, Penny, Shotton and West, 1973). The intervening glacial series can be either long or short and complicated or relatively simple, depending on location in the British Isles; this of course is also true to some extent of the various periglacial deposits. Most British authorities split the relevant glacial series between the earlier and later phases of the Last Glacial without even assigning separate status to the extensive spread of deposits caused by the last maximum British ice advance (e.g. West, 1968; Mitchell, Penny, Shotton and West, 1973). There are certainly local glacial series within the main periglacial episodes which are merely minor advances or readvances from local centres of glaciation (e.g. the Scottish readvances of Late Glacial Zones I and III, Sissons, 1967), but the last maximum advance is so extensive and severe in its impact on the British Isles that the present author prefers to follow the German practice of assigning such a separate status (e.g. Woldstedt, 1960); i.e. Full Last Glacial or Full Devensian. The Full Last Glacial (c.20,000–15,000 B.P.) with its maximum advances (British and Norwegian) and certain major readvances (Scottish, Irish and Northern Welsh) falls quite clearly between the Middle Last Glacial (c. 40,000–20,000 B.P.) and the Late Last Glacial (c. 15,000–10,000 B.P.) (Morgan, 1973; Penny, Coope and Catt, 1969; Saunders, 1968; Shotton, 1967). There is now no definite evidence for an earlier maximum advance within the Last Glacial, although there were probably minor advances in some of the highland zones.

The above and related Last Glacial evidence is summarized in Table 3. The age brackets given in the table are based on radiocarbon data discussed elsewhere (pp. 77–84) as is the case with the ecological data (pp. 84–134).

Basic British Upper Palaeolithic Stratigraphy: Earlier Field-work

As briefly described above (p. 36), the basic British Upper Palaeolithic sequence is most clearly demonstrated by Kent's Cavern, Soldier's Hole, King Arthur's Cave, Cae Gwyn Cave and Pin Hole. Proceeding in the clock-wise geographic order of this list of sites, beginning with Kent's Cavern, the stratigraphic evidence will now be considered in greater detail. As part of this research programme the author has visited all of these sites to be discussed and carefully examined and analysed their relevant collections and available records.

Kent's Cavern (Devonshire). Main periods of field study: 1865–80 and 1926–40.

Kent's Cavern lies in the lower northern slope of Lincombe Hill at Torquay. It consists of a large series of passages and chambers which have been formed by solution in the bedrock of Devonian limestone, and which have been partly filled with deposits at various times since the Middle Pleistocene or earlier. It at present has two main entrances, about 15 metres apart, lying at roughly the same level some 58 metres above the local mean tide and overlooking a dry valley known as Ilsham Valley. The northern-most (or "North") entrance faces south-east and the southern-most (or "South") entrance faces east-north-east. Buried additional entrances are known.

The only truly systematic excavations at Kent's Cavern, thus far, were those conducted by W. Pengelly, with some help from E. Vivian, from 1865 to 1880. The site is known to have yielded to him Lower, Middle and Upper Palaeolithic artifacts with contemporary faunal assemblages, as well as later material from Mesolithic to Medieval. Pengelly laid out a grid system of 1 foot (c. 30 cm.) parallels intersected at right angles by 1 yard (c. 90 cm.) lines which he employed throughout most of the site for horizontal control. For vertical control he dug in 1 foot (c. 30 cm.) levels, or spits, consistently to 4 feet (c. 120 cm.) beneath a granular stalagmitic floor, except in the Long Arcade where he dug to bedrock at about 9 feet (c. 275 cm.) below the same stalagmitic layer. He carefully recorded his observations on the composition, texture and colour of the various layers which he encountered, and he assigned find-numbers by prisms (or parallel, yard and level) to all artifacts, bones and teeth. He published monthly and annual reports on the progress of his excavations and a final summary (Pengelly, 1884). However, although he recorded in his Diary literally thousands of the appropriate measurements, he never found time to publish the detailed plans and sections of which he would have been capable. Also he never figured any of the artifacts, although at least a selection of these was reproduced by Sir John Evans shortly after their discovery (Evans, 1872).

Pengelly's five-volume Diary is preserved in the Museum of the Torquay Natural History Society. It has been studied and analysed in great detail during the course of research for this book, and the present author believes he is the first to have taken the trouble to extract a fair portion of the wealth of information which it contains. Garrod (1926) and Rogers (1955) apparently only glanced through the Diary in search of odd bits of information, e.g. artifact counts or brief summaries of stratigraphy.

Figure 4 shows a plan of the more important eastern part of Kent's Cavern according to the measurements given by Pengelly in his Diary. It is this area which has yielded most of the archaeological and environmental evidence. The positions and names of the various passages and chambers are given as well as the lines along which the sections in Figures 5 and 6 are drawn. Both sections are based on the relevant descriptions and measurements recorded by Pengelly in his Diary from 1865 to 1868, and on my interpretation of the relevant extant artifacts. The positions of these artifacts are determined by Pengelly's find-numbers, and hence his measurements. The Southern Section is drawn along Pengelly's data lines and projected on one plane. The Vestibule Section is drawn on both data and parallel lines and again projected on one plane, and it has Ogilvie's finds of 1926 to 1940 added according to what few details of depth he wrote on some of his specimens. The vertical scale on both sections is exaggerated in relation to the horizontal in order to show more clearly the occurrence of Later Upper Palaeolithic artifacts above Earlier Upper Palaeolithic artifacts, and the latter above Middle Palaeolithic artifacts. The deposits and finds from the entrance platforms are not recorded in great detail by Pengelly as he considered them to have been already largely disturbed by earlier diggers, although he eventually found this to be true only of their upper portions. The Sloping Chamber and the North and South Sally Ports (areas 5, 10 and 11 on Figure 4) are recorded within his grid system but are also considered largely disturbed.

What one might consider the "typical" stratigraphic sequence at Kent's Cavern is as follows (from top to bottom):

F/D. *Black Mould*, humic matter and sandy silt with Mesolithic, Neolithic, Bronze Age, Romano-British and Medieval artifacts.

C2. *Granular Stalagmite*, tufaceous flowstone with Mesolithic and Neolithic artifacts.

B2. *Stony "Cave Earth"*, sharp limestone fragments in light red silty sand with Later Upper Palaeolithic artifacts, and, in Vestibule only, with a palimpsest of L.U.P. hearths known as the "Black Band".

A2. *Loamy "Cave Earth"*, light red silty sand with some sharp and some rounded limestone fragments, and with Earlier Upper Palaeolithic artifacts above Middle Palaeolithic (Mousterian) artifacts. B2 grades into A2, thus B/A2 on Figures 5 and 6 denotes both.

C1. *Crystalline Stalagmite*, very hard flowstone.

B1. *Breccia*, sharp and rounded limestone fragments and dark red sand cemented together with masses of bear bones and teeth, with Lower Palaeolithic (Clactonian and Acheulian) artifacts.

A1. *Red Sand*, dark red sand with no or few artifacts.

C0. *?Earlier Crystalline Stalagmite*, recorded by Pengelly as fragments of very hard flowstone in B1 and traces on walls, some of which are in fact clearly traces of C1 but some of which are not.

A0. *?Laminated Silts*, recorded by Ogilvie (Dowie, 1928) as laminated sands, silts and clays on bedrock below Pengelly's excavation in the Gallery.

BEDROCK: Devonian limestone

Most of the above details of the "typical" sequence are taken from Pengelly's Diary (1865–68, pp. 1–600), except Ogilvie's "Laminated Silts" as cited. The classification of the artifacts is mine, but their positions are after Pengelly. The letter code (i.e. F/D, C2 etc.) is also added by me for convenience. In his publications Pengelly does not sub-divide the "Cave Earth" (B/A2), but in his much more copious Diary he clearly does.

Due to natural erosion at various times, one is also confronted with an "atypical" sequence at Kent's Cavern. In the Gallery it is basically as follows:

C1. *Crystalline Stalagmite*, suspended intact above an airspace with red sand adhering to its underside.

AIRSPACE

F/D. *Black Mould*

C2. *Granular Stalagmite*

A2. *Loamy "Cave Earth"*, with numerous fragments of C1 and B1, and with Earlier Upper Palaeolithic artifacts above Middle Palaeolithic (Mousterian) artifacts.

A1. *Red Sand*

This "atypical" sequence is according to Pengelly's notes in his Diary (1865, pp. 192–236). This sort of "atypical" arrangement of the layers may have been repeated at various times in antiquity in various parts of Kent's Cavern. Fragments of layers C1 and B1 do typically occur in B/A2, but they are only particularly numerous in the Gallery, the South West Chamber, the Wolf's Cave and the Long Arcade (layer B/A2 of the latter two chambers yielded the much discussed scimitar teeth of *Homotherium latidens* which seem most likely to have been derived in this fashion from layer B1 or some even earlier layer). For an excellent example of an eroded and complex cave sequence, excavated and reconstructed much more recently, see Henry de Lumley *et al.* (1972) on Grotte de l'Hortus.

The horizontal distribution of Middle and Upper Palaeolithic artifacts found by Pengelly is shown in Figures 7, 8 and 9. The plan of Later Middle Palaeolithic (or "Handaxe Mousterian") artifacts (Fig. 7) is included for comparison with the Upper Palaeolithic and because some of the artifacts which I have attributed to the Middle Palaeolithic may be Earlier Upper Palaeolithic, with of course the exception of the handaxes; there is no way of casting definite light on this in the present state of the evidence. Both extant and missing stone artifacts from Pengelly's excavations are included under stone waste, and extant tools found in disturbed areas are not plotted individually. Most of the Middle Palaeolithic artifacts are confined to the Great Chamber (see Figs. 4 and 7), with only an incidental scatter into neighbouring passages. A few Middle Palaeolithic artifacts also occurred in an undisturbed context in the Vestibule, but as they were found by Ogilvie in 1926 to 1940, their exact positions are unknown (see Fig. 6).

The distribution of Earlier Upper Palaeolithic artifacts (Fig. 8) is similar to but rather more extensive than the Middle Palaeolithic. It may be noted that all of the extant undisturbed leaf-points occur in the southern half of Kent's Cavern (i.e. the Great Chamber, Gallery and South West Chamber). Further, the extant disturbed leaf-points (not shown individually) are all from the South Sally Port (see Figs. 4 and 8). Scrapers, on the other hand, are scattered throughout the undisturbed area, with minor clusters at the North and South entrances. Awls and saws are also scattered over a fairly wide area. It is perhaps tempting to suggest that if the leaf-points were used as projectiles and knives, then any Earlier Upper Palaeolithic final butchering at Kent's Cavern was possibly carried out mainly in the Great Chamber with some discharge into the Gallery and South West Chamber (initial butchering was probably done at kill-sites away from the cave). This suggestion would seem to be supported by the horizontal scatter of possibly Earlier Upper Palaeolithic faunal evidence, as the total undisturbed large mammal tooth count of 1,519 (Pengelly's Diary, pp. 1–600) for the relevant levels shows the following distribution: 56.16% (853 teeth) in the Great Chamber, 42.07% (639 teeth) in the Gallery, Lecture Hall and South West Chamber, and only 1.78% (27 teeth) in the Vestibule. Scraping, cutting and sewing, on the other hand, may have taken place in all areas, perhaps with primary work near the butchering in the Great Chamber and secondary, or final, work well away from the butchering in the Vestibule.

The bifacial leaf-point in the Gallery (Fig. 8) now has a closely associated radiocarbon age estimate of 28,720± 450 B.P. (GrN–6202). The determination is from a humerus of *Ursus arctos* from the same grid and spit position. A unifacial leaf-point, two nosed scrapers and a saw in the northern middle of the Great Chamber (Fig. 8) also now have a closely associated age estimate: 28,160± 435 B.P. (GrN–6201). This determination is from a tibia of *Coelodonta antiquitatis* from the same grid and spit locality. An end-of-flake scraper and a scraper/saw in the middle of the Great Chamber (Fig. 8) now have an associated age estimate as well: 27,730± 350 B.P. (GrN-6325). This measurement is from a radius of *Bison* sp. from the same spit and about the same grid locality. Finally, two unifacial leaf-points and a saw in the south-western middle of the Great Chamber (Fig. 8) now have an associated age estimate of $38{,}270^{+1470}_{-1240}$ B.P. (GrN–6324). This is from a radius of *Equus* cf. *przewalskii* from the same spit and almost the same grid position.

These bone samples are from the Pengelly Collection at the Torquay Natural History Society and are marked with his appropriate find-numbers. They were selected by the present author for radiocarbon dating on the basis of their original close proximity with diagnostic Earlier Upper Palaeolithic tool-forms. The four age estimates they have provided are the most reliable obtained thus far for any British Earlier Upper Palaeolithic assemblage. They are certainly far less problematic than those from Badger Hole, Paviland Cave, Cae Gwyn/Ffynnon Beuno Caves and Robin Hood's Cave discussed below (see pp. 79–80).

The distribution of Later Upper Palaeolithic artifacts at Kent's Cavern (Fig. 9) is mainly confined to the Vestibule with slight extensions out the North Entrance and also into the North East Gallery and the Sloping Chamber (see Fig. 4 as well). The single backed tool in the South West Chamber is a fragment of a large tanged blade which is very likely to be Earlier Upper Palaeolithic. There are clearly two clusters of the other undisturbed backed tools, one at the eastern end of the Black Band (a palimpsest of hearths of wood charcoal and burnt bones) in the Vestibule and one at the North Entrance, and on certain stratigraphic evidence (Black Band group from 1st foot spit whilst North Entrance group is from 4th foot spit) from Pengelly's Diary (1866–67, pp. 287–359), they may be separate in time as well as space, the North Entrance group probably being a bit earlier (see Figs. 6 and 9). All of the bone/antler tools are either in or very near the Black Band, and most are at the western end with the exception of one uniserial harpoon which is at the eastern end and associated with that backed tool cluster. Burins are scattered to the east of the Black Band, and scrapers and saws are either in or near it. If backed tools were used as projectiles and knives, then perhaps any Later Upper Palaeolithic final butchering at Kent's Cavern was carried out mainly near either the eastern end of the Black Band or the North Entrance. This suggestion might be supported by the horizontal scatter of associated faunal evidence, as most of the relevant large mammal teeth recorded in Pengelly's Diary (1866–67, pp. 287–359) occur either in or near the Black Band or at the North Entrance, although the total count is small: 37 teeth. Scraping, cutting and sewing may have occurred in or around the Black Band only, with an emphasis at one time or another on the western end. Preparation of bone/antler tools by burins may have occurred more to the east, perhaps relying on greater daylight near the North Entrance.

The uniserial harpoon and backed tool complex at the eastern end of the Black Band (Fig. 9) now has an associated radiocarbon age estimate of 14,275 ± 120 B.P. (GrN–6203). This determination is from a tibia of *Ursus arctos* from the same grid and spit locality. The biserial harpoon and an end scraper just to the north-west of the Black Band now have an associated age estimate of 12,180 ± 100 B.P. (GrN–6204). This is from a metatarsal of *Megaloceros giganteus* from the same grid and spit position. These bones are also from the Pengelly Collection at the Torquay Natural History Society Museum and are again marked with his appropriate find-numbers. The two age estimates they have provided are among the few reliable ones obtained thus far for the British Later Upper Palaeolithic. It is now apparent that the eastern end of the Black Band at Kent's Cavern may be partly older than the western end, as one might suspect from the typological differences of the two harpoons concerned.

A summary analysis of the whole of the Palaeolithic and Pleistocene sequence at Kent's Cavern may be found in Campbell and Sampson (1971). Rogers' (1955) somewhat dubious views on the stratigraphy of the "Cave Earth" (layers B/A2) are now certainly open to serious doubt. He concluded that the "Cave Earth" had been largely or entirely disturbed before Pengelly's excavation simply because he thought the artifacts and faunas within each spit ought to be the same throughout Kent's Cavern, a most naive interpretation, to say the least. Despite the slight discrepancies pointed out by Rogers between the numbers of artifacts listed in Pengelly's publications and the numbers listed in his Diary, I think Pengelly is on the whole remarkably reliable and accurate for his day. In fact, it is really rather ironic (though also a fine tribute) that the best British Upper Palaeolithic radiocarbon dates should come from specimens excavated by Pengelly over a century ago!

Soldier's Hole (Somerset). Excavations of 1928–29.

Soldier's Hole is situated in the south-eastern face of Cheddar Gorge in the Mendip Hills. It consists of a single broad entrance and chamber with a small continuation to the north-east, and it has been formed by solution in a bedrock of Carboniferous limestone. All of the deposits which have filled it seem to be either Late Pleistocene or more recent. The entrance lies at about 46 metres above the immediate floor of Cheddar Gorge and at about 100 metres O.D. It is 5.8 metres across and the chamber inside immediately expands to 8.5 metres in breadth and

8.2 metres in length. The entrance faces north-west and the view from it is limited by the walls of the Gorge.

Systematic excavations at Soldier's Hole were carried out by R.F. Parry from 1928 to 1929 (Parry, 1931). He did not keep any horizontal control, but he excavated in 6 inch (c. 15 cm.) spits (his "working layers") which he adapted to the four distinct natural layers that he recognized. His published results include figures of most of the Upper Palaeolithic artifacts as well as detailed descriptions of their layers and associated fauna. He considered the artifacts from spits 4 to 9 to be of "La Madeleine type", and those from spits 10 to 17 to be of "Solutré type" (Parry, 1931, p. 52 and plate 12). Luckily, he recorded his spit numbers on the artifacts and bones themselves. However, I have not been able to obtain any radiocarbon age estimates, as the British Museum Research Laboratory, to which I submitted some of the bones, reports that the polyvinyl acetate which someone has added to them would make any determination unreliable (R. Burleigh, personal communication).

The stratigraphic sequence at Soldier's Hole, based on the findings of Parry, is as follows:

F/D. *Friable limey "cave earth" and leaf mould* with stalagmitic bosses, with Neolithic, Bronze Age and Romano-British artifacts, equals spit 1, thickness c. 15–75 cm.

C. *Buff-red "cave earth"* with many limestone fragments, equals spits 2–3, thickness c. 39–60 cm.

B. *Grey-red "marl"* with a considerable amount of limestone scree and calcitic fragments, with Later Upper Palaeolithic artifacts, equals spits 4–9, thickness c. 90 cm.

A2. *Dark red "marl"* of a much more clayey nature with many slightly rounded limestone fragments, with Earlier Upper Palaeolithic artifacts, equals spits 10–17, thickness c. 120 cm.

A1. *Ditto from fissure* in bedrock floor, with no artifacts, equals spits 18–21, thickness c. 60 cm.

BEDROCK: Carboniferous limestone.

The above descriptions of the layers at Soldier's Hole are as given by Parry (1931), but I have replaced his layer numbers by my letter code, changed his measurements to their metric equivalents, and used my terminology for his Upper Palaeolithic finds. The artifacts themselves are discussed in Chapter IV and V (see Figs. 91, 92 and 127). I have also drawn a hypothetical version of the section which Parry recorded for Soldier's Hole; this hypothetical section is based on my visit to the site and careful reading of Parry's observations. A copy of my hypothetical section of Soldier's Hole is on display at the new Cheddar Caves Museum, which I, in collaboration with Mr. L.V. Grinsell, helped reorganize during the course of this research programme.

King Arthur's Cave (Herefordshire). Excavations of 1871, 1925–27 and 1955.

King Arthur's Cave is situated at the foot of a low cliff at the north-western corner of Lord's Wood on the hill of Great Doward at Whitchurch near the River Wye. It consists of a broad entrance platform, a double interconnected entrance and two main chambers, and it has been formed by solution in a bedrock of Carboniferous limestone. The entrance platform lies at about 90 metres above the Wye and 122 metres O.D. It faces north-west and commands a good view of the broad saddle-back of Great Doward Hill. All of the deposits which have filled this cave seem to be either Late Pleistocene or more recent.

King Arthur's Cave was partly excavated by the Revd. W.S. Symonds in 1871, after some miners had removed some of the deposits the year before (Symonds, 1871). He found that considerable portions had been disturbed, but that in what remained of the undisturbed, there were two "cave earths", an upper and a lower separated by a thick stalagmite layer. Garrod (1926) has assigned his finds from the upper cave earth to the Upper Palaeolithic; those few from the lower cave earth might be Middle Palaeolithic.

From 1925 through 1927 members of the University of Bristol Spelaeological Society excavated the entrance platform and the passage connecting the two entrances (Hewer, 1926; Taylor, 1928). They found Earlier Upper Palaeolithic (or "Proto-Solutrean") artifacts and fauna in the upper cave earth of the passage, and three Later Upper Palaeolithic (or "Developed Aurignacian") assemblages followed by a Mesolithic assemblage in the platform deposits. Their diagrammatic section (Taylor, 1928, Fig.1) shows how the intervening deposits removed by Symonds in 1871 may have been related. They correlate the upper "cave earth" of their and Symonds' excavations with their exterior basal red-yellow clayey silt on the basis of a very similar fauna and matrix. If their correlation is correct, then the interior Earlier Upper Palaeolithic assemblage would be stratigraphically lower than the exterior Later Upper Palaeolithic assemblages.

In 1955 the same Society resumed work at King Arthur's Cave for a season (E.K. Tratman, personal communication). They found little new and so did not publish this field-work. However, A.M. ApSimon managed to produce a more accurate plan and sections, and these are now on display in the Society's Museum at Bristol. With the addition of my letter code and using my terminology for the archaeological occurrences, ApSimon's interpretation of the sequence is basically as follows:

- E. *Old Tip* (19th century, cave exterior and interior).
- F/D. *Humus*, with Neolithic, Bronze Age and Romano-British artifacts (exterior).
- C. *Weathered Limestone Scree* with hearths, Mesolithic artifacts (exterior).
- B. *Yellow Sharp Limestone Scree* with hearth at base, with at least two Later Upper Palaeolithic artifact assemblages (exterior).
- A3c. *Upper "Cave Earth"*, a reddish deposit with some sharp scree and a hearth at its base, with Earlier Upper Palaeolithic artifacts (interior).
- A3b. *Upper Red Loam*, a red-yellow clayey silt, with ?Earlier Upper Palaeolithic artifacts (exterior).
- A3a. *Upper Red Sand*, with no artifacts (interior).
- A2. *Stalagmite*, with no artifacts (interior).
- A1c. *Lower "Cave Earth"*, a reddish deposit with thin lenses of stalagmite, with ?Middle Palaeolithic artifacts (interior).
- A1b. *Lower Red Silt*, with no artifacts (interior).

A1a. *Lower Yellow Silt*, with no artifacts (interior).
BEDROCK: Carboniferous limestone.

The interface between layers C and B may have yielded yet another artifact assemblage, one perhaps transitional from Later Upper Palaeolithic to Mesolithic (E.K. Tratman, personal communication). Layer A3b may belong above rather than below A3c, but both belong below B and above A3a, A2 etc. The Earlier Upper Palaeolithic hearth in layer A3c was actually on a bedrock ledge in the passage. After visiting the site with a tracing of ApSimon's plan and sections, I have come to the conclusion that his account is probably the most accurate available for the King Arthur's Cave sequence. Although many of the Spelaeological Society's finds from before the Second World War were destroyed by bombing, those which survive, together with the published illustrations (Hewer, 1926; Taylor, 1928), are probably sufficient to form a rough picture of the original assemblages (see Chapters IV and V, pp. 146, 167–68).

Cae Gwyn Cave (Flintshire). Excavations of 1884–87.

Cae Gwyn Cave lies in the northern slope of a small valley just south of the village of Tremeirchion on the eastern side of the Vale of Clwyd. It consists of two entrances connected by a single narrow passage, and it has been formed by solution in a bedrock of Carboniferous limestone. One entrance faces south at about 19 metres above the floor of the small valley (Ffynnon Beuno Valley) and 120 metres O.D. The other entrance faces west and was completely buried before its excavation; it lies slightly higher than the southern entrance and about 40 metres from it through the cave. All of the deposits which have filled or covered Cae Gwyn Cave seem to be either Late Pleistocene or more recent.

Cae Gwyn Cave was excavated by H. Hicks and E.B. Luxmoore from 1884 to 1887 (Hicks, 1886 and 1888). The excavation was begun at the southern entrance before the western entrance was discovered. The sequence found at the southern entrance and inside the cave was as follows (Hicks, 1886):

4. *Recent Reddish Loam*, thickness c. 60 cm. at southern entrance.
3. *Laminated Clay*, with thin ferruginous and stalagmitic lenses in places, thickness c. 20 cm. at southern entrance.
2. *Reddish Sandy Clay*, with pebbles of felsite, granite, gneiss, quartz, quartzite, sandstone and limestone, with an Earlier Upper Palaeolithic-like flint end scraper, thickness c. 60 cm. at southern entrance (? partly disturbed by glacial meltwater, deposit may be partly outwash considering the sorts of pebbles).
1. *Sterile Gravel*, composed mainly of local materials, thickness c. 30 cm.

BEDROCK: Carboniferous limestone.

The above sequence was similar throughout the cave, although the deposits of course varied in thickness. At the supposed inner end of the cave, sands and gravels were found lying on the laminated clay (layer 3), and during the winter of 1886 a depression appeared in the field above that part of the cave. A sounding was excavated into what was obviously boulder clay or till, and it was found that this covered a buried limestone cliff with a second entrance to the cave. This entrance, facing west, proved to be more than 3.5 metres wide and as much as 2.4 metres high. A clear section was recorded, and the various layers were traced continuously across, both into this western entrance and over it. The sequence found here was as follows (Hicks, 1888; Garrod, 1926):

11. *Surface Soil*, thickness c. 15 cm.
10. *Brown Boulder Clay*, thickness c. 85 cm.
9. *Yellow Loamy Clay*, thickness c. 18 cm.
8. *Stiff Reddish Boulder Clay*, thickness c. 70 cm.
7. *Sand*, c. 5 cm.
6. *Purple Clay*, c. 25 cm.
5. *Sand with Boulders*, c. 50 cm.
4. *Gravelly Sand*, with boulders and lenses of purple clay, c. 65 cm.
3. *Sandy Gravel*, c. 60 cm.
2. *Finely Lensed Sand*, c. 40 cm.
1. *Red Laminated Clay and "Bone-earth"*, with angular fragments of limestone and a few boulders, c. 80 cm.

BEDROCK: Carboniferous limestone.

An unretouched flint blade, 51 mm. long, was found in the "bone-earth" (layer 1) just inside the western entrance and below the glacial deposits (Hicks, 1888). This blade is probably Earlier Upper Palaeolithic on the basis of its stratigraphic position. Its discovery clearly indicated that the cave had been used by man before being closed by glaciation. Layer 1 at the western entrance equals layers 2 and 3 at the southern entrance, the two being traced together as mentioned above (p. 45). Therefore the end scraper from Cae Gwyn is also older than the boulder clays. Some considerable opposition arose at the time against these discoveries, and this, as well as its conclusive refutation, has been summarized by Garrod (1926). Garrod has also reproduced the western entrance section (Garrod, 1926, Fig. 26).

The boulder clays at Cae Gwyn Cave are probably mostly of northern origin, although they do include some Welsh (Snowdonian) erratics. These tills are most likely due to the maximum ice advance of the Full Last Glacial (Full Devensian). Rowlands (1971) has summarized this evidence and has even suggested that not only Cae Gwyn Cave, but the neighbouring site of Ffynnon Beuno Cave, was originally sealed by the tills. Ffynnon Beuno Cave, excavated by Hicks and Luxmoore in 1885 (Hicks, 1886), is significant for its diagnostic Earlier Upper Palaeolithic unifacial leaf-point and typical "burin busqué", and it certainly seems reasonable to think that these tools are equally older than the last maximum ice advance. The entrance at Ffynnon Beuno Cave faces south and is slightly lower than Cae Gwyn at about 13 metres above the little valley and 116 metres O.D.

A carpal bone of *Mammuthus*? from the Hicks Collection from Cae Gwyn Cave has now been radiocarbon dated to $18{,}000^{+1400}_{-1200}$ B.P. (Birm-146, Rowlands, 1971). This age estimate may at least be tentatively accepted as an upper limit for access to the caves prior to the maximum ice advance, but one cannot be certain how closely contemporary it may be with the Earlier Upper Palaeolithic evidence, as *Crocuta*, or "Hyaena spelaea", is abundant in the faunal list (Hicks, 1886).

Pin Hole (Derbyshire). Excavations of 1874, 1924–38 and 1974.

Pin Hole is located in the northern face of the western end of the gorge known as Creswell Crags just to the east of the village of Creswell. It has a single entrance which faces south and leads to a long narrow inner passage, and it has been formed by solution in a bedrock of Permian limestone. Its entrance lies at about 4 metres above the floor of the Crags and at about 76 metres O.D. All of the deposits which have filled Pin Hole appear to be Late Pleistocene and more recent.

Pin Hole was first superficially excavated by J.M. Mello in 1874 (Mello, 1875). He dug at the entrance and found that that area had already been partly disturbed. The section he recorded was as follows:

3. *Surface Soil*, with recent artifacts, thickness c. 45 cm.
2. *Red Sand*, with limestone boulders, rolled quartzite pebbles and very abundant bones, thickness c. 90 cm.
1. *Lighter-coloured Sand*, consolidated by lime, no bones.

Pin Hole was then almost completely dug out by A.L. Armstrong from 1924 to 1938 (Armstrong, 1925, 1928 and 1931a; Kitching, 1963). Armstrong laid out a rough grid system of 8 foot "sections", beginning with Section A at 23 feet, or 7 metres, from the entrance barricade and ending with Section G at about 80 feet, or about 24 metres, from the same barricade. He recorded on most of his finds not only the appropriate section, but the actual distance in feet from the barricade. For vertical control, he excavated in spits of 6 to 12 inches (about 15 to 30 cm.) in thickness. This information he also recorded on most of his finds as a depth in feet and inches from a roughly horizontal level at the top of the deposits. The "typical" section of the stratigraphy published by him (Armstrong, 1931a) is as follows:

3. *Stalagmite or Brecciated Red Sand*, thickness c. 15 cm.
2. *Red Upper "Cave Earth"*, with "Developed Aurignacian" (my Later Upper Palaeolithic) artifacts overlying "Upper Aurignacian, Proto-Solutrean and Mousterian 3" (my Earlier Upper Palaeolithic) artifacts, thickness c. 200 cm.
1. *Yellow Lower "Cave Earth"*, with two thick lenses of limestone boulders and ?scree ("Slab Layers 1 and 2") near the top, with Middle Palaeolithic artifacts (his "Mousterian 1 and 2"), thickness c. 300 cm.

BEDROCK: Permian limestone.

I have added the number code to the above sequence and changed the measurements given by Armstrong to their metric equivalents. I have also brought his descriptions of the artifact assemblages into line with my suggested terminology, following my study of the original finds now in the Manchester University Museum. It is worth noting that Armstrong is quite definite and consistent in his reports of Later Upper Palaeolithic evidence being stratified above the Earlier Upper Palaeolithic at Pin Hole. However, I think his so-called "Mousterian 3" at the base of the Upper "Cave Earth" belongs to the Earlier Upper Palaeolithic rather than the Middle Palaeolithic (see Chapter IV.G. for further discussion). Also the Later Upper Palaeolithic finds may belong to at least two separate assemblages (see Chapter V.I.). If all of his artifacts which have grid and depth data are plotted on a diagrammatic section, one finds that there are a number of rather curious typological overlaps. For instance, two carefully decorated bone tools, which one would expect to be Later Upper Palaeolithic, seem to occur in association with Earlier Upper Palaeolithic side scrapers and other artifacts (see Chapter IV.G. and Fig. 102). Armstrong (1925) figures one of these bone tools with a Later Upper Palaeolithic-like shouldered point and says that they were found together at the same depth, as one might well expect, but his grid and depth records on the specimens themselves definitely do not correspond with this statement (see Chapter V.I.). Kitching (1963) also notes various discrepancies between Armstrong's records of the stratigraphy and his faunal finds. In fact, Kitching is strongly of the opinion that the average thickness of the two "Cave Earths" is less than that published by Armstrong, the Upper "Cave Earth" being perhaps about 120 cm. and the Lower "Cave Earth" about 200 cm. He also thinks the so-called "Slab Layers" are imaginary, and the Middle Palaeolithic evidence represents one assemblage, rather than two or three. As regards the "Slab Layers", Dr. J. Wilfred Jackson actually saw the field sections and says that boulders occurred at random, the only layer differences apparent to him being changes in colour, fauna and artifacts (Jackson, personal communication). But, of course, aside from all of the above and related problems regarding the Pin Hole stratigraphy, the basic archaeological sequence from Pin Hole would appear to remain the same: Later Upper Palaeolithic overlying Earlier Upper Palaeolithic which in turn overlies Middle Palaeolithic.

In 1974 S. Collcutt (personal communications) surveyed all of the caves at Creswell Crags and conducted small excavations at three of them: Pin Hole, a near-by slit-fissure (midway between Pin Hole and Robin Hood's Cave but higher up, see Map 14) and Mother Grundy's Parlour. His work at Pin Hole resulted in the collection of new faunal samples from a surviving section at the rear of the cave, although lacking artifacts it is difficult to correlate his results with those of Armstrong. But he has written up a comprehensive report (at the moment it is an Edinburgh M.A. thesis, but hopefully it will be published) on all of the stratigraphic problems at the Crags, a report which should prove quite useful, particularly with his new data on fish and small mammals at Pin Hole.

British Upper Palaeolithic Stratigraphy: Field-work in 1968–69

Following the above résumé of the stratigraphic evidence available for the general Upper Palaeolithic sequence in Britain, reports are now given on the sites selected by the author for excavation as part of this research project, in 1968–69, incorporating his original field observations together with a summary of any earlier work in each case. The sites are again discussed in a clock-wise geographic order from the south-west, beginning with Badger Hole in Somerset.

Badger Hole (Somerset). Excavations of 1938–53, 1958 and 1968.

Badger Hole is situated in the eastern face of the Wookey Hole Ravine not very far from the source of the River Axe at the cave of Wookey Hole. Map 10 shows its location in relation to present topography, water resources and the adjacent Upper Palaeolithic site known as the Hyena Den. The smaller map of southern Britain in the lower left-hand corner shows its position in relation to the approximate southern boundary of the Last Glacial (Devensian) maximum ice front, the latter being indicated by a dotted line.

Badger Hole at present has a single, very broad entrance which faces west-south-west and overlooks the Wookey Hole Ravine. This entrance lies at about 10 metres above the floor of the Ravine and at about 70 metres O.D. It leads through a short narrow passage to an inner chamber with various side passages and another opening which may possibly be the top of a concealed eastern entrance which has been blocked by a large, as yet unexcavated, talus cone. The cave has been formed by solution in a bedrock of Triassic dolomitic conglomerate, and it has been filled at various times by deposits which appear to be either Late Pleistocene or more recent.

The first known excavations at Badger Hole were conducted by H.E. Balch intermittently from 1938 to 1953. He recorded his work in a Diary (Balch, 1938–53) which is now kept at the Wells Museum, but he never properly published his results. Figure 10 shows a plan of the entrance platform and is based partly on a plan in Balch's Diary. As shown on this plan, Balch excavated by yard, or 0.84 metre, squares; for vertical control he employed 1 foot, or 30 centimetre, spits. Although most of the entrance was found to have been already greatly disturbed, he carefully recorded the appropriate yard and spit references on nearly all of his finds, and on this basis I have been able to show the horizontal distribution of definite and possible Earlier Upper Palaeolithic finds. Two of the four diagnostic unifacial leaf-points found by Balch are from definitely undisturbed contexts in the 5th foot spit of the southern portion of the entrance, more or less on bedrock, to which he generally excavated. Balch also excavated inside Badger Hole, leaving some rather curious tunnels, or what one might well term "burrows", through the deposits of the inner chamber and passages.

The next excavations at Badger Hole were carried out by C.B.M. McBurney in 1958 (McBurney, 1961). He dug two soundings, a northern square and a southern trench, as shown in Figure 10. Although his northern sounding revealed only disturbance, his southern sounding indicated the presence of some undisturbed Late Pleistocene deposits, as may be clearly seen in his published section drawing (McBurney, 1961, Fig.2). However, the only Upper Palaeolithic-like artifact he found, a large fragment of a struck flint blade, came from high up in a context that was almost certainly disturbed in grid square Bb (see Fig. 10).

Excavations by the present writer at Badger Hole were undertaken in 1968 with the hope of obtaining undisturbed Earlier Upper Palaeolithic artifacts in association with organic material for radiocarbon dating. The area chosen for sounding had been previously unexcavated, as shown on Figure 10, according to Balch's plan and McBurney's section. A 1 metre by 4 metre north-south trench was layed out parallel to and 50 centimetres from the ravine-side retainer wall, approximately along Balch's north–south grid line A/B from just in b to the end of f. Initially, a baulk was kept between each metre square, with the middle baulk removed first. The squares were labelled Bc in the south to Bf in the north for convenience, although by Balch's grid they were in fact "A/Bb/c", etc. The entire trench was excavated to bedrock and all faces were recorded, including those of the baulks. Most of the trench was found to have been already greatly disturbed (both by badgers and man, the latter presumably before Balch's work), even in places to bedrock (particularly in square Bf), and therefore only the most important sections have been reproduced in Figure 11. "Be NORTH" and "Bd SOUTH" are faces of the northern and southern-most baulks and more of the "EAST" is shown than of the "WEST".

The rather complicated stratigraphic sequence found at Badger Hole by the author is described under the layer key in Figure 11. In a preliminary report (Campbell, 1970) layers A2, B/A3 and E/C were simply coded I, II and III, and much briefer descriptions of them were given. Their coding for this present report has been brought in line with a general layer coding system begun by McBurney (1959) as a result of field-work at a number of British Upper Palaeolithic cave-sites, and found by me to be practical and useful to adopt. Provided the evidence for it is clear, this system in its simplest form generally runs from A on bedrock to F at the surface. A particularly sandy, silty and/or clayey undisturbed layer in a sequence is normally labelled 'A', whilst an undisturbed thermoclastic (or sharp freeze/thaw-formed) scree is labelled 'B'. A B-type layer would normally occur above an A-type, but they may be interstratified or blended together at some sites. A B-type layer never appears really weathered, but it may have a sandy matrix or be quite open (i.e. loose with air spaces). Overlying a B layer one often finds a partly weathered, sandy scree which is labelled 'C'. A C-type layer is then often overlain by a much more weathered scree with humic infiltration (a D layer), followed possibly by tip or other evidence of disturbance (an E layer) and almost certainly topped by modern or recent humus or topsoil (layer F) with living plants and animals. An E-type layer, i.e. tip, may sometimes overlie F or may, on the other hand, even penetrate its underlying layers to bedrock. When one is excavating a site, one normally sees the top layer first. If this top-most layer is humic and alive, it is labelled F, and if nothing else except bedrock is found below, the sequence would simply be F on bedrock. However, if other layers of a different nature as described above are also found, they are labelled accordingly in the sequence.

Although this layer coding system is certainly not foolproof, it seems far more applicable to British cave-site stratigraphy than the A to C horizon (from top to bottom) nomenclature of soil scientists in general (e.g. Butzer, 1964, pp. 76–77), which has no allowance for independent scree layers as such. Badger Hole itself is not a particularly good example of how well the A–F layer coding system can work; McBurney's sections from Bridged Pot in the Ebbor Rocks (McBurney, 1959, plate 18) offer much better examples, as does my section from Mother Grundy's Parlour at Creswell Crags (Campbell, 1969, Fig. 6; also Fig. 24 in this book).

Badger Hole has been thoroughly disturbed by badgers, as its name implies. In fact, open badger holes, or runs, occur in the sections shown in Figure 11, and layer E/D is generally quite loose with humic and similar débris. Layer E/C does not have the appearance of tip but does seem slightly disturbed and redeposited in places, possibly more by roots and worms than badgers, although it is clearly truncated by human and badger disturbance. Layer B/A3 appears generally less disturbed as does the underlying layer A2, although both are pierced and truncated in places by badger holes and disturbances. The upper portion of B/A3 has the most scree and it all appears to be thermoclastic in origin, although one cannot be absolutely certain as the parent material is a dolomitic conglomerate rather than a pure limestone. Layer A2 is mostly sandy and quite firmly packed but has some thermoclastic and weathered scree near its base on bedrock. The middle of layer A2 has a cluster of burnt bone fragments against the east face in square Bd, and against this I have plotted the stratigraphic range of the undisturbed unifacial leaf-point found by Balch in his adjacent square Cd (see Figs. 10 and 11). Part of this burnt bone cluster has yielded an infinite radiocarbon age estimate of >18,000 years B.P. (BM–497), and it seems quite reasonable that this "dating" might be interpreted as an upper limit for the Earlier Upper Palaeolithic occupation at Badger Hole. Inside the cave there is yet another layer on bedrock which is apparently a quite sterile yellow-red sand (Shackley, 1972 and personal communications), and this I am tentatively labelling layer A1. Sadly enough, the only possibly Upper Palaeolithic artifact found in 1968 was a derived, unretouched flint blade (for its location see Figs. 10 and 11); however, on the basis of the evidence now available it seems fair to assume that the level of the Earlier Upper Palaeolithic occupation was originally in layer A2. For the results of granulometric and pollen analyses, as well as fauna identified, see Chapter III.B.1, 2 and 3.

Sun Hole (Somerset). Excavations of 1927–28, 1951–53 and 1968.

Sun Hole is located in a northern face of Cheddar Gorge not far above the village of Cheddar and the source of the River Yeo. Map 11 shows its position in relation to present topography, water resources and neighbouring Later Upper Palaeolithic sites. The smaller map of southern Britain in the lower left-hand corner shows its situation in relation to the approximate southern boundary of the Last Glacial (Devensian) Scottish Readvance ice front, the latter being indicated by a dotted line as the most southerly possible extent of ice during the Later Upper Palaeolithic.

Sun Hole at present has a single entrance which faces south and overlooks the mouth of Cheddar Gorge and the Somerset Levels in the distance. This entrance lies at about 46 metres above the immediate floor of the Gorge and at about 100 metres O.D. It opens directly into a 4 metre wide passage which runs almost due north into the cliff and which is blocked at its inner end by a boulder-choke. The cave or fissure has been formed by solution in a bedrock of Carboniferous limestone, and it has been filled at various times by deposits which appear to be either Middle Pleistocene or more recent.

The first known excavations at Sun Hole were carried out by E.K. Tratman and the University of Bristol Spelaeological Society in 1927 to 1928 and resumed in 1951 to 1953 (Tratman and Henderson, 1928; Tratman, 1955). They excavated by a grid system of yard squares and 1 foot (c. 30 cm.) levels, or spits, and they recorded this data on nearly every find, although most of their finds made in 1927–28 were later destroyed by bombing during the Second World War. Figure 12 shows the entrance plan at Sun Hole with all of the area dug by Tratman marked in fine hatching. All of their extant Later Upper Palaeolithic finds with their appropriate measurements have also been plotted on this plan. These "Creswellian" artifacts appear to be clustered just inside the entrance close to the western cave wall with a thinner scatter over a much wider area. Although Tratman (1955) published a very much reduced plan and sections, as well as figuring a selection of the "Creswellian" artifacts, he did not illustrate either the horizontal or the vertical scatter of the artifacts, both of which would have been useful.

Excavations by the present writer at Sun Hole were conducted in 1968 with the hope of finding undisturbed Later Upper Palaeolithic artifacts in association with organic material for radiocarbon dating, as well as with the intention of making detailed section drawings of the faces of the deep sounding (area C to F and 3 to 5 on Fig. 12) which had been left partly open by Tratman. The latter objective was accomplished by removing what rubble had accumulated and cleaning back the faces of the deep sounding. In the search for an extant portion of the original "Creswellian" layers, a 1.5 by 3.5 metres north–south trench was then excavated in the entrance platform. The trench revealed almost entirely tip from the earlier excavations, except for a very small area of undisturbed thermoclastic scree and boulders in squares A and B4. This undisturbed area was at about the same depth as the original occurrences of Later Upper Palaeolithic artifacts and not very far from some of them, but the only finds were an unstruck flint pebble and some faunal remains. One bone, a complete humerus of *Ursus arctos*, has since yielded a radiocarbon age estimate of 12,378±150 years B.P. (BM–524).

Figures 13 and 14 show the east and north faces of the deep sounding, sections 2/3 and F/G, respectively. Part of the section in the entrance trench is also included along section 2/3, and the position of my radiocarbon sample, together with the approximate stratigraphic ranges of Tratman's finds, are projected on.

The positions of a column of 30 deposit samples are also indicated on section F/G (Fig. 14), and the results of their analyses are discussed in Chapter III.B.1 and 2. A layer key with descriptions of the various layers is as follows:

F/D. *Approximate original top of deposits*, humus and weathered stones (Tratman, 1955).
D. *Brown Sand*, with weathered stones, Beaker/Neolithic artifacts and hearth (Tratman, 1955).
E. *Twentieth Century Tip*, mostly scree and boulders.
C. *Sandy Thermoclastic and Weathered Scree*, buff-red with thermoclastic and weathered boulders, partly brecciated.
C/B. *Sandy Mostly Thermoclastic Scree*, buff.
B2-9. *Sandy Thermoclastic Scree*, buff-red with thermoclastic boulders, upper-most range of L.U.P. artifacts.
B2-8. *Very Open Thermoclastic Scree*, buff-grey, in L.U.P. artifact range.
B2-7. *Silty Sandy Thermoclastic Scree*, buff-grey with thermoclastic boulders, approximate centre of L.U.P. artifact range; radiocarbon sample from this layer.
B2-6. *Silty Sandy Thermoclastic Scree*, buff-red, approximate centre of L.U.P. artifact range.
B2-5. *Open Thermoclastic Scree*, buff-grey with thermoclastic boulders, in L.U.P. artifact range.
B2-4. *Silty Sandy Thermoclastic Scree*, grey-red with thermoclastic boulders, lower-most range of L.U.P. artifacts.
B2-3. *Slightly Open Thermoclastic Scree*, buff-grey.
B2-2. *Silty Sandy Thermoclastic Scree*, buff-grey.
B2-1. *Open Thermoclastic Scree*, buff-grey.
A2. *Silty Sand and Thermoclastic Scree*, light red with older stalagmitic fragments.
B1-3. *Silty Sandy Thermoclastic Scree*, grey with older breccia fragments.
B1-2. *Very Open Thermoclastic Scree*, grey with thermoclastic boulders and older breccia fragments.
B1-1. *Silty Sandy Thermoclastic Scree*, grey with thermoclastic boulders and older breccia, calcitic and stalagmitic fragments.
A1. *Silty Sand and Thermoclastic and Weathered Scree*, red with weathered boulders and older breccia, calcitic and stalagmitic fragments.
A0. *Sandy Calcitic Breccia*, grey-red with thermoclastic and weathered scree in a silty sandy matrix (layer survives as patches adhering to cave walls; hard stalagmitic shelves on cave walls well above the present top of deposits may have once covered this breccia and have almost certainly contributed the stalagmitic fragments found in layers A1, B1 and A2).
?BEDROCK: Carboniferous limestone (bedrock is certainly of this material, but its position vertically in the sequence is still uncertain).

As may be seen in Figures 13 and 14, the various layers of thermoclastic scree at Sun Hole are very much contorted and presumably soliflucted. The general slope of the deposits appears to be from north to south, down and out of the mouth of the cave rather than into it. Tratman (1955) has suggested that this sloping out of, rather than into, the present entrance is a result of the accumulation of deposits which have entered from some unknown entrance higher up at the back of Sun Hole, deposits which have then slipped out the present entrance and fallen into the Gorge below. He notes a slight trickle of silty water from the back of the cave in support of this hypothesis. However, it seems to me just as likely that only a portion of the finer component of the deposits may have trickled, or sludged, in from cracks, and not necessarily an actual entrance, at the rear of the cave, whilst the other portion of the finer component may have been wind-transported into the cave through the present entrance. As was observed in the field and quantitatively confirmed in the laboratory (see the author's granulometric analysis, Fig. 63 and p. 90), virtually all of the layers from A1 to C consist of a very substantial proportion of scree, as well as boulders (see Figs. 13 and 14). These screes and boulders which are generally quite sharp or angular, are probably most likely due to exfoliation of the bedrock walls and ceiling of Sun Hole, rather than introduction from the surface via some unknown entrance. As the ceiling of Sun Hole slopes down towards the Gorge, so might also the unexposed bedrock floor. Given a general inclination to the Gorge, any deposition of scree and boulders inside Sun Hole might have a tendency to roll out of the cave, a process which would be aided by solifluction, as well as possible permafrost sludging, under the periglacial conditions surrounding the site during the Last Glacial. The weathered scree and boulders of layers A0, A1 and C might possibly be due to chemical erosion *in situ* during the milder conditions of one or more earlier interglacials in the cases of layers A0 and A1, and the present "interglacial" or Post Glacial in the case of layer C. Layer A0 presents another example of the partial, or nearly complete, removal of earlier deposits from a cave, presumably by washing or sludging out of the cave mouth (this process presumably caused the "atypical" section at Kent's Cavern see pp. 39–40; as well as that at Grotte de l'Hortus in France, see Lumley, 1972; and that at Elder Bush Cave in Staffordshire, see Bramwell, 1964). Sun Hole thus provides a complex and interesting series of geomorphological and stratigraphic events, from perhaps as far back as sometime in the Middle Pleistocene to more or less the present.

Below layer B2-4 all of the preceding known layers appear to be archaeologically sterile, but from B2-4 up to B2-9 is the approximate stratigraphic range of the "Creswellian" flint artifacts found by Tratman (1955; his 4th, 5th and 6th foot levels below the top of the deposits). As shown by the projection of these artifacts on to my sections (Figs. 13 and 14), their vertical scatter is possibly wide enough to be the result of more than one Later Upper Palaeolithic occupation. However, the fairly dense concentration of artifacts in grid area C4-C5 (see Figs. 12 and 13) is morely likely to be the result of one occupation. The exact relationship between the radiocarbon sample obtained by me and the artifacts found by Tratman (1955, Fig. 10; and Fig. 128 in this book) is not certain, but as may be seen in Figure 13 they occur at about the same level. Furthermore, the flint pebble which I found in association with the radiocarbon sample was almost certainly brought in by man, as none occur naturally in the local bedrock. I am therefore tentatively accepting the radiocarbon date obtained as an age estimate for at least part of the Later Upper Palaeolithic activity at Sun Hole. There is, of course, the possibility that the bear represented by the bone in question died naturally in the cave, although its position at the entrance is not in a likely spot for hibernation. Also both this humerus and the broken bear teeth found near it would seem to be those of a fairly young, healthy individual. It seems on balance more likely to the author therefore that the occurrence of this and other bones and teeth found with it is the result of man's activity, and they were presumably brought to Sun Hole for meat and fur.

Layers C/B and C appear to be archaeologically sterile, but layer D, which occurred primarily just outside the entrance in the form of a slight talus cone and platform and was removed entirely by Tratman (1955, Fig. 8), included Beaker/ Neolithic artifacts and a hearth.

Cathole (Glamorganshire). Excavations of 1860s, 1958–59 and 1968.

Cathole is located in a rock exposure on the north-eastern slope of the presently dry portion of the little valley in Park Woods to the north-west of the village of Parkmill in the Gower Peninsula. It is not very far from the present source of the stream known as Penard Pill, which runs through Parkmill to the Bristol Channel. Map 12 shows the position of Cathole in relation to present topography and water resources. The smaller map of southern Britain in the lower left-hand corner shows its situation in relation to the approximate southern boundary of the Last Glacial (Devensian) Scottish Readvance ice front, the latter being indicated by a dotted line as the most southerly possible extent of ice during the Later Upper Palaeolithic (on subsequent maps this small British map is repeated with either the Last Glacial maximum ice front in relation to Earlier Upper Palaeolithic sites, or the Scottish Readvance front in relation to Later Upper Palaeolithic sites, accordingly; it is not meant to imply a definite knowledge of ice and site contemporaneity in either case).

Cathole at present has two entrances, a larger one which faces west-south-west and a smaller one which faces south-west. The smaller entrance is nearly choked up with deposits, but the larger entrance is at present quite open and commands a good view of the valley, or would do if the trees were not there. Cathole lies at about 10 metres above the immediate floor of the valley and at about 30 metres O.D. Both entrances open directly into a large chamber which then reduces to various side passages at the back. The cave has been formed by solution in a bed-rock of Carboniferous limestone, and it has been filled at various times with deposits which appear to be either Late Pleistocene or more recent.

The first known excavations at Cathole were carried out primarily inside the cave and at the mouth of the larger entrance by Col. Wood during the 1860s (Garrod, 1926). Although he did not record the stratigraphy, he did salvage a number of flint artifacts which Garrod (1926, Fig. 9) has assigned to the Upper Palaeolithic and attributed to either "Upper Aurignacian"or "Magdalenian".

The next known excavations at Cathole were conducted by C.B.M. McBurney in 1958 and 1959 (McBurney, 1959). He layed out a grid system of four foot squares and following these excavated at and outside the larger entrance. The position of his excavations and the grid are shown on Figure 15. McBurney dug in 6 inch spits which he adapted as nearly as possible to the natural layers as the latter became apparent. His published section drawings are from the work in 1958, and they show the natural sequence and its disturbances remarkably well (McBurney, 1959, plates 20 to 22). The sections recorded in 1959 have not as yet been published. McBurney (1959, plate 19) also published a plan of the 1958 excavations which includes the distribution of individual artifacts in his layer B, although I have reason to suspect that some of these artifacts may be from overlying layer C (this possibility is discussed below).

McBurney also published drawings of the principal tool-forms from his layer B (1959, Fig. 1), and diagnosed them as "Creswellian". Although some of these tool-forms are clearly Later Upper Palaeolithic in type, others appear typologically within the Mesolithic range. Having been kindly permitted by Dr. McBurney to examine his original field notes, section drawings and the finds themselves, I have found that several of the more "microlithic" tool-forms did in fact come from his layer C, and some even from the overlying layer D, with most apparently occurring at the interface of layers C and B. McBurney did indeed state in his published report (1959, p. 266) that the great majority of the 280 flint artifacts recovered in 1958 came from his layer B or the C/B interface (his "contact zone B/C"). He interpreted them however as the result of a single occupation even though he stated on the same page that they were found to occur throughout the thickness of layer B. His explanation of this vertical scattering is solifluction or other natural resorting. Although such resorting may certainly have occurred, I do not think it is the only reason for the fairly wide vertical scatter of artifacts at Cathole. Their surprising typological variation must raise the possibility that their vertical distribution also reflects a number of distinct occupations. In my illustration of selected artifacts from McBurney's 1958 excavations (see Fig. 131 in this book), I have therefore used some of McBurney's own drawings, but I have arranged them according to the layers and layer interfaces in which his notes suggest they were originally found.

Excavations by the present author at Cathole in 1968 were undertaken at the suggestion of Dr. McBurney with the hope of finding clearly stratified artifacts in association with materials suitable for radiocarbon dating. Two areas were chosen for sounding and these are shown in Figure 15. One area was adjacent to the southernmost portion of McBurney's excavations and followed his grid system. The other sounding was in the smaller entrance to the cave and was aligned with the orientation of that entrance towards the south-west; this second sounding yielded nothing of archaeological interest other than two very small nondescript flint flakes in the topsoil, but it did reveal a typical limestone cave sequence of weathered deposits overlying a thermoclastic scree which in turn overlay a silty sand on bedrock. This south-west entrance was found to be really more of a natural window than an actual entrance; as a ledge or wall of bedrock was noted running across the south-western end of the area dug above the level of the thermoclastic scree inside the cave. The small south-west entrance or window was thus possibly the result of a post-thermoclastic-scree collapse of that portion of the cave wall and ceiling; in other words, its formation may be Post Glacial (Flandrian), hence the lack of Upper Palaeolithic evidence in the area excavated there. As this sounding (grid area B to C7 on Fig. 15) proved sterile, its sections are not reproduced in this book.

The larger sounding (area G to H by 5 to 6 on Fig. 15), on the other hand, was found to be much more informative. Figures 16 and 17 show two of the sections recorded here. Section 6/7 is the south face of the sounding, and section 5/6 is a composite section, also facing south, across a line just to the north of the middle of my excavation and at the southernmost face of McBurney's excavations. The portion of section 5/6 which is based entirely on McBurney's as yet unpublished drawing (by his kind permission) is that above the solid line which cuts across layers USB, C and D, whilst the portion which is based on both of our records is that between this solid line and the dashed line further down which cuts across layers LSB, LOB and MSB. That portion below the dashed line is based entirely on my observations. Although McBurney recognized layer LOB, he did not notice the finer distinction between layers MSB and USB, nor did he find the discontinuous intervening layer UOB (see Fig. 16). My layers LSB to USB are thus really sub-divisions of his layer B, but they are generally quite distinct, particularly in section 6/7. All of the flint artifacts which I have projected on to these sections were found by me. As one may see, the artifacts occurred at three, if not four, separate levels. Artifacts of Later Upper Palaeolithic aspect were found in layers LOB and MSB, and they may represent either one or two occupations. Artifacts of Mesolithic aspect, on the other hand, were at or near the base of layer C and again at the base of layer D. All of these artifacts are described in much greater detail in Chapter V (see pp. 166–67), whilst the results of granulometric and pollen analyses of the deposit samples collected in 1968 are discussed later in this Chapter (see pp. 90 and 102).

A layer key with descriptions of the various layers observed by the present author at Cathole is as follows:

- F. *Modern Topsoil*, dark brown humus with weathered scree and boulders.
- E2. *Open Weathered Scree and Boulders*, brown-grey with many modern snail shells.
- E1. *Sandy Weathered Scree*, dark brown with boulders and snail shells.
- D. *Sandy Weathered Scree*, reddish brown with weathered boulders; Mesolithic artifacts at base.
- C. *Sandy Thermoclastic and Weathered Scree*, reddish with thermoclastic and weathered boulders; Mesolithic artifacts near base.
- USB. *Upper Sandy Thermoclastic Scree*, yellowish buff.
- UOB. *Upper Open Thermoclastic Scree*, yellowish buff with some thermoclastic boulders.
- MSB. *Middle Sandy Thermoclastic Scree*, very yellowish buff with thermoclastic boulders; Later Upper Palaeolithic artifacts.
- LOB. *Lower Open Thermoclastic Scree*, yellowish buff; Later Upper Palaeolithic artifacts.
- LSB. *Lower Sandy Thermoclastic Scree*, yellowish buff with thermoclastic boulders.
- A3. *Silty Sand*, medium grey with some thermoclastic scree and boulders.

A2. *Sand*, reddish with thermoclastic scree.
A1. *Sand*, buff with some weathered scree and in places at base, some fissured bedrock.
BEDROCK: Carboniferous limestone.

This reading of the sequence of natural layers at the site by me is basically in agreement with McBurney's field observations, even though my division of "layer B" is somewhat more elaborate. Also my interpretation of the "stone component" in layer A3 is thermoclastic rather than weathered (McBurney, 1959, plate 20), as I observed amongst the dispersed scree and boulders of that layer, only sharp elements. Although he does not mention the "stone component", if any, which he may have found in layers A2 and A1, I noticed as additional elements thermoclastic scree in the former and weathered scree in the latter. The stratigraphic gap between my Later Upper Palaeolithic finds and the overlying Mesolithic finds may, of course, be more apparent than real, as my total artifact samples number only 12 and 5, respectively, and they come from the previously unexcavated portion of this sounding (area G/H6 on Fig. 15), which was off to the south rather than directly in front of the main entrance at the site. In other words, McBurney's record of artifacts also occurring higher up in his layer B is not necessarily brought into question by the discoveries of the present author. The 1958–59 excavations took place directly in front of the main entrance in an area which could easily have been the main focus of quite a few Later Upper Palaeolithic and Mesolithic activities at almost any stage within their respective time ranges; such greater activity is certainly implied by his much larger number of finds. My few finds may simply be the result of peripheral activity during more intense periods of occupation. I have therefore not included with this account plans of the horizontal distribution of my individual finds, although this information was of course recorded in my field notes. Finally, the British Museum Research Laboratory has concluded that the bone fragments and specks of charcoal collected by me from layers LOB and MSB at Cathole are insufficient for radiocarbon measurements (R. Burleigh, personal communication).

Long Hole (Glamorganshire). Excavations of 1861 and 1969.

Long Hole is located in a low escarpment well above the sea on the south-western coast of the Gower Peninsula. It is about 2 kilometres west of the village of Port Eynon, and it has a single entrance which faces south-south-east and commands a very good view of the Bristol Channel and the opposing coast of northern Devonshire. The entrance lies at about 58 metres above the low water rock platform and at about 55 metres O.D. It opens directly into a single passage which extends inwards for about 15 metres and slopes upwards towards the inner end. The site has been formed by solution in a bedrock of Carboniferous limestone, and it has been filled at various times by deposits which appear to be either Late Pleistocene or more recent. Its situation is shown on Map 13 in relation to present topography, water resources and the not too distant Earlier Upper Palaeolithic site known as Paviland Cave (or the "Goat's Hole").

The first excavations at Long Hole were carried out in 1861 by Col. Wood, who discovered the cave earlier in that same year, and H. Falconer (Falconer, 1868). When first found the cave entrance was almost completely blocked by its talus cone, but after this had been removed the deposit inside the cave was found to consist of a "ferruginous cave earth" which had been mixed with angular limestone fragments. The "cave earth" was recorded as about 7 feet (or approximately 2.1 m.) thick and resting directly on bedrock. A small assemblage of flint and chert artifacts was found in this "cave earth" at a depth of 4.5 feet (or about 1.4 m.) and a distance of approximately 6 feet (or roughly 1.8 m.) from the entrance. Garrod has since suggested, on the basis of a well-made burin/end scraper in chert and a rough keeled scraper in flint (Garrod, 1926, p. 69 and Fig. 9), which were included in the finds of 1861, that the assemblage may be "Middle or Upper Aurignacian" in age. However, she noted that the fauna supposedly found associated with it seemed rather mixed, some of the mammals being interglacial in character whilst others were of the more usual glacial, or rather periglacial, species so often found at Upper Palaeolithic cave-sites. She therefore suggested that either there had been some disturbance of the "cave earth", or that two levels were present but not detected. This faunal evidence has been considered in greater detail elsewhere in the present study (see pp. 121–22), as have the artifacts found by Wood and Falconer (see pp. 144–46 and Fig. 97).

The next excavations at Long Hole were undertaken by the present writer in 1969. A preliminary visit was made to the site in 1968 when it was concluded that some of the entrance platform might still be undisturbed, even though the interior of the cave appeared almost entirely dug out, there being only a very small section of deposits left in the innermost part. It was hoped that new excavations might clarify the problem of the apparently mixed faunal assemblage and obtain diagnostic Upper Palaeolithic tool-forms in association with materials suitable for radiocarbon dating.

Figure 18 shows a plan of the entrance area at the site including the metre grid and excavations of 1969. The approximate boundary between Wood's entrance trench and what I found to be previously undisturbed deposits is also indicated. The 1969 excavation consisted of a 2 metre wide by 7 metre long north–south trench, with a 0.5 metre wide by 3 metre long strip added to the north–west to retard the possible collapse of Wood's tip. The sections recorded in 1969 are shown on Figures 19 to 22, and a description of the stratigraphy is given under the layer key on Figure 19. The positions of the undisturbed flint and chert artifacts found are indicated on section 2/3 (see Fig. 21). They occurred no further than 50 centimetres to either side of that section, and they appear most likely to be Earlier Upper Palaeolithic, although they only include struck waste. They are further described in Chapter IV (see pp. 144–46), where my reasons for considering them Earlier Upper Palaeolithic are given. As they are very few in number, plans of their horizontal distribution according to their two layers are not included in this account, although the information was of course recorded in my field notes. Two possibly Middle Palaeolithic artifacts were also found; these occurred in layer A2b in square D2. They are an apparently struck flake of limestone and a bone tool. The bone tool is a medial fragment of a metapodial of a large herbivore which has been split in half and crudely chipped to a point at one end. The very tip of the point has been worn smooth, and the tool as a whole is not unlike the two French Mousterian ones figured by Bordes (1961, plate 108, no. 1 and 2), although apart from its stratigraphic position, it is of such generalized design that it could belong to any period.

In Figure 21 the vertical positions of deposit samples collected for granulometric and pollen analyses may be seen. Reports on the results of these studies, as well as the various fauna found, are given in parts III. B.1, 2 and 3 of this present Chapter (see pp. 90, 102, 121). Although reasonably large samples of bone fragments from the layer complexes of A2 and A3 were submitted to the British Museum Research Laboratory for radiocarbon measurements, they have been found by that laboratory to have insufficient collagen for present methods of 14-carbon age determination (R. Burleigh, personal communication). This is disappointing but nothing can be done about it at present.

Mother Grundy's Parlour (Derbyshire). Excavations of 1874–79, 1924, 1959–60, 1969 and 1974.

Mother Grundy's Parlour is located at the north-eastern end of the gorge known as Creswell Crags which lies to the east of the village of Creswell. It has a single broad entrance which gives way to what is really a deep-set rock-shelter rather than a cave, although there is a small side passage off to the north-east at the rear of the shelter. It has been formed by solution in a bedrock of Permian limestone, and it has been filled at various times by deposits which appear to be either Late Pleistocene or more recent. The entrance faces south-east at about 5 metres above the floor of the Crags and 74 metres O.D., and there is a broad platform of deposits in front of it. Map 14 shows the position of the site in relation to present topography, water resources and neighbouring Later Upper Palaeolithic sites. Creswell Crags itself straddles part of the present border between Derbyshire on the north and Nottinghamshire on the south.

The positions of excavations which have taken place at Mother Grundy's Parlour are shown on Figure 23, with the exception of the sterile 1974 sounding by S. Collcutt outside and to the east of the entrance (i.e. in grid area 8/9–D/E). The first known excavations were those conducted by W.B. Dawkins and J.M. Mello intermittently between 1874 and 1879 (Dawkins and Mello, 1879). As may be seen on my plan of the entrance, they worked primarily inside the shelter, although some of the disturbance which I found just beyond the overhang may also have been due to them. In general their record of the stratigraphy from top to bottom is basically as follows:

5. *Surface Soil,* thickness c. 13 cm. at entrance.
4. *Red Sandy "Cave Earth",* with rudely chipped quartzite pebbles (of "Mousterian" aspect according to Garrod, 1926, p. 136) and bones of woolly rhinoceros, etc.; thickness c. 75 cm. at entrance and c. 105 cm. in rear passage.
3. *Stiff Red Clay,* no artifacts but bones of *Hippopotamus,* etc.; thickness c. 15 cm. with deposit only occurring in rear passage.
2. *Ferruginous Sand,* no artifacts but with bones as in layer 3; thickness c. 30 cm. with deposit only occurring in rear passage.
1. *White Calcareous Sand,* no remains; deposit occurring both at entrance and in rear passage.

Mother Grundy's Parlour is the only site at Creswell Crags which is known to have yielded what almost certainly appears to be a Last Interglacial (Ipswichian) fauna. However it is worth stressing that Dawkins and Mello found neither this fauna nor the relevant layers (2 and 3) at the entrance to the site, nor have subsequent investigators found any such evidence in the entrance platform. The artifacts from their overlying layer 4, which Garrod considered possibly "Mousterian", are thought by the present author probably to have been at least as late as Later Upper Palaeolithic, if not even younger (see 1969 excavations below and Chapter V, pp. 171–75). After Dawkins and Mello had finished at the site, the remaining deposits inside the shelter are thought to have been dug out by a Dr. Laing of Newcastle in or about 1887 (Armstrong, 1925).

The next excavations at the site were undertaken by A.L. Armstrong in 1924 (Armstrong, 1925). He dug primarily in the north-western corner of the entrance platform, and working by 6 to 12 inch (or about 15 to 30 cm.) spits, which he termed "zones", he recorded the following section from top to bottom:

5. *Recent Throw-out* (i.e. tip), thickness c. 15 cm.
4. *Old Surface of Dark Sandy Humus,* Romano-British and later artifacts, thickness c. 15 cm.
3. *Red Sandy "Cave Earth" and Stones,* a concretion of limestone scree cemented compactly together by the sandy "cave earth"; Upper spit ("upper middle zone") c. 15 cm. thick with Later Mesolithic (his "Azilio-Tardenoisian") artifacts; Middle spit ("middle zone") c. 22 cm. thick with Earlier Mesolithic ("more Azilian", according to Garrod, 1926, p. 145) artifacts; Lower spit ("lower middle zone") c. 22 cm. thick with Mesolithic/Later Upper Palaeolithic (his "Developed Aurinacian") artifacts; Base spit ("base zone") c. 30 cm. thick (including all of layer 2 below) with Later Upper Palaeolithic (his "mixed Aurignacian and Mousterian") artifacts; total layer 3 thickness c. 75 cm.
2. *Yellow "Cave Earth" and Stones,* merging almost imperceptibly into overlying layer 3 and included in base spit with Later Upper Palaeolithic artifacts as above although supposedly "more Mousterian" at very bottom; with L.U.P. prepared hearth sunken in top; thickness c. 15 cm.
1. *"Basement Bed" of Yellow Calcareous Sand,* sterile.

The above description of Armstrong's sequence is a combination of his "typical section" (Armstrong, 1925, Fig. 4) and his lengthier descriptions of the layers and finds (1925, pp. 149–74). As I have studied his numerous original finds which are now housed at the British Museum according to their "zone", or spit, I have not hesitated to give priority to my own terminology for them, but Armstrong's original descriptions are also quoted. They are further described in Chapter V (see pp. 171–75 and Fig. 144–49).

In 1959 and 1960 C.B.M. McBurney conducted new excavations at Mother Grundy's Parlour, working mainly on the north-eastern side of the entrance platform (McBurney, personal communications). The aim of his work was to check Armstrong's results and obtain suitable samples for radiocarbon dating. I have examined his as yet unpublished section drawings and finds (by his kind permission) and discussed them with him, and the following is a summary of his reading of the stratigraphy from top to bottom:

F. *Modern Turf.*
E2. *Tip of Armstrong's Excavation.*
E1. *Tip of Earlier Excavations.*
E/D. *Subrecent Humic Deposit with Weathered Boulders*, Romano-British artifacts.
C. *Reddish Sand with Sharp Elements*, some weathering; artifacts of Mesolithic aspect.
SB. *Small Scree with Sharp Elements* in a Sandy Matrix, artifacts of Mesolithic/Later Upper Palaeolithic aspect.
LB. *Large Airhole Sharp Scree*, unweathered; artifacts of Mesolithic/Later Upper Palaeolithic aspect.
A. *Decayed Fissured Surface of Bedrock* with Sterile Silty Infilling.

The total number of undisturbed recognizable tool-forms found by McBurney in an excavation area as large as Armstrong's was only 28, whereas Armstrong found at least 209 (see Fig. 23 to compare areas). Of Armstrong's tools, 107 are possibly Later Upper Palaeolithic, whilst McBurney's include only 7 which might conceivably be Later Upper Palaeolithic, 3 of the latter being from the C/SB interface and 4 from within SB/LB. The immediate impression one gets from this difference in their tool samples is that greater Mesolithic and Later Upper Palaeolithic activity took place in the north-western corner of the entrance platform. In the case of Later Upper Palaeolithic activity in particular, this may be because of the potential sheltering effect of a bedrock buttress and extended overhang on that side. In support of this observation one may cite Armstrong's discovery of a well prepared hearth which may well have been of Later Upper Palaeolithic age (Armstrong, 1925, p. 152, Fig. 3 and plate 22).

Four radiocarbon age estimates were obtained by Godwin and Willis (1962) from samples collected by McBurney in 1960. A "date" of only 8,800 ± 300 years B.P. (Q–551) was read for charcoal from within layer B (or SB/LB), and another of 7,602 ± 140 years B.P. for charcoal from the C/B interface. By comparison with age estimates since obtained by the present author for Later Upper Palaeolithic samples from Kent's Cavern (p. 42) and Sun Hole (pp. 52–55), these from Mother Grundy's Parlour certainly seem rather too young. The other two age estimates were both read from the same bulked sample of charcoal from within layer C and their published average was 6,760 ± 140 years B.P. (Q–553/4). This latter "date" is not very unreasonable for material which is probably Mesolithic. However, it is worth noting that amongst the charcoal of all of the samples were found recognizable remains of shells of hazel nuts (*Corylus avellana*, Godwin and Willis, 1962, pp. 61–62), a shrub characteristically abundant in Post Glacial

Zone VI, but not present, or very rare, before that. As layers C and B are comprised mainly of scree and sand, it may of course be that some of these nut-shells slipped down quite naturally in the passage of time from C into B.

Excavations by the present writer were therefore undertaken in 1969 with the hope of finding diagnostic tool-forms in association with more reliable radiocarbon samples, as well as with the intention of collecting deposit samples for granulometric and pollen analyses. The area selected for these excavations was a narrow strip of probably undisturbed deposits between the areas dug by Armstrong and McBurney (see Fig. 23). Figure 24 shows the south-western face of the trench (the north-eastern face is not shown in this account as it intersects McBurney's excavations, although it has been useful in correlating with his sections). A description of the stratigraphy recorded in 1969 is given under the layer key on the same figure. In a preliminary report (Campbell, 1969, Fig. 6) the undisturbed flint artifacts found within 50 cm. of this section were projected on to it, but in this present account they have been left off for greater clarity of the section itself, and they are instead shown horizontally with other finds on Figures 25 and 26. My reading of the stratigraphy is remarkably similar to McBurney's, and I think it may be taken that our combined observations are the most accurate thus far for Mother Grundy's Parlour, although I also think Armstrong's records may be accepted with a certain amount of reservation. Even though the charcoal specks and bone fragments collected in 1969 from layers LB to D have proved insufficient for radiocarbon measurements (R. Burleigh, personal communication), the deposit samples have yielded rather informative results (see pp. 104–05). Also I found layer D to be less disturbed than McBurney did, and it together with layer C yielded no less than 110 undisturbed Mesolithic flint artifacts. These artifacts occurred primarily in squares D to F (see Fig. 26), but this off-set distribution may be more apparent than real, as squares A to C were almost entirely disturbed. Furthermore, Armstrong (1925, Fig. 3) plotted in a very general way a dense occurrence of his "Azilio-Tardenoisian" artifacts just to the north-west of my squares A to C and extending more or less alongside. On the other hand, the distribution of bone and teeth fragments in layers D and C is confined to squares D to F, with the exception of one tooth in square C, layer D, which might suggest that their users discarded them together with some of the artifacts down the slope towards the floor of the Crags. Although, of course, these Mesolithic hunter/gatherers may have actually sat on the slope and knapped some of their stone artifacts there, as well as eaten some of their kill, not to mention hazel nuts. How many successive Mesolithic occupations might be represented is uncertain, but as the majority of these finds occurred near, although not precisely at, the interface of layers D and C, there may have been only one main Mesolithic occupation. A subsequent occupation was probably Neolithic, as two backed tools of the so-called "petit tranchet dérivatif" (or "transverse arrowhead") form were found near the top of layer D, and three sherds of crude pottery were found at the base of the immediately overlying layer F (see Fig. 26).

By comparison with the Mesolithic evidence, material which might be Later Upper Palaeolithic was rather scarce (see Fig. 25), only 40 flint artifacts, including 13 from the C/B interface which might be Earlier Mesolithic instead. In fact, one got the impression that one really was only on the eastern periphery of Armstrong's dense cluster of Later Upper Palaeolithic finds. The absence of any finds from layer LB in squares E and F, and layer SB and interface C/B in square F, was due to their extending below the bottom of the excavation in those squares (see Fig. 24 as well). The little cluster of charcoal specks found in square C of layer LB might well have been scattered from the above mentioned hearth found by Armstrong, but as already stated, they were *in toto* insufficient for radiocarbon measurements, as were the "bones", or large bone fragments, found and plotted in layers LB and SB. The undisturbed artifacts from the 1969 excavations are further described in Chapter V (see pp. 171–75 and Fig. 150), but as regards those in quartzite, most of which would be better termed "manuports", it should be noted that as they occurred predominantly in layer D in a context which was Mesolithic and later, many of the similar quartzite finds from the "cave earth" of Dawkins/Mello and Armstrong might also have been as young, rather than "Mousterian". This possibility is underlined by the fact that McBurney (personal communication) also found absolutely no evidence for either Middle Palaeolithic or Earlier Upper Palaeolithic at the site.

The 1974 sounding by S. Collcutt (personal communication) was undertaken to clarify further the stratigraphic problems at Mother Grundy's Parlour. His section is similar to McBurney's, but shallower and completely sterile.

Robin Hood's Cave (Derbyshire). Excavations of 1874–76 and 1969.

Robin Hood's Cave is located just west of centre on the northern side of Creswell Crags. It has two adjacent entrances which both face south-south-east and which lead to a complex series of parallel and interconnected chambers, passages and more or less vertical shafts. As with the other sites at Creswell Crags, it has been formed by solution in a bedrock of Permian limestone, and it has been filled at various times by deposits which again appear to be either Late Pleistocene or more recent. Its position may also be seen on Map 14 in relation to the usual data. The two entrances open out on to a broad entrance platform with a fairly extensive overhang. This platform lies at about 5 metres above the floor of the Crags and 75 metres O.D.

The first recorded excavations at Robin Hood's Cave were those carried out by W.B. Dawkins and J.M. Mello from 1874 to 1876 (Mello, 1875, 1876 and 1877; Dawkins, 1876 and 1877). They worked entirely at and inside the westernmost entrance, the easternmost entrance having been dug out by some unknown excavator at some unknown time (as Dawkins and Mello do not mention the eastern entrance, it may be that it was buried and therefore unknown to them at the time). Figure 27 shows a plan of these two entrances and their exterior platform as found by the present author in 1969. The maximum possible extent of the 1874–76 excavations has been plotted according to the superficial and deeper disturbances uncovered in 1969, although it may well have been that Dawkins and Mello were not solely responsible for these earlier disturbances, particularly in the case of the platform itself.

When Mello began digging at the western entrance to what he initially termed "Cavern B" (i.e. Robin Hood's Cave), he found teeth of woolly rhinoceros and hyena in a surface soil which was only about 1 inch, or 2.5 cm., thick (Mello, 1875, pp. 682–83), and if I have read him correctly, this was probably an indication of some previous disturbance of the platform, perhaps by early nineteenth century occupants. As he proceeded further into the cave mouth, he deepened his cutting from about 3 feet (c. 90 cm.) to 8 feet (c. 245 cm.) where he found limestone blocks which he interpreted as bedrock (Mello, 1876, p. 240). Dawkins then joined Mello and guided his efforts, during and after which he reported in detail on the fauna and artifacts found (Dawkins, 1876 and 1877). Finally, Mello reported that the site was "worked out" (Mello, 1877, p. 587), although in 1887 someone found a well-made side scraper which Armstrong (1925, p. 175 and Fig. 19) subsequently discovered in the collection of a W.F. Jackson of Dore, near Sheffield. As Dr. Laing was thought to have ransacked Mother Grundy's Parlour in about 1887 (Armstrong, 1925, p. 146), it could be that he was the one who dug out the eastern entrance at Robin Hood's Cave and that Jackson's side scraper was his only find ever eventually published, although Armstrong does not even suggest such a possible connection. I have now traced the original side scraper to the British Museum and found a red-buff silt adhering to some parts of its surface.

The artifacts found by Dawkins and Mello are described in detail in Chapters IV and V (pp. 147–49, 171–75). They found what appear to be both Earlier and Later Upper Palaeolithic tool-forms, but the exact stratigraphic relationship of these remains uncertain. The general sequence recorded by them inside the cave is as follows (Mello, 1877; Dawkins, 1877 and 1880):

F/C. *Surface Soil and Stalagmite*, with Roman and later artifacts, thickness c. 5–60 cm.

B3. *Stalagmitic Limestone Breccia*, with Upper Palaeolithic artifacts including possibly both E.U.P. and L.U.P. types (see Dawkins, 1880, Fig. 47, 48, 50 and 51); thickness 0 to c. 90 cm.

B2. *"Cave Earth"*, light-coloured, sandy and very calcareous; upper portion with Upper Palaeolithic artifacts said to include a rib segment with an incised head of a horse (see Dawkins, 1880, Fig. 53); middle and lower portions with Middle Palaeolithic artifacts including handaxes of "Mousterian" aspect (e.g. Dawkins, 1880, Fig. 45); upper "Cave Earth" and Breccia thought by excavators to be roughly contemporaneous formations as they observed an inverse thickness ratio between the two; thickness of "Cave Earth" c. 40–120 cm.

B1. *"Mottled Bed"*, deposit of "cave earth" with numerous small angular fragments of limestone and possibly Middle Palaeolithic artifacts; thickness 0 to c. 65 cm.

A2. *Red Sand*, with laminated red clay at base and possibly Middle Palaeolithic artifacts;

thickness c. 60–120 cm.
A1. *Light-coloured or White Sand*, with limestone blocks; sterile; thickness c. 60 cm.
BEDROCK: Permian limestone.

The above sequence is apparently quite variable, and the excavators themselves appear somewhat inconsistent in their reports on it. One enigma is the supposed occurrence in the upper portion of the "Cave Earth" (or layer B2, as I have coded it) of a canine of *Homotherium latidens* (or "Machairodus"). This tooth was apparently found at the inner end of their "Chamber F" (Dawkins, 1877), and as that part of the cave has a vertical shaft above it, it seems to me quite conceivable that the tooth, which is more likely to be of Early or Middle Pleistocene age, was derived naturally from some earlier deposit in the shaft to the "Cave Earth" in the passage below.

As it was thought that some deposits in the entrance platform might still be undisturbed, excavations by the present author at Robin Hood's Cave were undertaken in 1969. The area excavated is indicated on Figure 27 together with the author's grid system of metre squares, and the sections recorded are shown on Figures 28 to 31. A description of the layers observed is given under the layer key on Figure 28 and continued on Figure 31. As may be seen, most of the upper portion of the area excavated consisted of nineteenth century tip, whilst most of the lower portion consisted of more or less undisturbed thermoclastic screes (layers B/A to USB) and a red sand (layer A) on bedrock. Layer A yielded an undiagnostic assemblage of three flint flakes and a crude end scraper on a thermally fractured pebble of quartzite, and these might be Middle Palaeolithic in age. Layers B/A to USB yielded four successive assemblages of diagnostic backed tools and other artifacts of Later Upper Palaeolithic type, the uppermost assemblage possibly also incorporating a Mesolithic component. For convenience I have labelled these backed tool assemblages from bottom to top as L.U.P.1 (layer B/A), L.U.P.2 (layer LSB), L.U.P.3 (layer OB) and Meso./L.U.P. (layer USB). In a preliminary report (Campbell, 1969, Figs. 3 and 4) some of these artifacts were projected on to sections 0/1 and 2/3, but in this present account they have been left off for greater clarity of the sections, only the vertical positions of deposit samples, radiocarbon samples, a Later Upper Palaeolithic hearth (H1 on section 2/3) and a human frontal bone being plotted.

Figures 32 to 36 show the horizontal distribution by layer of the finds made in 1969. It should be noted that layers OB and USB had already been partly or largely removed by earlier excavations, as is indicated on Figures 35 and 36. In the case of layer USB, this disturbance would doubtless help to explain the unexpected age estimate which the British Museum Research Laboratory obtained from a radiocarbon sample that I at the time of excavation thought I had extracted from an undisturbed context (see Figs. 28 and 36 and relevant statements thereon). Although I found no definite Earlier Upper Palaeolithic evidence whatsoever, the "date" in question (c. 28,500 B.P.) would certainly be a more likely age for such an assemblage than for my Meso./L.U.P. of layer USB. It may therefore be that sample BM-602 was unwittingly transferred by Dawkins and Mello from an original association with their Earlier Upper Palaeolithic finds inside the cave to their entrance trench where it got trodden by them into layer USB.

Radiocarbon sample BM-601, which was collected as two large lumps of wood charcoal from layer B/A and bulked for the measurements, also yielded an unexpected, although not stratigraphically inconsistent, age estimate (> 40,000 B.P.). In this case the material "dated" has quite likely been naturally derived by erosion from the underlying layer A, as it could not possibly be the minimum age of the backed tools (including two "Creswell points") from layer B/A (see Figs. 29, 32 and 33 and relevant statements thereon). Although basal layer A appeared *in situ*, it is quite possible that its uppermost portion and any other deposit overlying it at the time (e.g. such as Dawkins and Mello's "Mottled Bed", see p. 65) were mixed by cryoturbation and a partial sludging-out of the cave's interior at some point during the Full Last Glacial (Full Devensian), thus leaving a composite layer B/A as a veneer over A on the platform outside. This hypothesis would help to explain the complete lack of any definite Earlier Upper Palaeolithic evidence in the area excavated in 1969 (such evidence may possibly still survive somewhere between the entrance platform and the floor of the Crags).

The quartzite scraper found in layer A (see Fig. 32) appeared to the present author to have been burnt, and it was therefore submitted to the Research Laboratory for Archaeology and the History of Art at Oxford University as a possibility for thermoluminescent dating. It was subsequently found that this tool carries an "equivalent radiation dose" of 60 kilorads, which indicates that either it has not been burnt at all, or it was burnt longer ago than around 100,000 years (M.J. Aitken, personal communication). Thus the finds from layer A (including such fauna as *Coelodonta antiquitatis*) might even belong to some part of the Penultimate Glacial (Wolstonian), rather than the earlier part of the Last Glacial (Devensian). But for the moment the "real age" of layer A remains uncertain, although by extrapolation from radiocarbon sample BM-601 of layer B/A, layer A is at least older than 40,000 years ago.

Radiocarbon samples BM-604 (L.U.P.2, layer LSB, Figs. 30 and 34) and BM-603 (L.U.P.3, layer OB, Figs. 29 and 35) have proved much more satisfactory, as their age estimates are finite, stratigraphically consistent and not too unreasonable for the Later Upper Palaeolithic. However, on the basis of the results of pollen and small mammal analyses (see pp. 105 and 125-26 and Figs. 77 and 82), I suspect these two "dates" may each be about 500 to 1,500 years too young.

As may be seen on Figures 33 to 36, the main clustering of Later Upper Palaeolithic artifacts from L.U.P.1 to Meso./L.U.P. falls consistently in square A1, with a variable scatter spilling over into adjacent and/or more distant squares. This recurring distribution pattern suggests that the most intense activity took place at or inside the adjacent (i.e. westernmost) entrance to the cave, a place which although narrow is well sheltered (aside from a draught that issues at present from this mouth of the cave, but which may not have been as strong or even present during the Late Last Glacial; properly hung mammal skins would reduce such in any case). Bones, on the other hand, are scattered pretty well everywhere, although they also seem to have a tendency to concentrate or increase in numbers in or near square A1. The occasional fall of large boulders from the retreating overhang may also have had a reduction effect on potential activities beneath the "dripline" (see sections and layer plans). But obviously, Later Upper Palaeolithic activities of one sort or another took place at various times more or less everywhere about the site, and the site may have been used repetitively (but not necessarily continuously) during the Late Last Glacial as a kind of "home base" for exploitation of surrounding fauna and presumably flora. Changes in fauna and flora are apparent from layer to layer (see this Chapter, pp. 105, 125), as are, perhaps in response, changes in backed tool-forms and other artifacts (see Chapter V, pp. 171–75).

A noteworthy discovery at the base of layer LSB in square C3 was what appeared to be a prepared hearth (see Figs. 30 and 34). This hearth consisted of thinly scattered charcoal specks and burnt bone fragments in a matrix of what appeared to be burnt sand, surrounded and partly underlain by five small, weathered and worn boulders, and right up against a low-lying, fissured vertical buttress of bedrock. These boulders were presumably carried in and intentionally placed, as their nature was quite distinct from the thermoclastic boulders and scree of the layer at large. As these little weathered boulders about the hearth were also of Permian limestone, they were probably obtained from near-by, perhaps the floor of the Crags. Five pebbles of quartzite were also found in the hearth, but as they appeared neither burnt nor thermally fractured at all, they had presumably been tossed into the hearth after it had grown quite cold, whether or not they had originally been intended as so-called "pot-boilers". Material collected from this hearth for radiocarbon dating proved insufficient (R. Burleigh, personal communication), but a humerus of *Equus* found only a few centimetres above in square C2 has become BM–604 and yielded the age estimate quoted on Figures 30 and 34. As may be clearly seen on Figure 34, in addition to the cluster of artifacts and bone fragments in squares A0 and A1, there is another concentration of mostly bone fragments around the hearth. It may therefore be in the case of layer LSB that whilst most tool manufacture and use took place over in A0–1, eating (and possibly sleeping) happened nearer the hearth, perhaps mainly in squares B2 and B3 which would have been well under the overhang. The suggestion of sleeping outside the cave during the Late Last Glacial is not as absurd as it may at first seem. In addition to any warmth provided by the hearth, one could have easily erected a rough shelter of large mammal skins on a frame of perhaps willow or birch branches, which might then have been partly supported by any of the numerous bedrock ledges occurring between the –1 metre and −2 metre bedrock contours in squares A4 to C4 (see Fig. 34). Of course, given the screey nature of the deposit excavated, no evidence of possible stake-holes was found, although if stakes had been pushed down as far as the basal layer of sand (layer A), their traces in one form or another would probably have survived (compare the fairly convincing evidence for a much earlier cave-site structure at Grotte du Lazaret in Nice, Lumley *et al.*, 1969).

Another noteworthy find is a frontal bone of *Homo sapiens sapiens* (identification verified by K.P. Oakley, personal communication) found in layer OB and recently studied by Rosemary Powers at the British Museum of Natural History, together with some possibly related human bones and teeth from the overlying tip. The find-spot of this frontal is marked on Figures 30 and 35. It is presumably contemporary with the L.U.P.3 assemblage of layer OB, but its almost isolated occurrence (see Fig. 35 in particular) probably indicates it had been simply discarded. It is unabraded and there are no apparent cut marks, so although it may have been the result of a meal, there is no conclusive evidence for such. The human skeletal material is reported on in detail in Appendix 6 by Rosemary Powers and the author.

Hengistbury Head, Sites C1 and C2 (Hampshire). Excavations of 1957 and 1968–69.

Hengistbury Head is a peninsula which juts out into the English Channel to the east of the town of Bournemouth, forming the southern shore of Christchurch Harbour. The higher part of the headland is known as Warren Hill and consists of a core of Eocene sand and clay beds, the latter having large ironstone concretions. This hill is overlain by a capping of gravels and sands which appear to have been deposited at various times since the Late Pleistocene or earlier. The hill is connected with the mainland to the west by a plain of alluvium which appears to be both recent and possibly Late Pleistocene. At the eastern end of the hill there is a slight depression which has become partly filled with peat and wind-blown sand since at least the very latest part of the Pleistocene. In this depression a number of flint tool-forms and other artifacts of Later Upper Palaeolithic aspect have been uncovered since 1913. The find-areas of these artifacts are indicated on Map 15 as Sites A, B, C1 and C2. The few finds from Sites A and B have already been described by Angela Mace (1959) and are further described by the present author in Chapter V (pp. 177–81). What I have labelled, for convenience, as Site C1 was found and excavated by her (Mace, 1959), and its adjacent Site C2 was found and excavated by me. It will be noticed on Map 15 that there are a number of rock outcrops of ironstone to the south of these sites in the Channel itself, the largest group of which is known as the Beerpan Rocks. These rocks are remnants of the approximate southern limit of Hengistbury Head prior to the disastrous effect of ironstone mining operations in the nineteenth century (Pepin, 1967). There is a local legend that numerous remains of horse were found somewhere between these rocks and the present headland, but whether these were ancient or modern will probably never be known as their present whereabouts are unknown (J.H. Lavender, personal communication).

The excavations by A. Mace in 1957 were undertaken to determine whether there might be any undisturbed Later Upper Palaeolithic evidence surviving under a narrow strip of unploughed heath known to stretch along the seaward edge of the above-mentioned depression at the south-easternmost part of Hengistbury (Mace, 1959). The positions of some of her initial soundings and her subsequent main excavation are indicated on my plan of Sites C1 and C2 (Fig. 37). The area of her main excavation is what I have termed Site C1, and it was here that she recovered no less than 2,263 flint artifacts which she interpreted as a single Later Upper Palaeolithic assemblage. She dug by yard (or 0.84 metre) squares, allowing the direction of their spread to be dictated by the density of the artifacts and eventually removing a total area of 85 square yards (or 71.1 square metres). For vertical control she presumably dug by spits, but she provides no precise record of the vertical distribution of her artifacts, stating only that they came from a depth of 9 inches to 2 foot, 3 inches (or c. 23 to 68 cm.), with most between the 1 and 2 foot levels (or c. 30 to 60 cm.). Also, interpreting her exposed sections as insignificant because of their podzolic appearance, she only published a "typical section" which was drawn by J.H. Lavender (Mace, 1959, Fig. 2). She illustrated the horizontal distribution of her finds diagrammatically by yard squares, showing total finds per square and total individual artifact-types per square (Mace, 1959, Fig. 3). She also developed an elaborate morphological typology, the basic concepts of which have influenced my own typological approach (see Chapter II, pp. 20–23), even though I consider her excavation itself inaccurate. She applied her typology to her finds and figured a fair selection of them (Mace, 1959, Figs. 4 to 8), but it should be noted that she had probably only salvaged those artifacts greater than 15 millimetres in size (1959, p. 238). Thus, despite her excavation standards, her published report, for its details on the 1957 and earlier finds, remains of great value as one may study her figures, counts and percentages and see how she arrived at her conclusion of possible typological affinities with north-western German assemblages such as some of those from Rissen, near Hamburg.

Most of the material found by Mace at Site C1 is housed in the British Museum, and the extant earlier collections from Sites A and B are mainly in the Red House Museum at Christchurch. I have examined all these flint artifacts in detail, and on the basis of my excavations at Site C2 (see below) as well as the results of metrical and statistical analyses of all the available finds (see Chapter V, pp. 177–93), I suspect that only a few of Mace's 2,263 finds (namely, 9 of her backed tools) and a few of the finds from A and B are possibly Later Upper Palaeolithic, whilst most are more likely Mesolithic and younger. As regards Site A, Mace herself (1959, p. 238) says most of them are of Neolithic or Bronze Age character, only about 60 finds being possibly Later Upper Palaeolithic. The facts that most of Mace's finds cluster within her main area of excavation (Site C1) and many of them fit together (e.g. spalls on burins and flakes on cores) help to support her conclusion that they represent a single assemblage, but I would suggest that what she uncovered was a predominantly Mesolithic assemblage (generally unpatinated) which probably overlaid a thin scatter of unrelated Later Upper Palaeolithic artifacts (generally patinated).

With an eye on the possibility of new excavations at Hengistbury Head, I corresponded with Mrs. Mace early in 1968, who replied that she thought further investigation would be well worthwhile. She recommended digging between the cliff-side footpath and the cliff edge near her excavation site, and for that suggestion I am forever indebted to her as the evidence which came to light during my excavations there in 1968 and 1969 was far more than initially expected, and would probably have otherwise soon disappeared into the sea below, given the high rate of natural and pedestrian-caused erosion to which the remnants of the headland are now subjected.

The area excavated in 1968–69 by the present author is indicated on Figure 37, where it is coded as Site C2. The heavy lines mark both the outer limits of the excavation and the lines along which sections were recorded. Initially, a grid system of metre squares was laid out, and work was begun with alternate metre-wide trenches near the active stream gully to the east and proceeding west at right-angles to and along the cliff edge. Although flint artifacts of predominantly Mesolithic aspect were found scattered and/or clustered pretty well everywhere in the turf and just below, flint artifacts of Later Upper Palaeolithic aspect were only found deeper down and to the west in what appeared to be a fairly well defined cluster. In order to retrieve as much of this apparently Later Upper Palaeolithic cluster as possible, the trenches were accordingly extended and connected together, a total area of 136 square metres eventually being dug.

A selection of the sections recorded in 1968–69 is given in Figures 38 to 47, and a description of the layers observed is given under the layer key on Figure 38. Although the appropriate information is available for nearly all of the finds, for clarity only classifiable flint tools occurring within 50 centimetres of a given section are projected on to the section. A key to the symbols used for these tools is also given on Figure 38. In order to avoid the necessity of awkward fold-outs, the longer sections shown are divided into separate lengths, which in every case are kept on the same page, except section E/F (Figs. 38 and 39). Section E/F is from a cleaned-up natural exposure of the deposits in the upper portion of the cliff-face adjacent to the excavation site, as is the southernmost part of section 16/17 (Fig. 44), whilst all of the other sections shown are from the excavation itself. The metre heights marked on the sides of the sections are O.D., the "0 depth datum" for the excavations having been established at 13.5 metres O.D. The positions of deposit samples collected are indicated on Figures 39, 41 and 42, whilst those of radiocarbon samples are given with their results on Figures 39, 41 and 45. The results of granulometric and pollen analyses are described in detail later in this chapter (see pp.92, 105), but it is worth noting at this point that Hengistbury provided some of the richest pollen samples collected by the present author during the course of this research. In fact, Hengistbury is the only sequence for which I am able to present a proper "tree and shrub pollen diagram" (Fig. 79) in addition to my usual, more generalized "total pollen diagram" (Fig. 78). Mace (1959, p. 236) apparently wrongly assumed there was no hope of extracting pollen from the deposits concerned, but rightly assumed that any attempt at radiocarbon dating would probably prove unsatisfactory.

In a partial departure from my cave-site layer coding practice, as Hengistbury is a series of open-sites with some peat formation and some soil (in the sense of intermixed mineral and plant matter) formation, as well as some apparent podzolic features, I have employed the soil science practice (e.g. Butzer, 1964, pp. 76–77) of lettering layers, or "horizons", from the top down as A, B and C. The soil science concept is that C is more or less unaltered parent material such as sediment or rock, which may be overlaid by B which is partly altered parent material with perhaps some humus washed into it from above, B being overlaid by A which is commonly characterized by an admixture of decomposed or partly decomposed organic matter, i.e. humus, the lowermost portion of which may be washed down to B, leaving mainly mineral matter in the bottom of A. But in the case of Hengistbury I think such an interpretation would be only partly true, as in fact separate layers and sub-layers do actually seem to exist with distinct interfaces (and distinct flora as represented by pollen, see Figs. 78 and 79) and in the case of layers B and A, distinct artifacts. I have therefore numbered the sub-divisions of A, B and C from the bottom upwards. Also, as with the cave-sites, I have coded any tip or other apparent disturbance with the letter E. As may be seen from the radiocarbon results (see Figs. 41 and 45), layer B2 has probably suffered from considerable subsequent humic infiltration, although a very small proportion of its organic matter may be much older. Layer B2 also includes a discontinuous series of fairly hard iron-pans, but rather than being mostly iron derived from subsequent humic formations above, most appears to have come from the highly ferruginous layers and bedrock below and elsewhere on Warren Hill by a process of wind-derivation of fine particles and solifluction of coarser elements. The cementation of these iron-pans within layer B2, many of which may be seen exposed on the surface of Warren Hill, has of course taken place primarily during the humid, milder conditions of the Post Glacial (Flandrian), perhaps particularly under a Sub-Atlantic climate as the radiocarbon "dates" might suggest (the "dates" being perhaps the approximate time of the cessation of most of the humic contamination of layer B2, i.e. the end of the podzolic effect).

Layers C1 to C3 appear to be alluvial deposits, and according to their height of about 11 metres O.D. (and their contained pollen, see Figs. 78 and 79), they are probably part of a Last Interglacial (Ipswichian) river system. Layers B1 and

B2 appear to be aeolian and soliflucted deposits with what might be an ancient soil formation at their interface. The interface between layer B1 and the underlying C3 and C2 is often intensely involuted, and this is presumably the result of cryoturbation during the Last Glacial (Devensian). Layer A1a appears to be entirely aeolian, whilst its overlying, discontinuous layer A1b appears to be a mixture of aeolian deposits and peaty lenses. Layer A2 is a discontinuous peat which is generally thin about the site, but particularly thick where it occurs as the fill of a drainage ditch of apparently Bronze Age/Neolithic date at the eastern end of the site (see Fig. 39). Layer A3 is the modern, living turf with principally grasses, heather and gorse; it is discontinuous as shown on Figure 37. The discontinuity of layers A1b, A2 and A3 is due to natural "sheet" and subsequent down-cutting erosion, as well as frequent pedestrian traffic, and it doubtless affects the distribution of Mesolithic and later finds. Military activities during the World Wars, previous archaeological soundings, rabbit burrows and, to some extent, ancient erosion (in the case of the latter see particularly section M/N on Fig. 42) have also probably slightly affected the distribution of Later Upper Palaeolithic finds. Layers A1b to A3 would appear to be entirely Post Glacial, but layers B2 and A1a might be Late Last Glacial (contained pollen also suggests this, see Figs. 78 and 79), whilst layer B1 is certainly Last Glacial and perhaps mostly Full Last Glacial in age. The top of layer B1 also has a number of what appear to be ice wedges (in particular see section J/K, square 17, Fig. 41; section 13/14, square F, Fig. 43; section 17/18, square J, Fig. 44; and section 26/27, square F, Fig. 47).

The horizontal distribution of finds made in 1968–69 is shown in Figures 48 to 55. As Later Upper Palaeolithic material occurred mainly at or just above the interface of layers B1 and B2, its level is referred to on these plans as "layers B". Similarly, as Mesolithic material occurred mainly at or just above the interface of layers A1a and A2, or within layer A1b where such was present, its level is referred to as "layers A". In addition to the detailed plans (Figs. 49, 50, 51, 53, 54 and 55), in order to convey a clear impression of the general pattern of Later Upper Palaeolithic and Mesolithic flint artifacts, their distributions are shown diagrammatically in Figures 48 and 52 by plotting the total number of artifacts, in each level, per 0.5 metre square. Neighbouring higher counts are connected together to form "contours" and empty areas are left blank. One may easily see from these general plans that the Later Upper Palaeolithic artifacts are less widely scattered than the Mesolithic ones. Also, unlike the Mesolithic, the Later Upper Palaeolithic has a singular, particularly dense concentration in and about square G23 with only two lesser peaks in squares F24 and J16 (see Fig. 48). The Mesolithic evidence, on the other hand, has no comparatively dense cluster, but instead four small peaks in squares E22, H22, L23 and M23, the latter two being the densest grouping (see Fig. 52). The total number of Mesolithic artifacts found is 2,350, whilst the total Later Upper Palaeolithic number is only 1,161. From these basic differences in distribution and sample size, it is apparent that the Later Upper

Palaeolithic activity is the more limited of the two.

Figure 49 shows the distribution of individual Later Upper Palaeolithic finds, as well as major disturbances. It is of interest that most of the tools and struck waste occur inside a rough, incomplete circle or oval formed by cores weighing more than 200 grammes (with some in fact over 1,000 grammes), as well as two cobbles and two large, possible anvils all weighing well over 200 grammes. 200 grammes is an arbitrary limit, but in fact all of the cores weighing less than that amount fall just inside the oval (also none of the Mesolithic cores found higher up weigh more than that amount). The anomaly to the oval is the core in square F13, but of course not all cores weighing over 200 grammes need have been used for what I am about to suggest. Closely similar stone rings, sometimes of a pear-shaped plan, have been already recognized at a number of Hamburgian and Ahrensburgian sites in north-western Germany (e.g. Ahrensburg-Bornwisch, Borneck and Poggenwisch, Rust, 1958 and 1962) where they have been interpreted as edge-weights for tents which had most likely been made from reindeer skins. Other stone settings, generally oval in plan, have been recognized further afield at Upper Palaeolithic sites in central and eastern Europe (e.g. Pavlov in Moravia, Klíma, 1955; and Gagarino on the Don near Lipsetsk, Zamyatnin, 1934). All of these stone settings are generally 4 to 6 or 7 metres in diameter, and they recall, in a general way, the rings of stone weights left behind, for example, by the Iglulik Eskimo (Mathiassen, 1928) and the Netsilik Eskimo (Balikci, 1970), when they abandon a summer camp-site. The Iglulik and the Netsilik of the Canadian Arctic make their summer tents from the skins of caribou (or reindeer) by stretching the skins over a few wooden poles and weighting down the edges by a ring of stones. If the Later Upper Palaeolithic cores found at Hengistbury were initially abandoned as waste-products, there seems no good reason why some of the heavier ones might not be subsequently employed, together with any other available heavy stone, as edge-weights for some sort of shelter against the rigours of the Late Last Glacial. As the heavier cores, etc. were in fact found more or less encircling most of the rest of the assemblage, and the approximate diameter of that open ring was about 6 metres east-west and 5 metres north-south, the rough outline of a temporary structure seems a reasonable interpretation. Although one could perhaps produce a similar roughly circular distribution pattern of cores by sitting approximately in the centre to knap the flint and tossing the "finished" cores about oneself, I think the smaller semi-circle of tools in area F–I by 21–19 stands against such an interpretation. This semi-circle of recognizable tool-forms inside the ring of possible weights gives me the impression of tools, perhaps both used and unused, which have been tossed or placed against the inside of the hypothetical structure. This distribution pattern of tools and possible weights is perhaps somewhat clearer in Figures 50 and 51, where only the relevant information is plotted.

It may be significant that the only evidence for possible Later Upper Palaeolithic hearths occurred either at the gap(s) in or within the ring of heavy stones (see Figs. 50 and 51). These were three in number and consisted of little clusters of thermally fractured flint pebbles which appeared to have been altered by fire rather than frost, as they were reddened and "calcined". These clusters occurred at the interface between layers B1 and B2, and that in square G23 has been projected on to section 23/24 (see Fig. 46). A pebble from this larger cluster has been given to the Department of Physics at the University of Birmingham as a possibility for thermoluminiscent dating, but the study of it (as well as a selection of large mammal teeth from all of the present author's cave-site excavations) is still in progress (C. Christodoulides, personal communication). An interesting fact in relation to this larger possible hearth and the hypothetical "shelter" is that the greatest concentration of waste flakes and blades occurred at or near this "hearth" in square G23, but no cores occurred to the south-west of it, their having occurred in every other direction (see Figs. 48, 49, 50 and 51). Thus most of the flint knapping at the site probably took place near the "hearths" at what might have been the main entrance to the "shelter", whilst the heavier cores were probably removed to serve as "weights" around the "shelter", perhaps particularly on the southern side, the "anvils" being employed as "weights" on the east and north-east (perhaps a cluster of lighter cores to the north-west in conjunction with a heavy one also served as "weights").

Other noteworthy discoveries in layers B1 and B2 included two possible pits and some minute specks of bright red ochre. The small oval depressions which may have been intentional pits occurred in the top of layer B1, were each about 15 to 20 centimetres deep and filled with normal layer B2 deposits, the easternmost depression (square F18) also having a backed tool in it (see Figs. 49, 50 and 51). As the interface of layers B1 and B2 is often somewhat contorted, these little depressions may of course have been quite natural, although it is curious that these quite likely-looking "pits" occurred just outside the ring of "weights", raising the possibility that they are faint traces of major, exterior "stake-holes" from the hypothetical structure, rather than "pits" or otherwise. However, the evidence was too scanty to allow a firm conclusion. The occurrence of red ochre specks was confined to the northern part of the excavation (squares K23, M23 and M21) beyond the ring of "weights" (see Figs. 49 and 51), and may also have been natural, as ochre of various forms occurs in the alluvium below (in the clayey portion of layer C1 in particular). However, these bright red specks were the only traces of ochre encountered within the area excavated in layers B1 and B2. If one wished to push conjecture to the limit, one might suggest that their occurrence outside the "structure" was due to some sort of outdoor "ceremony", or perhaps scraping activity (cf. Rosenfeld, 1971). On the negative side, no bones or traces of bones whatsoever were found in any of the layers during the excavation; they had presumably been long since leached away. Much more convincing structural evidence was found higher up in association with Mesolithic artifacts at the interface between layers A1a and A1b at the eastern end of the main excavation area. This included an oblong feature with fairly clear stake-holes, partly surrounded by roughly parallel depressions or channels which may have been "gutters" for catching and draining away any precipitation (see Figs. 53, 54 and 55). These "gutters" left the area with the stake-holes somewhat higher on a sort of platform composed of the undisturbed sand of layer A1a below. A possible outline of the inner structure, or presumably "tent-base", has been suggested by dashed lines connecting the stake-holes to the south-west and leaving an opening to the north-east where the traces of the stake-holes ceased. A section from north to south through the "gutters" is included in section 13/14, with projected sections of the two nearest stake-holes, their width being exaggerated twice for the sake of their clarity (see Fig. 43). All of the stake-holes were found to be filled with what appeared to be the ordinary peaty sand of the overlying layer A1b.

Additional Mesolithic structural evidence, as well as better Later Upper Palaeolithic structural evidence, may eventually be uncovered to the north of Site C2, or to the north-east if that area survives the down-cutting effect of the stream gully long enough (see Fig. 37). An active peat-bog running from just north of the footpath to further north between Sites A and B (see Map 15 as well) might well preserve much useful information, and it should be a worthwhile possibility for any new excavations at Hengistbury.

In comparing the Mesolithic and Later Upper Palaeolithic evidence found in 1968–69, it is of interest that no apparent "weights" were found in association with the Mesolithic structure. However, the Mesolithic structure is obviously much smaller than the suggested Later Upper Palaeolithic one and may therefore not have needed weights as such, its stakes being sufficient to hold down whatever sort of tent might have been employed. On other lines of evidence, the settlement patterns of the two Stone Age assemblages again seem quite different. For example, the Mesolithic hearths found (see Figs. 39 and 41 for sections, and Figs. 53 and 54 for horizontal positions) occurred well away from the structural evidence rather than near it, and as already shown by the diagrammatic plans (Figs. 48 and 52), Mesolithic artifacts occurred almost everywhere with many little clusters (see Fig. 53), rather than in a more confined area with generally only a thin scatter beyond (see Fig. 49). These differences are doubtless due to different patterns of activity, the Mesolithic "occupants" perhaps spreading themselves all about the place in the warmth of the Post Glacial, whilst the Later Upper Palaeolithic "occupants" stayed near their "hearth/shelter" when at the site. Fanciful reconstructions of the Later Upper Palaeolithic and Mesolithic camps are offered in Figure 56.

The artifacts found in 1968–69 at Hengistbury are further described in detail in Chapter V (see pp.177–93), where metrical and statistical differences between the Later Upper Palaeolithic and Mesolithic evidence are also given.

2. Radiocarbon Measurements and Other Age Estimates

All available British Upper Palaeolithic radiocarbon age estimates, whether from reliable or problematic samples, are listed in Table 4. They are divided into the relevant groups, Earlier and Later Upper Palaeolithic, according to their apparent degree of reliability, with question marks against problematic or unreliable groups. Available age estimates, whether reliable or problematic, are also listed for "bracketing" British industries, i.e. Middle Palaeolithic and Earlier Mesolithic. The relevant additional data given in Table 4 includes laboratory sample reference number, material "dated", deposit from which material came or probably came, site name, county, National Grid reference and bibliographic source. An indication of whether a sample is likely to have been contaminated in one way or another is also given, as is an indication of whether the sample was measured by the now obsolete "solid carbon" method. Most of the Earlier and Later Upper Palaeolithic age estimates are plotted in Figure 57. It is therefore considered unnecessary to go into a lengthy and repetitive listing and discussion here in the text, particularly as most of these "dates" are discussed by site, either under the section on stratigraphic evidence (pp. 36–76), or below under various other sections, often that on floral evidence (pp. 96–107). However, as regards the most reliable British Upper Palaeolithic age estimates obtained thus far, it might be stressed here that all of them are based on the measurement of residual 14-carbon from bone collagen, and that the present author is directly responsible for submission of all of them, except those from Dead Man's Cave (Yorkshire).

It is directly relevant that opinions vary on the reliability of "bone dates" and that better methods for the extraction of collagen are still being developed (e.g. see Longin, 1971). Dr. McBurney (personal communication) is of the opinion that "dates" based on charcoal are often more consistent stratigraphically and slightly older than "dates" based on bone from the same levels (e.g. the Molodova sequences, Klein, 1969b and references therein). But Prof. Bordes (personal communication) is of the converse opinion, particularly as regards the numerous "dates" now available for the south-western French Upper Palaeolithic (e.g. the Abri Pataud sequence, Vogel and Waterbolk, 1963 and 1967; also see Bordes, 1973; Movius, 1975). Also the laboratories (British Museum and Groningen University) which have carried out what I consider the most reliable British Upper Palaeolithic "datings", are both of the opinion that "good" samples of bone collagen are generally just as reliable as "good" samples of charcoal (R. Burleigh and W.G. Mook, personal communications). Thus the least problematic British Upper Palaeolithic "bone dates" can reasonably be accepted for the moment as reliable age estimates, although in the present author's opinion five of the Later Upper Palaeolithic "dates" (namely BM–439, 440A, 440B, 603 and 604; Dead Man's Cave and Robin Hood's Cave; Table 4) appear slightly too young according to their associated floral and faunal evidence (see pp. 103, 105 and Figs. 75, 77 and 82). Further age measurements are obviously necessary to obtain a more complete picture of British Earlier and Later Upper Palaeolithic age ranges.

It is perhaps not strictly relevant, but certainly of interest, that variations in the concentration of radiocarbon in the earth's atmosphere at different periods have been recognized (Willis, Tauber and Münnich, 1960; Suess, 1967), and that these fluctuations are thought to occur on a world-wide scale (Olsson, 1970; Renfrew, 1970; Stuiver, 1971; and references therein). The discovery of this source of error in radiocarbon dating has come about by comparison of tree-ring, or dendrochronological, calendar years with radiocarbon age estimates of selected groups of tree-rings. Thus provisional correction charts have been constructed for calibrating the approximate calendar age of radiocarbon "dates" back to about 7,000 year B.P. (Suess, 1967; Olsson, 1970; Renfrew, 1971; the Ottaways, 1972). These charts are based on the tree-ring sequence from the remarkably long-lived bristlecone pine (*Pinus aristata*), and they clearly indicate a tendency for radiocarbon age estimates to be increasingly too young from about 2,000 calendar years B.P. to the present tree-ring limit of about 7,000 calendar years B.P., being perhaps as much as 1,000 years too young at 7,000 B.P. (also see Michael and Ralph, 1974). If this difference is eventually found to continue right into the Last Glacial, then any "reliable" Upper Palaeolithic radiocarbon age estimates would obviously be affected. However, it is just possible that the difference between calendar years and radiocarbon years might decrease rather than increase during the Last Glacial. I suggest this because:

(a) radiocarbon, tree-ring and historical years seem to agree fairly well from about 500 to 2,500 B.P., thus implying that radiocarbon is not always too young and that its least error might correlate with recent cooling-down, its greatest known error having been associated with the Post Glacial climatic optimum (Suess, 1967; Olsson, 1970);

(b) the glacial varve analyses of de Geer (1912, 1934 and 1940) and Sauramo (1923) from Scandinavian outwash series seem to agree reasonably well with radiocarbon age estimates for the retreat of Last Glacial north-west European ice, commencing at about 17,000 B.P., with the Last Glacial itself interpreted by varve counting as ending at about 10,000 B.P. and by radiocarbon assays at about 10,200 to 10,500 B.P. (Hammen, Maarleveld, Vogel and Zagwijn, 1967; and see also British Late Last Glacial Zone III radiocarbon age estimates listed in Table 5 of this book, as well as Figs. 59 and 60); this point is further discussed by Stuiver (1971);

(c) there is the possibility of 14-carbon enrichment in wood types growing at high altitudes in environments such as that inhabited by *Pinus aristata*, the source of the bristlecone pine tree-ring "corrections" (Burleigh, Switsur and Renfrew, 1973; Harkness and Burleigh, 1974).

For the present one must simply rely on what few reasonable radiocarbon age estimates are available for the British Upper Palaeolithic, without attempting to extrapolate any sort of tree-ring calibration back into their range. All of the radiocarbon age estimates quoted or newly presented in this present study are therefore simply based on Libby's 14-carbon half-life of 5,570 ± 30 years, as this is the standard practice of all radiocarbon laboratories and the international journal *Radiocarbon*. These age estimates may of course be easily brought into line with what is currently thought to be the best value for the half-life of 14-carbon, 5,730 ± 40 years, by multiplying each one by 1.03. The standard laboratory and *Radiocarbon* journal practice of presenting age estimates as the number of years Before Present, or simply B.P., is also employed, rather than the cumbersome, ethnocentric and irrelevant "A.D./B.C." and "b.p./b.c." systems of later prehistorians and historians (e.g. see Renfrew, 1974). Thus the B.P. age estimates in this book are all "uncorrected dates".

Most available British Upper Palaeolithic radiocarbon age estimates older than 8,500 years B.P. are projected at one and two standard deviations on to a linear scale in Figure 57 (the reader is referred to Table 4 for a complete list of actual measurements, including the relevant readings younger than 8,500 B.P., which are definitely incorrect). The age estimates which the present author considers the most reliable thus far are plotted accordingly under either E.U.P. or L.U.P., whilst those "ages" from more problematic samples, layers and/or sites are plotted accordingly under either "?E.U.P." or "?L.U.P.". With the single exception of Badger Hole (BM–497), all of the age estimates shown are finite; that is, they have been determined as a number of years B.P. plus or minus so many years. The bulked Badger Hole sample had only sufficient collagen remaining to determine that its age was probably greater than 18,000 B.P.

The only four reasonably reliable finite age estimates available for the British Earlier Upper Palaeolithic are those from Kent's Cavern (GrN–6201, –6202 –6324 and –6325; see p. 41). Badger Hole is nevertheless included in the E.U.P. column as I consider the close association with a diagnostic leaf-point sufficiently certain (see p. 51 and Figs. 10 and 11). Whether the Earlier Upper Palaeolithic only "dates" from about 38,000 to 28,000 B.P. is still uncertain, but it would seem on the basis of the less clearly associated age estimates for the "Red Lady" of Paviland Cave and the "cave earth" of Cae Gwyn/Ffynnon Beuno Caves, that it might possibly have a longer "life-span". However, it is certainly rather curious that a derived bone from Robin Hood's Cave 1969 (a site with definite, but poorly stratified E.U.P. evidence) has conveniently yielded an estimate of about 28,500 B.P. The associations of the Ogof-yr-Ychen and Lea Valley age estimates with local "?E.U.P." evidence are even more tenuous and are merely offered as possibilities.

All ten finite Earlier Upper Palaeolithic "dates" are also plotted in Figure 57 on a quartile dispersion bar after the method of Ottaway (1973). Assuming that this dispersion might represent most or all of the "life-span" of the Earlier Upper Palaeolithic, and allowing it to be weighted by the best "dates" (i.e. Kent's Cavern), then as shown on Figure 57 the Earlier Upper Palaeolithic was possibly current in Britain, however sporadically, from about 38,000 to 18,000 B.P., but mostly round about 28,000 B.P. in the lower end of the statistically more reliable inter-quartile range. Finally, the onslaught of the Full Last Glacial maximum ice advance after about 20,000 B.P. seems the most likely termination time to me for the Earlier Upper Palaeolithic, if indeed it survived that long at all.

Thus far there are eight reasonably reliable finite radiocarbon age estimates for the British Later Upper Palaeolithic (see Table 4) and these are shown on Figure 57 under L.U.P. The less certain possibly Later Upper Palaeolithic age estimates range from about 12,200 to 2,800 B.P., those younger than 8,500 B.P. definitely being incorrect (see Table 4). Seven of these possibly associated "dates" older than 8,500 B.P. are plotted under "?L.U.P." on Figure 57. To these could be added the four age estimates from Sproughton, although they have been left off this diagram as their relationship to the Sproughton "?Meso./L.U.P." artifacts is still tenuous (see Table 4), and there are now already a fair number of "L.U.P." dates from other sites.

If one accepts the general premiss that the British Palaeolithic formally ceases with the known end of the Last Glacial (and the so-called "End of the Pleistocene"), then any British Later Upper Palaeolithic radiocarbon "dates" younger than about 10,000 B.P. ought to be considered unacceptable. This is borne out by the fact that eight reasonably reliable age estimates for British Earlier Mesolithic material from Thatcham and Star Carr range from about 10,400 to 9,500 B.P. (see Table 4). Therefore the best fitting radiocarbon age, for a change or possibly "transition" from Later Upper Palaeolithic to Earlier Mesolithic, if anyone is determined to draw a sharp dividing line between them, would still seem to be about 10,000 B.P., if not slightly earlier. Assuming that the available Later Upper Palaeolithic age estimates which have measurements older than 10,000 B.P. represent, however sporadically, most of the "life-span" of the Later Upper Palaeolithic, then as suggested by the inter-quartile dispersion bar on Figure 57 it was probably current from about 12,000 to 10,000 B.P., although typologically "Later Upper Palaeolithic" industries may sometimes have survived alongside frankly "Mesolithic" ones until as late as 9,000 B.P., whilst the Later Upper Palaeolithic itself may have begun by 14,000 B.P. or perhaps even earlier.

As may be clearly observed on Figure 57, there is an apparent gap of at least 3,000 radiocarbon years between the Earlier and the Later Upper Palaeolithic in Britain, and this gap may well be as much as 5,000 years or even more. As will now be demonstrated, this apparent archaeological hiatus is most likely the result of the climatic deterioration and maximum ice advance of the Full Last Glacial or Full "Devensian".

All available radiocarbon age estimates for the British Last Glacial, or "Devensian" as it is now termed by Mitchell, Penny, Shotton and West (1973), are listed in Table 5. They are arranged in groups in stratigraphic order beginning with the so-called "Chelford Interstadial" of the Early Last Glacial and ending with the various sub-divisions of the Late Last Glacial. Within these groupings the "dates" are arranged in chronological order from oldest to youngest. The same background information as before is given for each sample. It is considered unnecessary to go into a lengthy and repetitive listing and discussion here in the text.

Figure 58 shows the dispersion bars and inter-quartile ranges of those British Last Glacial age estimates which fall in the approximate span 45,000 to 5,000 B.P., including for comparison those available so far for the Upper Palaeolithic and Earlier Mesolithic, however problematic the estimates may be (again, see Tables 4 and 5 for actual measurements). On the basis of careful stratigraphic studies and radiocarbon "dating" it is now fairly certain that the maximum advance of ice in the Last Glacial (or "Devensian") occurred somewhere between 20,000 and 16,000 B.P., reaching its height between about 18,000 and 17,000 B.P. (see particularly Penny, Coope and Catt, 1969; Rowlands, 1971). As one might expect, this agrees with age estimates for the last ice maximum already obtained in Germany and North America which also range from about 20,000 to 16,000 B.P. (see particularly Woldstedt, 1960 and 1967; Turekian, 1971). A major readvance of the ice in Britain seems then to have occurred between about 16,000 and 15,000 B.P. (see particularly Saunders, 1968). Although it has many local names, I have termed this major readvance the "Scottish, Irish and Welsh Readvance", a phrase which describes where it was prevalent (see pp. 84–87 as well). This Scottish, Irish and Welsh Readvance may have been more or less synchronous with the "Pomeranian Readvance" of Woldstedt (1967) in northern Germany. Both the Last British Ice Advance Maximum and the subsequent Scottish, Irish and Welsh Readvance, for their extent, severity and combined duration of about 5,000 years, are considered by the present author to represent the Full Last Glacial. They are apparently preceded in Britain by a long cold phase from about 40,000 to 20,000 B.P. (see Fig. 58), the Middle Last Glacial, during the latter part of which ice was presumably beginning to accumulate, as is suggested by the age estimates for glacially derived material (the youngest for material caught up by the maximum advance are round about 25,000 B.P.; the isolated one at about 17,000 B.P. is from material caught up in the Welsh Readvance; see Fig. 58 and Table 5). Prior to somewhere between 35,000 and 40,000 B.P. the sea level on a world-wide scale may have actually been near or just below its present level before falling as much as 130 metres during the Full Last Glacial (Milliman and Emery, 1968); such may be suggested by purely British evidence as well (see Donovan, 1962).

The changing environments associated with the Middle, Full and Late Last Glacial in Britain doubtless had an effect on Upper Palaeolithic hunter/gatherers, but how strong that effect was is still uncertain, although it would seem that the Earlier Upper Palaeolithic inhabitants came generally after sea level had begun dropping, thus they were probably able more or less to walk in, and that they were eventually driven out, presumably by a general lack of game, during the onslaught of the Full Last Glacial maximum ice advance. The subsequent Later Upper Palaeolithic people then entered during the Late Last Glacial, perhaps evolving *in situ* into the British Earlier Mesolithic.

The last definite interstadial, in the sense of relative warming-up, in Britain before the maximum ice advance was the so-called "Upton Warren Interstadial *s.s.*" of about 43,000 to 40,000 B.P. (see Fig. 58 and Table 5), so it also seems likely that the Earlier Upper Palaeolithic was unable to enjoy the mildest conditions possible during the British Last Glacial (in contrast to the L.U.P. which went through both mild and severe conditions), although there does appear to be some Dutch evidence for "interstadials" there at about 32,000 to 30,000 B.P. (Hammen, Maarleveld, Vogel and Zagwijn, 1967), and again possibly at about 18,600 B.P. (Vogel and Waterbolk, 1972).

Figures 59 and 60 show available radiocarbon age estimates for the English and Welsh Late Last Glacial and the Irish and Scottish Late Last Glacial, respectively. These are again presented as dispersion diagrams with their inter-quartile ranges clearly indicated. Here the relevant evidence is divided into separate illustrations because regional variations are more apparent in the Late Last Glacial than in the earlier phases of the Last Glacial. Figure 59 also includes Later Upper Palaeolithic and Earlier Mesolithic radiocarbon dispersion bars and inter-quartile ranges. There is as yet no definite either Upper Palaeolithic or even earliest Mesolithic (i.e. older than 9,000 B.P.) evidence from Ireland and Scotland. Figure 60 also includes data on the apparent local changes in sea/land level and the ice readvances of the Scottish Late Last Glacial (Bishop and Dickson, 1970; Donner, 1970; Sissons, 1967); the Scottish sea/land level changes also affected the north-eastern part of Ireland and possibly the northern and western coasts of the Isle of Man (Stephens and Synge, 1966; Donner, 1970). The floral information on Figures 59 and 60 is modified after West (1968, particularly his table 12.10) with amendments from my own pollen studies (see pp. 96–107). The interpretation of the temperature, or so-called "thermal", requirements of beetles given on Figure 59 is based on the studies of G.R. Coope (Coope, 1969b and personal communications; Coope, Morgan and Osborne, 1971). The apparent discrepancy between floral and coleopteran evidence in Zones Ic and II may be partly due to different rates of migration by various plants and beetles on the one hand, but Coope (personal communication; but also see Coope, Morgan and Osborne, 1971) is of the opinion that it is largely due to a partial misinterpretation of the requirements of the relevant flora, notably *Betula* and *Juniperus*. If July average temperatures really were higher in Zone Ic than in Zone II (the so-called "Allerød Interstadial"), then *Juniperus communis*, with its somewhat greater range of climatic and soil tolerance than *Betula pendula* and *B. pubescens*, the three species one is probably dealing with (Godwin, 1956; Clapham, Tutin and Warburg, 1962), may have been able to take better advantage of the Zone Ic extremes of comparatively cold winters and hot summers (Coope, personal communication, estimates a July average temperature of about 17 °C. for Zone Ic which then gradually fell during Zone II, reaching a low point near 10 °C. in Zone III; also see Coope, Morgan and Osborne, 1971). High peaks of *Juniperus* during Zone Ic are apparent in many British pollen profiles (e.g. Tadcaster in Yorkshire, Bartley, 1962; Llyn Dwythwch in Caernarvonshire, Seddon, 1962), and, interestingly enough, they are again apparent at the transition from Late Glacial Zone III to Post Glacial Zone IV, particularly in the more northerly and/or mountainous parts of Britain and Ireland (e.g. Scaleby Moss in Cumberland, Godwin, Walker and Willis, 1957). It may be significant that the maximum warmth of Zone Ic as indicated by beetle remains in southern Britain appears to be contemporaneous with the height of the marine transgression during the Late Last Glacial "isostatic recovery" of northern Britain (compare Figs. 59 and 60). In other words, although the world-wide sea level was already beginning to rise (Milliman and Emery, 1968), as were those regions such as Scotland which had been substantially depressed or down-warped by the great weight of Full Last Glacial ice caps (Stephens and Synge, 1966; Donner, 1970), in response to a general melting of the ice sheets caused presumably by a general rise in temperature, a sudden increase of warmth towards the middle or end of Zone I may have aided the sea in quickly inundating the continental shelves and present coasts of the only slowly rising down-warped regions. Cooling down during the latter part of Zone II and minor ice readvances in Zone III (see Fig. 60) may then have allowed Scotland and other down-warped regions to catch up with a still generally rising sea, so that by Post Glacial Zone V the coastline became temporarily about the same as at present. Regions which had not been generally down-warped during the Full Last Glacial, such as Wales and most of England, remained above the rising sea throughout the Late Last Glacial and Early Post Glacial, although their previously exposed continental shelves became more and more inundated.

The above-mentioned environmental and topographic changes, most of which have now been radiocarbon "dated", appear to have been fairly rapid during the Late Last Glacial and Early Post Glacial, and they doubtless had a profound effect on the Later Upper Palaeolithic and Earliest Mesolithic hunter/gatherers who ventured into southern Britain during and after the various glacial retreats. Scotland and Ireland were probably for the most part inaccessible and/or undesirable to these people(s), even though some flora and fauna managed to reach those further regions long before the first definite traces of man (Mesolithic, younger than 9,000 B.P., or really most often younger than 8,000 B.P.). Of course, this is not to say that some of the more adventurous Late Last Glacial hunter/gatherers did not attempt to follow whatever migration routes there were into Scotland and Ireland for the larger land mammals such as *Rangifer* and *Megaloceros*, but if they did they left few, if any, discernible traces (Movius, 1942; Lacaille, 1954; Coles *et al*., 1971; Mitchell, 1971; Mercer, 1974). Differences between and amongst Later Upper Palaeolithic and Early Mesolithic tool kits might well be partly explained by the necessity for different responses to different environments at different times; in other words, chronologically aligned functional changes might be implied by apparent variations in tool-forms both within and between these two archaeological groups.

However, although many radiocarbon age estimates are available for the relevant environmental evidence, much more "dating" of the archaeological material is required as well as further extensive field-work and artifact analyses, before one may really make reasonably sound generalizations about morphological/functional variability as a possible indicator of ecological interaction with varying environments during any part of the Upper Palaeolithic or Mesolithic in Britain. We are simply on the threshold of the "absolute dating" and "ecosystematic" era in archaeology, or rather in what is becoming "palaeoanthropology/palaeoecology".

B. ECOLOGICAL FRAMEWORK FOR THE UPPER PALAEOLITHIC AND THE LAST GLACIAL IN BRITAIN

1. Geomorphological Evidence

Maps 16 and 17 show the main ice fronts of the British Last Glacial as generally agreed upon by the late D.F.W. Baden-Powell and the present author. They have been drawn by me after a careful review of the available literature and Mr. Baden-Powell's own field maps (as yet unpublished, but kindly made available a few years before his death in 1973), and they are, as far as I am aware, the most up-to-date maps which could possibly be produced by me at the time of writing. The ice fronts of the Late Last Glacial (Map 16) are shown first for two reasons: (a) the present outline of the British Isles is less obscured than it is on Map 17; and (b) the maximum ice advances of the Full Last Glacial (Map 17) presumably had their origins in the Scottish Highlands and Southern Uplands as well as the corries of mountainous regions further to the south and south-west. However, our discussion of the main lines of evidence along which these two maps are drawn will be in stratigraphical/chronological order from bottom to top, or maximum to least ice, beginning with just a brief word on Middle Last Glacial ice fronts.

Middle Last Glacial, about 40,000 to 20,000 B.P.

If any ice fronts were already in existence in Britain during the Middle Last Glacial, their precise geographical positions are still quite unknown. The ice sheets which attained their maximum during the Full Last Glacial (Map 17) presumably grew from ice centres such as those which survived into the subsequent Late Last Glacial (Map 16), but whether they grew rapidly or slowly is still uncertain. The "Weertman Theory of Glacial Nonequilibrium" deduces from available world-wide evidence that an ice sheet requires from 30,000 to 15,000 years to form, but that it can be dissipated in only 4,000 to 2,000 years (Weertman, 1964). If this be true, then some glaciers, whether corrie glaciers or slightly larger, were probably present and gradually growing in Britain during the Middle Last Glacial. And by the latter part of the Middle Last Glacial these hypothetical glaciers might well have reached what became the later limits of the Scottish, Irish and North Welsh Readvances (see Map 17), or they may have been only somewhere between those limits and the still later lines of the Dinnet-Perth-Armoy Readvance (see Map 16). As has already been suggested on the basis of radiocarbon age estimates from glacially derived materials (see Fig. 58 and p. 81), some sort of ice advance was most likely underway by at least about 25,000 B.P. and was gathering up earlier marine shells, land mammal bones and plant remains; materials which could of course be of most any age earlier than the glaciers concerned. Depths and radiocarbon ages of sea-level indicators from continental shelves throughout the world, but particularly from the reasonably stable Atlantic shelf of the United States, suggest a world-wide drop in sea level from at about or just below the present level at about 35,000 B.P. to at least −50 metres by about 20,000 B.P. (Milliman and Emery, 1968). This change in sea level was presumably directly linked to the expansion of glaciers in many parts of the world, including Britain.

Full Last Glacial, about 20,000 to 15,000 B.P.

The glaciers, whose existence before the end of the Middle Last Glacial may be assumed, appear to have attained their maximum extent over most of Ireland, more than half of Britain and large adjacent areas of what are now sea beds (Map 17) by about 18,000 to 17,000 B.P. during the Full Last Glacial (see pp. 80–82; as well as Fig. 58 and Table 5). More or less simultaneously, great Scandinavian glaciers spread out covering the Baltic plains of Russia, Poland, Germany and Denmark, and almost certainly extending a huge lobe of Norwegian ice across the northern bed of the present North Sea (Woldstedt, 1967, Figs.1 and 2). This Norwegian ice sheet most likely would have helped the deflection of north-eastern British glaciers in a principally south-eastern direction so that the latter eventually impinged on the present coasts of Yorkshire, Lincolnshire and Norfolk (see Map 17). Perhaps about 1,000 years after the British Last Glacial maximum ice had halted its advance and melting had begun, a series of important but localized readvances occurred over most of Scotland, the northern half of Ireland and separately, but probably simultaneously, the northern half of Wales (Map 17). World-wide sea-level indicators suggest that the sea carried on dropping from about −50 metres at approximately 20,000 B.P. to about −130 metres by about 16,000 B.P. (Milliman and Emery, 1968), if not to as much as −175 metres (Jongsma, 1970). The −100 metre sea level suggested on Map 17 is what I consider to be the most cautious estimate for the lowest sea level reached in relation to the British Isles. This clearly illustrates a "land-bridge" (i.e. the exposed, ice-free, or unglaciated, −100 metre portion of the continental shelf) between southern Britain and the European continent, while if the sea were in fact lowered to −130 metres or more, there would certainly have been a substantial "land-bridge" between southern Britain and southern Ireland as well, with perhaps only a broad river or two dividing these dry-land connections.

Late Last Glacial, about 15,000 to 10,200 B.P.

During the Late Last Glacial at least two ice readvances occurred, one at some point early within Zone I and the other, the lesser of the two, at some point within Zone III. In terms of ice sheets and valley glaciers these readvances affected only Scotland, except for a broad lobe which probably just impinged on the northernmost coast of Ireland (see Map 16). I have termed the Zone I readvance the "Dinnet-Perth-Armoy Readvance" after three of the villages or towns which it just managed to reach in Scotland and northern Ireland. The Zone III readvance has been commonly called either the "Highland Readvance", the "Loch Lomond Readvance" or a combination of both names; I prefer to call it the "Loch Lomond Readvance" as that is the name in current general usage. A short name for the Zone I readvance is often simply "Perth Readvance". Beyond these readvance ice fronts there were a number of widely scattered, but localized reappearances of corrie (or "cirque") glaciers during both Zones I and III. Their principal occurrences are indicated on Map 16. World-wide sea-level indicators suggest that the sea was still below −100 metres at about 15,000 B.P., but that by 14,000 B.P. it had begun rising (Milliman and Emery, 1968). By about 12,000 B.P., or the transition from Zone I to II, the sea had risen to about −70 metres, and had begun inundating glacially down-warped regions which were rising at a slower rate (West, 1968; Bishop and Dickson, 1970; Mörner, 1969). Finally, by 10,200 B.P., or the approximate transition from Zone III to Post Glacial Zone IV, the sea level had recovered as far as about −50 to −40 metres (Milliman and Emery, 1968; Mörner, 1969). The −100 metre sea level shown on Map 11 allows for what I consider the maximum possible exposure of the British continental shelves during Zone I. This excludes the present sea beds adjacent to mainland Scotland, as those areas were almost certainly transgressed by the latter part of Zone I (see Fig. 60 as well). Various narrow "land-bridges" probably connected Ireland to southern Britain during the earlier part of Zone I but were probably severed by Zone II. Some sort of "land-bridge", initially quite broad but gradually decreasing, connected southern Britain to the European continent throughout the Late Last Glacial.

There are well over one hundred bibliographic references for the many minutiae which support, or in a few cases contest, the information on Last Glacial ice fronts which I have presented in Maps 16 and 17. As it is mainly the synthesis and interpretation of such studies which is relevant to this present work, it is considered reasonably sufficient to state only the major sources: Bowen, 1966 and 1970; Charlesworth, 1929, 1956 and 1963; John, 1971; Penny, 1964; Penny, Coope and Catt, 1969; Rowlands, 1971; Saunders, 1968; Shotton, 1967; Sissons, 1967; Stephens and Synge, 1966; Synge, 1956; Synge and Stephens, 1960; Turekian, 1971; Watson, 1967; West, 1968; Woldstedt, 1967; as well as personal communications from D.F.W. Baden-Powell, G.R. Coope and D.T. Donovan. Where these sources or their references are in disagreement on points of detail, I have attempted to employ the most reasonable and/or up-to-date interpretation of the evidence thus far available. In some cases I have offered new correlations of previously discontinuous ice fronts. When such correlations have been across present sea beds they are obviously largely speculative, but field studies are now in progress on tills, etc. found off the present coasts of Britain (Donovan, personal communication; but also see Donovan, 1968 for a preliminary account). New correlations across present land surfaces, on the other hand, have been based mostly on the as yet unpublished field maps and notes of the late Mr. Baden-Powell (by his kind permission). Studies by the latter were still in progress, particularly on the Scottish, Irish and North Welsh Readvance, just before his untimely death in 1973. Many minor known ice-free areas, including principally "nunataks", have not been included on Map 17 for reasons of general clarity. These minor ice-free areas, as well as the major ones, may have served as "glacial refugia" for various cold-tolerant flora and fauna. Some ice-free areas, such as Kent, East Anglia and the Peak District in England and various parts of south-western Wales, were almost certainly covered by fine wind-blown deposits of "loess" grade (see Weir, Catt and Madgett, 1971; Catt, Corbett *et al.* 1971; Bryan, 1970; Watson, 1967; respectively), probably mainly during the latter part of the Full Last Glacial. Vast spreads of glacial outwash also covered some of the areas adjacent to the various Last Glacial ice fronts (West, 1968), not to mention the many morainic features, etc. which were left behind by these ice fronts. Melt-waters formed "proglacial" lakes and cut various overflow channels, whilst advancing or retreating ice diverted various river courses. In brief, the face of Britain changed rapidly and drastically during the latter half of the Last Glacial.

The above summary of British Last Glacial ice fronts, sea levels, ice-free areas and "land-bridges" has been included in this section on geomorphological evidence for Upper Palaeolithic environments as it is absolutely essential background information for any study of those hunter/gatherers who attempted to exploit whatever resources were available, within the limits of their own technologies.

There now follows a brief review of the principal results and possible interpretations of granulometric analyses which have been carried out by the present author on samples collected during the course of his field-work in 1968 and 1969 at various British Upper Palaeolithic sites. The relevant sites are considered in the same clock-wise geographic order as before beginning with the south-west, and the results from each site are presented on linear diagrams as percentages of the different components: scree, sand, silt and clay. These diagrams are based on the laboratory observations recorded in full in Tables 6 to 13, where information on the colour of the various samples, when wet and when dry, may also be found (see pp. 10–11 for description of author's method of granulometry). The vertical scale on each of these diagrams is in "decimetres", or 10 centimetre units.

Badger Hole, 1968: Granulometric Analysis

Figure 61 shows the results of the analysis of five deposit samples which were collected during the present writer's excavations in the entrance platform at Badger Hole (Somerset) in 1968 (see Figs. 10 and 11 for sampling positions at the cave-site). The interesting aspects of this diagram are higher scree frequencies with some weathering at the bottom and top of the sequence, and an intervening, unweathered peak of mainly sand grades, just above the level of the occurrence of Balch's near-by, undisturbed Earlier Upper Palaeolithic leaf-point. As this rise in mainly medium and fine sand is immediately above the Earlier Upper Palaeolithic level and the latter has a radiocarbon age estimate of > 18,000 B.P. (BM–497), it seems not unreasonable to interpret the sand peak as due to intense periglacial winds during the Full Last Glacial, particularly as the associated scree component is entirely thermoclastic. This may appear an unusual interpretation to some, but it has been my field and laboratory observation that some scree of one kind or another nearly always forms at a hill-side site beneath a rock exposure of limestone or similar material in Britain. Such scree production may be reduced at times by milder conditions and abundant plant growth, but it is not stopped by them, the scree itself simply becoming weathered to some extent before, during and/or after its detachment from a given rock face. However, sometimes during mild, damp conditions with frequent, comparatively minor temperature changes, the production of weathered scree may be quite high. Also during cold, damp conditions with frequent temperature changes the production of thermoclastic scree may be quite high. But as long as the rock face remains exposed, formation of some scree generally continues. This scree production is then apparently only greatly reduced or even sometimes temporarily halted by what is assumed to be maximum cold, dry conditions such as probably existed for a time during particularly the Full Last Glacial and resulted in the appearance of proglacial and periglacial "Arctic desert cover-sands", definitely in the Netherlands (Hammen, Maarleveld, Vogal and Zagwijn, 1967) and most likely in parts of Britain as well (Bryan, 1970; Watson, 1967). Such a halt, or near halt, in scree production is generally associated at hill-side sites with a high rise in sand, silt and possibly some clay, any of the scree that is formed being entirely thermoclastic. At Morfa-bychan near Aberystwyth there is a 37 metre high succession of Last Glacial screes, interrupted only within the upper portion by a true loess grade deposit of 70% fine sand and silt with just a few lenses of thermoclastic scree, this loess being interpreted as Full Last Glacial (Watson, 1967). Against the south-facing limestone cliff at Brean Down (Somerset) there is a 35 metre high succession of thermoclastic screes followed by a very thick accumulation of blown sand with intercalated thermoclastic screes and topped by thermoclastic and weathered screes and soils. I have visited the Brean Down section, and both the authors of the published account (ApSimon, Donovan and Taylor, 1961) and I agree that it is probably mostly Last Glacial, but whereas they attribute the massive blown sand layers to the Late Last Glacial, I would strongly suggest these layers are mostly Full Last Glacial, the nearest maximum ice front having been just across the Bristol Channel at Cardiff (see Map 17). They only found faunal evidence below and above these main sand layers, the sands themselves being quite sterile. Thus on this growing, although still problematic, body of evidence it would seem that Britain probably experienced some sort of "Arctic desert" conditions during the Full Last Glacial, and that this was associated with temporary, but substantial, falls in thermoclastic scree formation with substantial rises in sand and silt at at least the more exposed sites; and this is one of my reasons for interpreting such falls in many cases as quite likely due to the Full Last Glacial.

Hyena Den, 1968: Granulometric Analysis

Figure 62 shows the results of an analysis of deposit samples collected from a new exposure at the Wookey Hole Hyena Den (Somerset) in 1968. These samples were collected by the present author from the south-eastern face of excavations which were being conducted by Professor E.K. Tratman (for sampling positions see Tratman, Donovan and Campbell, 1971, Figs. 40 and 41B). The exact levels of the Middle Palaeolithic and Earlier Upper Palaeolithic tool-forms found by Dawkins in 1859–74 remain uncertain, but it seems quite likely that they were within the stratigraphic range of the bone fragments and identifiable fauna found in 1966–70 and plotted on Figure 62 (also see Tratman, Donovan and Campbell, 1971, Figs. 40 to 44B, Table 1). It is therefore of interest that the granulometric analysis indicates a peak of principally sands in sample 6 at the top of the overlying layer A3. As may be seen, this sand peak is associated with some thermoclastic scree as well as a slight rise in silt, and it has been interpreted as possibly representative of part of the Full Last Glacial. It may therefore be of some relevance that the interface of layers A3 and A2b was found to have been quite severely contorted, presumably by solifluction and cryoturbation (Tratman, Donovan and Campbell, 1971, Fig. 41B). The bone scatter itself is associated with two peaks of scree, the lowermost partly weathered and the upper entirely thermoclastic, with an intervening minor peak of sand (see Fig. 62). These varying phenomena are thought indicative of changing climatic conditions during the Early and Middle Last Glacial, the Earlier Upper Palaeolithic artifacts having presumably been associated with the latter climatic phase.

Sun Hole, 1968: Granulometric Analysis

Figure 63 shows the results of the analysis of thirty deposit samples which were collected during the present writer's excavations at Sun Hole (Somerset) in 1968 (see Figs. 12 and 14 for sampling positions, although it should be noted here that these samples came from not outside the cave-site, but 5 metres inside from the present overhang). The interesting aspect of this diagram is a variable, and perhaps continual, series of abundant scree grades from bottom to top of the entire, long sequence. The lack of any substantial rise in the sand grades may, however, be more apparent than real, as the samples are from a location that even in winter today is very sheltered (my final season at Sun Hole was actually in December, when it was sufficiently warm at and inside the entrance under just a low rising sun to work comfortably without a shirt!). There are clearly various minor fluctuations in the deposition of sand, silt and clay, with two possible peaks in layers B1−1 and B2−2. The stratigraphic range of the Later Upper Palaeolithic tool-forms found by Tratman (1955, Fig. 10) is indicated on Figure 63, with two small boxes with hatching to denote the levels of the apparent concentration of these artifacts. Their range is associated with two brief, but exceedingly high, peaks in thermoclastic scree formation, and as shown these peaks have been interpreted as possibly indicative of the rapid climatic alterations of the Late Last Glacial.

Cathole, 1968: Granulometric Analysis

Figure 64 shows the results of the analysis of thirteen deposit samples which were collected during the present author's excavations in the entrance platform at Cathole (Glamorganshire) in 1968 (see Figs. 15 and 16 for sampling positions at the cave-site). The possibly significant aspects of this diagram are very high proportions of sand, silt and clay in layers A1 to A3, followed by high frequencies of screes in layers LSB to F, the uppermost screes being quite weathered and clearly Post Glacial. The particularly high peak of medium and fine sand, as well as silt and clay, in layer A3 is thought to be due to the almost proglacial situation of Cathole during the Full Last Glacial as the maximum ice reached the Swansea Bay (see Map 17). Artifacts of Later Upper Palaeolithic aspect occurred above this level in layers LOB and MSB, which both have abundant thermoclastic scree although the latter of the two layers has slightly more sand. These thermoclastic screes are thought to have been formed during the Late Last Glacial.

Long Hole, 1969: Granulometric Analysis

Figure 65 illustrates the results of the analysis of twenty-two deposit samples which were collected during the present author's excavations in the entrance platform at Long Hole (Glamorganshire) in 1969 (see Figs. 18 and 21 for sampling positions at the cave-site). The possibly significant aspects of this diagram are three clearly distinguishable peaks in scree formation with two intervening peaks of considerable magnitude of sand and silt deposition, particularly in the case of layer A3c. The rise of sands which comprises layer A3c consists almost entirely of medium and fine sand and silt, with some material of clay grade and no or very little scree, all of which is thermoclastic. Layer A3c is thought to be almost certainly due to the deposition of wind-transported particles during the Full Last Glacial, and it is therefore of some importance that artifacts of Earlier Upper Palaeolithic aspect occurred immediately below in layers A3a and A3b, in deposits which seem quite likely due to the cold, but not maximum cold, conditions of the Middle Last Glacial. Possible interpretations of the other layers are also offered on Figure 65, but it is worth noting here that the interface of layers A3b and A3c was found to have been very contorted, presumably by solifluction and cryoturbation (see Figs. 19 and 21). Also the general sequence at Long Hole from the geomorphological point of view does not seem very unlike that already described from Morfa-bychan (Cardiganshire) and interpreted as covering the whole of the Last Glacial (Watson, 1967).

Mother Grundy's Parlour, 1969: Granulometric Analysis

Figure 66 shows the results of the analysis of eight deposit samples which were collected during the present writer's excavations in the entrance platform at Mother Grundy's Parlour (Derbyshire) in 1969 (see Figs. 23 and 24 for sampling positions at the rock-shelter). The possibly significant aspects of this diagram are very high proportions of medium and fine sand, with relatively abundant silt in layer A, followed by high frequencies of thermoclastic screes in layers LB and SB, and capped by lower proportions of partly or entirely weathered screes with principally coarse and medium sands in layers C and D, these upper two layers being quite clearly Post Glacial and containing abundant Mesolithic and possibly some Neolithic artifacts. The basal sand peaks with their intervening slight rise of entirely thermoclastic scree in sample 2 may be due to the wind-transport of fine particles during the Full Last Glacial which had maximum ice advances which nearly, but not completely, surrounded Derbyshire and Nottinghamshire (see Map 17). However, if Dawkins and Mello's layer 1, sterile "white calcareous sand", equals my layer A, then A may possibly be due instead to the Penultimate Glacial as their layer 1 was directly overlain inside at the rear of the site by two discontinuous layers with a Last Interglacial fauna (see pp. 60–61). This alternative interpretation of the origin of layer A would not be impossible, as Creswell Crags itself, where the site is situated, was quite likely formed in part as a meltwater overflow channel at some point during the Penultimate Glacial, presumably towards the end (D.F.W. Baden-Powell, personal communication). But as the correlation of layers 1 and A remains uncertain, I am retaining my tentative assumption that layer A, as observed outside the shelter, is of Full Last Glacial origin. Being of a somewhat less problematic nature, layers LB and SB seem quite likely to have been formed during the Late Last Glacial, and this probability is of direct interest as artifacts of Later Upper Palaeolithic aspect have been found in those layers.

Robin Hood's Cave, 1969: Granulometric Analysis

Figure 67 demonstrates the results of the analysis of eight deposit samples which were collected by the present author during his excavations in the entrance platform at Robin Hood's Cave (Derbyshire) in 1969 (see Figs. 27 and 28 for sampling positions). The important aspects of this diagram are two clear peaks of thermoclastic scree formation, including mostly coarse scree in the peak of layer OB, and an intervening slight peak of sand grades, silt and clay as suggested by sample 4. The basal peak of principally coarse and medium sand in layer A is also of some interest, as it is associated with some medium and fine scree which is mostly weathered but hardly any silt and clay. My interpretation of layer A is that it is probably largely hill-wash, or "colluvium", and that it may be due to the comparatively mild conditions of the Early Last Glacial or some even earlier mild or cool climatic phase. The overlying layer B/A has a scree component which is entirely thermoclastic and therefore probably its own, but its finer elements appear to be a mixture of what might be its own with some from layer A. I would suggest that this apparent mixture which may have formed B/A might be due in part to a sludging out from the site of most of any Middle and Full Last Glacial deposits which may have previously accumulated (see pp. 64–67). Such sludging out might have occurred either towards the end of the Full Last Glacial or in the beginning of Zone I of the Late Last Glacial, when permafrost was presumably starting to lose its hold on Britain. Layers LSB to USB, on the other hand, are probably entirely Late Last Glacial, and as implied on Figure 67, their two peaks of thermoclastic scree may be due to the milder parts of the Late Last Glacial when there would probably have been more frequent freezing and thawing, particularly during springs and autumns. Layers B/A to OB yielded Later Upper Palaeolithic tool-forms, whilst layer USB yielded tool-forms which appear to combine Later Upper Palaeolithic and Mesolithic aspects.

Hengistbury Head, 1968–69: Granulometric Analysis

Figure 68 shows the results of the analysis of eight of the seventeen deposit samples which were collected during the present author's excavations at Site C2 on Hengistbury Head (Hampshire) in 1968–69 (see Figs. 37 and 39 for sampling positions at this open-site; samples 7, 8 and 11 to 17 were not included in the granulometric studies because of their peaty structure). This diagram clearly demonstrates the high proportion of medium sand which constituted samples 2 to 10, as well as the comparatively high frequencies of fine sand, silt and clay in those same samples. It also shows the proportion of gravel in the basal sample 1, a gravel which appeared generally water-worn and was presumably deposited by some ancient river, perhaps in the Last Interglacial, but which had been subsequently partly frost-shattered, presumably during the Last Glacial. The overlying sands of layers C2 and C3 are also probably of fluvial origin and Last Interglacial, but the sands of layers B1 to A1a include slightly higher frequencies of fine sand, silt and clay which suggest in part a wind-blown origin, together with possibly some soliflucted redeposition, perhaps during the latter portion of the Last Glacial. Sample 6, at the level of the main concentration of Later Upper Palaeolithic tool-forms and other artifacts, has a slight rise in coarse thermoclastic gravel, which is presumably due largely to soliflucted redeposition of a thin spread of gravel on the surface of layer B1 from the higher exposures of the underlying layer C1 to the west as well as the east of the site. Samples 4 and 10, on the other hand, seem quite likely the result of almost entirely wind-transported materials. It is clearly significant that the interface of layers C3 (or C2 where C3 was absent) and B1 was found to be intensely involuted (see Figs. 38, 39 and 44); this is most likely due to cryoturbation during the Full Last Glacial after some of layer B1 had been deposited; the interface itself seems to represent a gap in the depositional sequence. Layer A2 at the top is definitely a Post Glacial peat, and Mesolithic tool-forms and other artifacts were found often at its interface with the underlying layer A1a or within an intervening, discontinuous peaty layer A1b. Layer A1a itself seems most likely to be composed of Late Last Glacial wind-blown sands, into the top of which Mesolithic artifacts became subsequently trodden.

The above results of the granulometric analyses are included as lines of essential background information. However, it must be stressed that their interpretation is often very difficult, and the suggested origins and ages of deposition and erosion are simply those considered by the present author to be the most reasonable interpretations; others would be possible. Linear diagrams have been employed rather than cumulative curve diagrams as the former are considered to be the clearest method of presenting a stratified column of samples, although for soil scientists, the latter may often be preferable for the detailed presentation of one or possibly a very few samples (e.g. Cornwall, 1958, Figs. 1 and 14 to 16). In France at least, similar linear diagrams are now in common use for stratified Palaeolithic sites, both cave and open-air, and where minutiae are desired as well, they are presented together with various cumulative curve diagrams (e.g. Grotte du Lazaret, Miskovsky, 1969, Fig. 2; "la station de plein air du Basté", Chauchat and Thibault, 1968, Fig. 7). The aim of the present study is simply to determine the general nature of samples from British Upper Palaeolithic sites.

In summary, it would appear from the granulometric analyses of the author's samples that Earlier Upper Palaeolithic artifacts at Badger Hole, Long Hole and possibly the Hyena Den, are associated with layers of silty sand containing variable proportions of thermoclastic scree, the suggested age of which is Middle and possibly early Full Last Glacial, whilst Later Upper Palaeolithic artifacts at Sun Hole, Cathole, Mother Grundy's Parlour and Robin Hood's Cave are more often associated with higher frequencies of thermoclastic scree, the suggested age of which is Late Last Glacial. The single open-air site considered, Hengistbury Head, would appear to have Later Upper Palaeolithic artifacts in association with soliflucted and wind-blown silty sand, the suggested age of which is Late Last Glacial. These conclusions are in no way at variance with the broader picture presented in the section on chronology.

2. Floral Evidence

The known flora of the British Last Glacial is that of a generally open vegetation of tundra to steppe type, interrupted by perhaps only two brief episodes of presumably localized cool woodland development: the "Chelford Interstadial" of about 60,800 ± 1500 radiocarbon years B.P. with a Boreal coniferous forest; and the "Allerød Interstadial", or Late Last Glacial pollen Zone II, of about 11,950 to 10,750 radiocarbon years B.P. with a Sub-Arctic/Boreal birch "park tundra" (West, 1968). Macroscopic plant remains and preserved pollen and spores are the supporting evidence for these reconstructions, in conjunction with a rapidly growing body of radiocarbon age estimates. Macroscopic plant remains, unlike most pollen, often permit specific identification but do not usually permit the calculation of percentages. Also, although macroscopic remains can be transported and redeposited to some extent (e.g. as in "peat erratics"), they are not normally susceptible to the long-distance wind-transport which some pollen types are, particularly tree pollen (e.g. especially *Pinus* with its "air-sacs" or "bladders"). Therefore, prior to a discussion of the results of pollen analyses carried out by the present author on samples from selected British Upper Palaeolithic sites, it is essential at least briefly to consider some of the characteristic macroscopic plant remains which have been reported by various authorities from British Middle and Late Last Glacial deposits. The sites which have yielded these remains are chiefly concentrated in lowland England with British highland extensions mainly in the Late Last Glacial. Some of these floral sites would undoubtedly have been within the general ranges of territories exploited by Upper Palaeolithic hunter/gatherers, although no direct association between the two has come to light as yet. Table 14 lists a selection of the plant species identified from macroscopic material, and the species named are essentially from Godwin's extensive lists for what he termed "Full-glacial" and "Late-glacial" (Godwin, 1956), checked against those mentioned by Pearson (1964) and West (1968). Additional information is included on whether a species reaches what Polunin (1959) considers an Arctic occurrence today and similarly whether it achieves a circumpolar distribution; on what its upper altitudinal limit is if it survives in Britain today, according to Clapham, Tutin and Warburg (1962) if they give such information; and on what it might be used for today by the Nunamiut Eskimos, if such data is given by Gubser (1965), or by other peoples if such data is given by Dimbleby (1967) and Harrison, Masefield and Wallis (1969).

This additional information is given for its often neglected value as a guide to the variable severity of the British Last Glacial and to the potential uses for certain plants during the Upper Palaeolithic. It should be noted that the list is unbalanced by a bias which I have given to the trees and shrubs because of their economic potential, there being in fact a considerably greater number of specific identifications of herbs available, particularly in Godwin (1956), Bell (1969), Morgan (1969), Walker and West (1970) and Deacon (1972; really a reference to the new data bank at Cambridge). Herbs in general may of course have been of some economic importance to the Upper Palaeolithic, for Gubser (1965) says that the Nunamiut find what they call "ivaroq" (general term for tundra growth, i.e. herbs, mosses, etc.) useful in house construction and hygiene (see also pp. 28 and 33). Also some herbs such as *Rumex acetosa* and *Urtica dioica* are considered quite edible plants by many peoples, e.g. the Scottish today who prepare "nettle pudding" (Harrison, Masefield and Wallis, 1969). However, aside from pollen whose presence is doubtless quite natural in the cases thus far, the only known close archaeological association of the British Upper Palaeolithic with plant remains is with wood charcoal, which in most cases when found is very difficult for even generic identification. A good example of such charcoal which has been identified is that of *Betula* from an association with Later Upper Palaeolithic artifacts at Mother Grundy's Parlour ("base and lower middle zones", Armstrong, 1925, p. 175). A more problematic example comes from the same site and approximately the same layer: charred nut-shells of *Corylus avellana*, normally an interglacial rather than glacial species (layer B, Godwin and Willis, 1962, pp. 61–62; see also pp. 62–63 in the present study). In the hopes of identification of wood charcoal from the Later Upper Palaeolithic "Black Band" found by Pengelly in Kent's Cavern (see pp. 38-39 and 41–42, Figs. 6 and 9), I borrowed some of this charcoal from the Torquay Natural History Society Museum (by their kind permission) and submitted it to the Forest Products Research Laboratory at Princes Risborough in 1969. Surprisingly, this has been identified as *Quercus, Ulmus* and *?Rhamnus catharticus*, normally interglacial wood types (see Appendix 1 by the author with identifications by J.D. Brazier). In addition to being potential sources of fuel some plants, such as *Juniperus communis* and *Betula pendula* may have been used for colouring in the manufacture of clothing (Dimbleby, 1967).

Macroscopic remains of the aspen, *Populus tremula*, are known from the British Late Last Glacial Zone II (Godwin, 1956) but are still rare and are therefore not included in Table 14. On the basis of those remains listed in Table 14, the presence of pollen referable to *Juniperus, Betula, Salix* and *Empetrum* in Middle and Late Last Glacial deposits may be considered reasonably reliable as indicating the presence of those trees and shrubs. However, the pollen of *Pinus*, which is sometimes present, is generally thought to be due to redeposition from earlier beds and/or long-distance wind-transport, with the possible exception of south-eastern England in Zone II where fairly high frequencies are encountered (Godwin, 1956; West, 1968). Pollen of *Hippophaë* and *Artemisia* is often present with some considerably high frequencies of the latter, and these two genera are thought to give a steppe character to the Last Glacial flora. Generally, however, the most abundant pollen types are those of sedges and grasses, which only permit assignment to their respective families, Cyperaceae and Gramineae. Spores of the more montane clubmosses, *Lycopodium* and *Selaginella*, are often present, sometimes in large numbers. Some of the British flora may well have survived south of the ice fronts during what I consider the "true" Full Last Glacial (i.e. the time of the last maximum advance of ice, see p. 85, Figs. 58 to 60 and Map 17), as has already been suggested by Godwin (1956) for his "Full-glacial" which corresponds mainly to my Middle Last Glacial and partly to my Full Last Glacial. Some of the hardier flora may have survived in any ice-free "refugia", such as "nunataks", that may have been available during the Full Last Glacial (see Lindroth, 1970, on ice-free refugia today and during the Pleistocene).

There now follows a brief review of the principal results and possible interpretations, or "zonations", of pollen analyses which have been carried out by the present author on samples collected during the course of his field-work in 1968 and 1969 at various British Upper Palaeolithic sites. The relevant sites are considered in the same clock-wise geographic order as before beginning with the south-west, and the results from each site are presented on linear diagrams as percentages of the different components: trees and shrubs, herbs, ferns. These are "total pollen diagrams" as the pollen counts are generally comparatively low and such diagrams are considered by some authorities to be the best method of illustration of what one assumes to be tundra spectra (Faegri and Iversen, 1964). Such diagrams tend to suppress misleading rises of tree pollen such as *Pinus* and they present a better general picture of how open or closed the vegetation may have been at a given time. The only "tree pollen diagram" presented is that for Hengistbury Head and is based on usually high pollen counts, i.e. above 150 grains, which may probably be taken as reliable for apparent changes of individual tree and shrub genera. All of these diagrams are based on the laboratory observations recorded in full in Tables 15 to 26, where counts and percentages for individual families or genera may be found (see also pp. 11–13 for description of author's method of pollen analysis).

Kent's Cavern, 1969: Pollen Analysis

Table 15 gives the results of the analysis of three samples which were collected by the present writer from deposits surviving between the Lecture Hall and Gallery and against the north-western wall of the Vestibule inside Kent's Cavern (Devonshire) in 1969 (see Figs. 4 to 6 for sampling positions). A diagram has not been constructed for these results as there are only three samples and number 1 yielded only two pollen grains. Sample 2 is from layer A2 (loamy "cave earth", see pp. 38–41) in the surviving section between the Lecture Hall and the Gallery, and it is therefore from the same layer and, being 30 to 60 centimetres below the "granular stalagmite", at about the same level as an Earlier Upper Palaeolithic bifacial leaf-point found by Pengelly in 1865 (see Figs. 4, 5 and 8). This sample yielded about 28% tree and shrub pollen and 66% herb pollen. Notable trees/shrubs are *Juniperus* and *Salix*, and the impression one gets from the spectrum as a whole is of quite open vegetation. Three presumably derived tree types, comprising about 1% each, are *Pinus, Tilia* and *Quercus*. Harder to interpret is the possible presence of *Menyanthes*?, and aquatic or bog plant which is normally insect pollinated. If this identification be correct, then it is probably the single species, *M. trifoliata*, which does in fact reach the Arctic today (Polunin, 1959), but how did the two grains get into Kent's Cavern? Presumably an insect or some cave-dwelling mammal would probably be the best answer, although it could have been derived from some earlier deposit as the type is known from both glacial and interglacial sites in Britain (Godwin, 1956).

Sample 3 is from layer B2 (stony "cave-earth", see pp. 38–42) in a surviving section or veneer against the north-western cave wall in the Vestibule, and it is therefore from the same layer and, being 15 to 45 centimetres below the "granular stalagmite", at about the same level as the Later Upper Palaeolithic "Black Band" and its diagnostic artifacts found by Pengelly in 1866 (see Figs. 4, 6 and 9). This sample yielded about 33% tree and shrub pollen and 61% herb pollen. Notable tree/shrubs are *Juniperus* (about 16% of total pollen and spores), *Salix* and *Empetrum*, and the impression one gets from the entire spectrum is again that of open vegetation. The possible occurrence of *Populus*? in both samples 2 and 3 is of immense interest as its pollen is fragile and difficult to identify. It is unlikely to have been derived and may therefore suggest the presence, however sparse, of *P. tremula* somewhere in the neighbourhood. Neither sample 2 nor 3 is inconsistent with the Last Glacial age suggested by radiocarbon measurements (see pp. 41–42, 79–80 and Fig. 57).

Badger Hole, 1968: Pollen Analysis

Figure 69 shows the results of the analysis of five deposit samples which were collected during the present author's excavations in the entrance platform at Badger Hole (Somerset) in 1968 (see Figs. 10 and 11 for sampling positions). The only thermophilous tree/shrub in this sequence is *Corylus* in sample 5, which together with a sharp rise in *Pinus* and *Betula*, suggests an Early Post Glacial age for at least the bottom of layer E/C, a layer which was greatly but not completely disturbed in other parts of the excavations. The underlying other four samples, with their higher percentages of herb pollen and Sub-Arctic to Boreal trees/shrubs, all appear to represent various stages of the Last Glacial. Sample 3 has only 6% trees/shrubs, about 4% of which is *Salix*, and may therefore indicate Full Last Glacial conditions, an interpretation which would not be inconsistent with its stratigraphic position above the range of Balch's unifacial leaf-point of Earlier Upper Palaeolithic type and the level of the radiocarbon age estimate for a burnt bone cluster (see Fig. 69). Samples 1 and 2 might be indicative, on the other hand, of slightly milder conditions. The presence of from 4 to 7% *Juniperus* in samples 1, 2 and 4 is of particular interest as its pollen is fragile and liable to rupture. However, it normally breaks up in a characteristic manner with two portions of ellipsoid shape which often remain attached, and it is therefore not always difficult to identify (see Beug, 1961, plate 3, no. 1–7, pp. 22–23), although a reference slide is essential.

In Figure 69 and subsequent "total pollen diagrams", it will be noted that I have grouped the trees and shrubs into sub-groups which normally include more than one genus. This has been done because: (a) the samples are usually small; and (b) these groupings, rather than isolated symbols for individual genera, make the general patterns of continuity or change easier to read and may reflect ecological communities. In order to find the exact counts and percentages for individual genera, the reader is again referred to Tables 15 to 26. Temperate or thermophilous trees and tree-shrubs such as *Alnus, Corylus, Quercus,* etc. are plotted on the extreme left when present and indicated by dark shading; the Boreal trees and/or shrubs, *Pinus* and *Betula*, are plotted next to the right when either is present and indicated by vertical hatching; the Boreal to Arctic tree-shrubs and/or low shrubs, *Juniperus* and *Salix*, are next to the right when either is present and indicated by no shading; and finally any other shrubs from Temperate to Arctic in requirements and/or tolerance are grouped on the extreme right of the "tree and shrub" category when any are present and indicated by dotted shading. Rare or unusual genera such as *Carpinus* and *Picea* when encountered in quantity are normally given their own symbolic shading pattern and positioned accordingly near the left on either side of the "thermophilous tree" group, *Carpinus* being shown against the extreme left and *Picea* somewhere to the right before the "pine-birch" group. Thus these "total pollen diagrams" are designed to allow a clear, although necessarily generalized, reading of how "closed" or "open", and how "hot" or "cold", the flora suggested by a given pollen spectrum may have been. In other words, apparent afforestation progresses from the left, whilst apparent deforestation progresses from the right. This system is much simplified, and therefore apparent changes in flora should only be read with due caution. Of course, any marked shift to open or partly open conditions in the flora of samples which are known to be from the more recent portion of the Post Glacial is most likely due to man's farming activities, etc. rather than a loss of Temperate conditions (see Godwin, 1956; Dimbleby, 1967; West, 1968).

Gough's Cave, 1969: Pollen Analysis

Figure 70 shows the results of the analysis of six deposit samples which were selected by the present writer in 1969 as the largest available amongst a series excavated by R.F. Parry from inside Gough's Cave (Somerset) in 1927–31. The series preserved by Parry consists of flint artifacts of Later Upper Palaeolithic aspect embedded in portions of their original deposits, with each artifact having its relevant spit number recorded directly on it. This material is housed in the Cheddar Caves Museum and was temporarily loaned to me (by the kind permission of the owner, the Marquis of Bath) for the palynological section of this research. The outer portion of the deposit samples selected was removed and discarded, and the inner portion was then treated by the method of pollen isolation described on pages 11–13. Sample 1 was a mixture of rounded limestone pebbles and sand cemented together by calcite (Parry's "gravel", 1929; Donovan's "conglomerate", 1955), which I have labelled layer A2. What I would term layer A1 is a "clayey loam and sand" (Parry, 1929) or a "laminated coarse sand" (Donovan, 1955) resting on limestone bedrock. Samples 2 to 6 were a sandy thermoclastic scree which had been cemented together by calcite (Parry's "cave-earth with small angular limestones", 1929; Donovan's "Creswellian cave earth and breccia", 1955), which I have labelled layer B. According to Parry (1929) there was Early Iron Age disturbance of layer B down to his spit 9 (represented by pottery and recent fauna), and this I have indicated on Figure 67 under "associated artifacts" as "E.I.A.". Artifacts of Later Upper Palaeolithic aspect, including many diagnostic tool-forms, were found principally in or close to his spit 13, with a wider scatter running from beyond spit 18 at the bottom to beyond spit 9 at the top of the sequence shown in Figure 70. Artifacts of what I would term "Mesolithic/Later Upper Palaeolithic" aspect apparently also occurred from spit 11 upwards.

It would appear from the results of the pollen analysis of samples 3 to 5 that what was presumably the main or part of the main Later Upper Palaeolithic occupation at the site took place during a time of fairly open vegetation, perhaps Late Last Glacial pollen Zone III as suggested on Figure 70. Sample 2 from spit 16 below, on the other hand, includes about 21% *Betula* pollen and may, perhaps together with sample 1, belong to Zone II, which might represent the beginning of the main Later Upper Palaeolithic occupation. Sample 6 seems to indicate milder conditions than those implied by any of the other samples, and it may represent some early part of the Post Glacial, although its higher proportion of tree pollen could just be due to the downwashing of a later pollen rain into an otherwise Late Last Glacial thermoclastic scree. The finer matrix of the scree would probably prohibit such downwashing from penetrating very far, but the presence of the scree itself might well permit some "vertical displacement of pollen after primary deposition" (see Faegri and Iversen, 1964, p. 123 on downwashing as a potential source of error). Of course, as these samples are simply stratified according to Parry's spits they may be from almost anywhere within the vast horizontal range of his excavations at the site (see plan in Donovan, 1955, Fig. 12). However, as Parry (1929) claims to have adapted his spits, or "working layers", to the slope of the natural layers as he observed them, this "horizontal" source of error may not be too great. Again, I stress due caution in reading the diagram.

Hyena Den, 1968: Pollen Analysis

Figure 71 shows the results of the analysis of nine deposit samples which were collected by the present author from a new exposure at the Wookey Hole Hyena Den (Somerset) in 1968, during the course of excavations by Prof. E.K. Tratman (for sampling positions see Tratman, Donovan and Campbell, 1971, Figs. 40 and 41B; also see discussion of stratigraphy and granulometry in this book, p. 89). From its general appearance this pollen sequence may well span the whole of the Last Glacial. The attribution of these few samples to such a long period in the absence of radiocarbon evidence is of course not beyond doubt. However, the higher frequencies of trees and shrubs in samples 1 to 3 and 7 to 9 certainly suggest that 4 to 6 represents an intervening rather cold phase. Further, the open heathland with some thermophilous trees and comparatively abundant *Betula* indicated by sample 1 might well belong either to the very end of the Last Interglacial (compare West, 1968, Fig. 13.9) or the beginning of the Early Last Glacial, perhaps before or just after the so-called "Chelford Interstadial" (Simpson and West, 1958) which itself does not appear to be clearly represented in this sequence. A Late Last Interglacial/Early Last Glacial age for the base of this sequence would certainly agree well with the presence of *Dicerorhinus hemitoechus* in the fauna found by Dawkins (1863a and 1874). Samples 7 to 9, on the other hand, seem to reflect the better known floral changes of the Late Last Glacial and Early Post Glacial, although only in a very general way (compare diagrams in Godwin, 1956; West, 1968). The substantial rise in *Juniperus* in sample 3 to about 22% of the total pollen before the apparent Full Last Glacial rise in herbs might be indicative of some sort of "interstadial", perhaps part of the so-called "Upton Warren Interstadial" (Coope, Shotton and Strachan, 1961) even though *Juniperus* is not listed for Upton Warren itself. The occurrence of the thermophilous tree, *Alnus*, in sample 3 is probably due to derivation as the samples are from a sloping, partly soliflucted series. The possible presence of *Dryas* in samples 5 and 6 (see Table 18) is extremely interesting as its pollen, which is difficult to identify, is apparently seldom encountered in British deposits, although macroscopic remains of *Dryas* (see Table 14) have been recognized from various British Last Glacial sites (Godwin, 1956; West, 1968).

As may be seen in Figure 71 it would appear that the occurrence of faunal remains at the Hyena Den ceased before the Full Last Glacial, as indeed the human occupation presumably did, since only artifacts of Later Middle Palaeolithic and Earlier Upper Palaeolithic aspect are represented at the site. The maximum scatter of bone fragments is according to those found in the deposit samples, as is the occurrence of the hyena, *Crocuta*, in sample 2 as indicated by an incisor. The association of *Crocuta* with sample 1 is inferred from the results of the 1966–70 excavations by Tratman. It is tempting to suggest that the Later Middle Palaeolithic tool-forms found by Dawkins (see Tratman, Donovan and Campbell, 1971, Figs. 42 and 43) came from the lower portion of layer A2b, whilst the

Earlier Upper Palaeolithic tool-forms found by him (see Fig. 90 in this book) came from the upper portion of layer A2b, both assemblages thereby helping to account for the total vertical scatter of bone fragments and falling into their expected stratigraphic positions. Such a suggestion is of course almost entirely speculative.

Sun Hole, 1968: Pollen Analysis

Figure 72 shows the results of the analysis of thirty deposit samples which were collected during the present author's excavations inside Sun Hole (Somerset) in 1968 (see Figs. 12 and 14 for sampling positions). From its general appearance this pollen sequence may not only span the whole of the Last Glacial for the Mendip Hills, but cover it in somewhat greater detail than the sequence described above from the Hyena Den. The occurrence of *Carpinus* in layer A1 is quite acceptable for the proposed Last Interglacial age of that level, but if that really is the age of that layer, then the presence of *Picea* may be due to either derivation from some much earlier deposit (fragments of an older breccia and stalagmite occurred in layer A1, see pp. 52–54) or downwashing from some slightly younger deposit. *Picea* supposedly does not occur in the British Last Interglacial (West, 1968), but it did, however, reach Britain during the "Chelford Interstadial" of the Early Last Glacial (Simpson and West, 1958). The floral spectra from at least layer B1–2 up to layer C/B are thought to be almost entirely Last Glacial, with only slight traces of probable derivation of such thermophilous forms as *Corylus*. The various minor peaks in trees and shrubs are mostly thought due to sampling and statistical factors, but the major peaks probably do represent actual fluctuations. The lowest percentage encountered for trees and shrubs is represented by only about 2% *Salix* in sample 14, and that sample together with some of the samples immediately below and above should indicate Full Last Glacial conditions. The substantial rise in ferns/club mosses in layers B2–2 to B2–4, if also partly Full Last Glacial, may be the result of over-representation by the microspores of the clubmosses concerned, *Lycopodium* and *Selaginella*, two "cosmopolitan" genera which may well have sheltered and flourished inside the south-facing cave mouth at that time. Samples 19 to 30 appear to show the changes in trees, shrubs and herbs characteristic of the Late Last Glacial and Early Post Glacial. I have suggested a tentative zonation for these from Zones I to VI, the assumed span of I and II being not inconsistent with the intervening radiocarbon age estimate and the stratigraphic range of the Later Upper Palaeolithic artifacts from the site (see Fig. 72). The two apparent main concentrations of these artifacts are plotted as little boxes with oblique hatching, and it may be that some of these artifacts actually come from fairly early in Late Last Glacial Zone I, whilst others come from about the beginning and/or later in Zone II.

Cathole, 1968: Pollen Analysis

Figure 73 shows the results of the analysis of thirteen deposit samples which were collected during the present writer's excavations outside Cathole (Glamorganshire) in 1968 (see Figs. 15 and 16 for sampling positions). This pollen sequence would appear to cover Last Glacial and Post Glacial changes in the southern Welsh flora right up to the present time, layer F being the modern topsoil. The changes within the Post Glacial are compressed and obscured by the very nature of the samples, the latter being from somewhat loose deposits and yielding only rather small pollen counts, but a general pattern of natural afforestation and subsequent deforestation by post-Mesolithic man seems reasonably clear. What I have interpreted as Early to Full Last Glacial also seems somewhat compressed, considering the known length of that time span, but a general trend from open woodland in sample 1 to herbaceous tundra in sample 4 seems fairly clear, although the apparent sharp peak in herbs in sample 4 is really caused by a sharp rise in the representation of the "ferns", i.e. the clubmosses *Lycopodium* and *Selaginella*. On the other hand, Late Last Glacial Zones I to III appear comparatively drawn out; this is largely due to the thickness of the deposits concerned. As the apparent indication of Zone II is represented by a total count of only 98 pollen grains and spores, the rise of *Betula* to about 30% of that count may be simply a statistical effect, but its increase is nevertheless very suggestive when considered in conjunction with the associated small mammals from layer UOB (see Fig. 81 and Table 35). If this interpretation of samples 5 to 9 be correct, then the artifacts of Later Upper Palaeolithic aspect found in 1968 should belong entirely to Late Last Glacial Zone I. The artifacts of Mesolithic aspect, on the other hand, are clearly associated with the earlier part of the Post Glacial. It may of course be that some of the artifacts found by McBurney (1959) at about the interface of layers C and USB (his "C/B contact zone") or just below it actually "date" from Late Last Glacial Zone III, whether one regards them as Later Upper Palaeolithic or Mesolithic, his interpretation being "Creswellian".

Long Hole, 1969: Pollen Analysis

Figure 74 shows the results of the analysis of twenty-two deposit samples which were collected during the present author's excavations outside Long Hole (Glamorganshire) in 1969 (see Figs. 18 and 21 for sampling positions). Although the total counts are all rather low, this pollen sequence would appear to be the best thus far available for the whole of the British Last Glacial. Not only is a prolonged phase of Arctic tundra development fairly clearly represented by samples 4 to 15, but there is also a reasonably clear implication of Boreal coniferous forest development, including *Picea, Pinus* and *Betula*, in sample 3, which when taken together with its stratigraphic position is probably indicative of the "Chelford Interstadial" of Simpson and West (1958), West (1968) and Mitchell, Penny, Shotton and West (1973). Of course, the total pollen count in sample 3 is only 99 and the total tree and shrub count numbers only 70, or about half the minimum count recommended by Faegri and Iversen (1964) for any sort of reasonable statistical reliability. Pollen is very sparse in this sample, and it was therefore exceedingly difficult to concentrate it and obtain a count even as high as 70 tree and shrub grains, let alone an overall total of 99. But considering the proportion of trees and shrubs and those types most commonly represented (even though they are genera known to over-produce pollen), an equivalency with the Early Last Glacial "Chelford Interstadial" seems the most reasonable interpretation of sample 3, particularly as there is about 23% *Pinus*, or more than double the minimum 10% recommended by Faegri and Iversen (1964) for acceptance of the true presence of *Pinus* in a local flora. It is therefore of interest that a possible Middle Palaeolithic occurrence was observed in the same layer, layer A2b (see p. 60).

The occurrences of artifacts of Earlier Upper Palaeolithic aspect were found higher up in layers A3a and A3b, and from their associated pollen it would appear that they "date" from a time of comparative abundance of *Juniperus* and *Salix*, presumably during the Middle Last Glacial. Also these artifacts are stratigraphically below the apparent tree and shrub minimum indicated by samples 9 and 10, which I have attributed to the Full Last Glacial. A noteworthy specimen in sample 10 is a single grain of what appears to be *Dryas* pollen.

Dead Man's Cave, 1969: Pollen Analysis

Figure 75 shows the results of the analysis of four deposit samples which were collected by the present author with the assistance of Dr. P.A. Mellars and Mr. G.F. White from a section at the entrance to Dead Man's Cave, also known as "Anston Stones Cave" (Yorkshire) in 1969 (see Figs. 1 and 2 in White, 1970, for sampling positions). What I have termed layer A is a clay and stalagmitic breccia on bedrock (the lower half of White's spit 10); layer LSB is a buff silty sand with some thermoclastic scree, partly White's spits 9 and 10 containing his main occurrence of "Creswellian" backed tools; layer MSB is a hard, brecciated sand and silt with a small amount of thermoclastic scree (partly White's spits 8 and 9) and with no artifacts; layer USB is a buff silty sand, brecciated towards its bottom, with some thermoclastic scree and boulders (partly White's spits 6 to 8) and with one "Creswell point" of Later Upper Palaeolithic type, as well as some localized disturbance (see White, 1970, pp. 2, 4 and Fig. 2); and layer C is a series of stalagmitic breccias intercalated with weathered scree and boulders in a sandy to silty matrix (White's spits 3 to top of 6).

The pollen sequence obtained from these four samples is suggestive of conditions not unlike those which occurred during the latter part of the Late Last Glacial, but the precise zonation of this series is uncertain owing to its isolated occurrence and the smallness of the counts which it yielded. The presence of *Alnus* in sample 2 is presumably due either to derivation from some much earlier deposit or to long-distance wind-transport. The presence of what might be *Populus* in samples 2 to 4, on the other hand, would not necessarily be out of place, and if the fragments thought to represent it had been properly identified, Dead Man's Cave would be the second cave-site to have yielded this fragile and rare type from a Last Glacial context. The other site was Kent's Cavern (Devonshire), and here to the north it is again presumably indicative of the aspen, *Populus tremula*, which has already been found elsewhere in the north of England at Windermere where it was represented by a catkin scale in a lake deposit assigned to Zone II (Godwin, 1956, p. 217). Pollen of *Populus* sp. is also known in small frequency at the transition from Zone III to IV in southern Scotland (Godwin, 1956, pp. 216–17).

It will be noted on Figure 75 that the tentative attribution of samples 1 to 3 to Late Last Glacial Zone II is quite inconsistent with the radiocarbon results obtained by Mellars (1969; White, 1970). However, these radiocarbon age estimates are on bone fragments attributed to *Rangifer* a herbivore which one would normally consider more at home in the Late Last Glacial than in the Early Post Glacial (BM–439 and –440A; BM–440B is from a less well stratified position but is also on *Rangifer*). Incidentially, no *Rangifer* is included in the Boreal type faunal lists from the Earlier Mesolithic sites of Thatcham (Wymer, 1962) and Star Carr (Clark, 1954), prolific sites with radiocarbon age estimates of the same order as those from Dead Man's Cave, but carried out on plant material rather than bone (see Table 4). I would myself suggest that the Dead Man's Cave age estimates are too young by about 1,000 radiocarbon years (also see section on radiocarbon measurements, pp. 77–80).

Mother Grundy's Parlour, 1969: Pollen Analysis

Figure 76 shows the results of the analysis of eight deposit samples which were collected during the present author's excavations outside Mother Grundy's Parlour (Derbyshire) in 1969 (see Figs. 23 and 24 for sampling positions). This pollen sequence would appear to represent in a general manner some of the floral changes which occurred in the Creswell Crags region from the Full Last Glacial to about Zone VII in the Post Glacial. Samples 1 to 3 have a very high herbaceous content which is most likely the result of "full glacial" conditions, whether Full Last Glacial as assumed, or near Full Penultimate Glacial as might just be possible (see pp. 60, 91 on granulometry and possible stratigraphic correlations with the finds reported by Dawkins and Mello). Samples 4 to 5 see a slight increase in trees and shrubs, including principally *Betula* and *Salix*, and are quite likely the result of Late Last Glacial environmental improvements. Sample 6 shows a sharp rise in trees and shrubs, including mainly *Corylus, Pinus* and *Betula*, which is indicative of the Early Post Glacial climatic amelioration and not at all inconsistent with the radiocarbon age estimate obtained earlier for material, including *Corylus* nut-shells, by McBurney for about that level (Godwin and Willis, 1962). However, as layer

SB is a thermoclastic scree, it is conceivable that both some of my pollen and some of his radiocarbon sample Q–551 were the product of long-term downwashing. Pollen samples 7 and 8 suggest first afforestation and then deforestation. The other two radiocarbon age estimates seem to agree reasonably well, although they appear somewhat too young.

Robin Hood's Cave, 1969: Pollen Analysis

Figure 77 shows the results of the analysis of eight deposit samples which were collected during the present author's excavations outside Robin Hood's Cave (Derbyshire) in 1969 (see Figs. 27 and 28 for sampling positions). This sequence has been truncated at the top by earlier excavations and partly disjointed towards the base by assumed periglacial erosion of the surface of layer A (see pp. 66–67). Sample 1 would appear to be at least as old as Early Last Glacial, if not indeed somewhat earlier. Sample 2 suggests much more herbaceous conditions and may be indicative of late Full Last Glacial or early Zone I recovery of a fairly open vegetation from the High Arctic conditions of the main Full Last Glacial maximum ice advances. It implies an early, but not unreasonably early, age for the associated Later Upper Palaeolithic artifact assemblage, "L.U.P.1". Samples 3 and 4 appear to be more into Zone I with a rise of *Juniperus* to about 15% in sample 3, and they suggest that radiocarbon sample BM–604 from near the top of layer LSB may be about 1,500 years too young, whilst the "L.U.P.2" artifact assemblage is probably late Zone I. Samples 5 and 6 indicate an increase in *Betula* to about 37% in sample 6, and imply, despite their small counts, a local example of the well known Zone II amelioration. Again the associated radiocarbon sample (BM–603) may be too young by about 500 to 1,000 years. The "L.U.P.3" artifact assemblage seems on these bases most likely Late Last Glacial Zone II in age. Samples 7 and 8 suggest a return to herbaceous tundra and are thought the result of the climatic deterioration associated with Zone III, but whether the age of the "Meso./L.U.P." artifact assemblage from layer USB is in fact Zone III remains uncertain, as that layer was found to be already greatly disturbed by previous excavators. The occurrence of small amounts of *Alnus* pollen (1 grain each) in samples 6 and 8 is most likely due to derivation or some other source of contamination.

Hengistbury Head, 1968–69: Pollen Analysis

Figures 78 and 79 show the results of analyses of seventeen deposit samples which were collected during the present author's excavations at an open-air site, Site C2, at Hengistbury Head (Hampshire) in 1968–69 (see Figs. 37, 39, 41 and 42 for sampling positions). Figure 79 is a "tree pollen diagram" and demonstrates the apparent changes in trees and shrubs only. The "tree sum" for each sample is the total tree pollen counted, and percentages of both trees and shrubs are based on this amount in each case; thus the percentage calculated for some shrub pollen is greater than 100%. This is the fairly standard method for presentation of tree and shrub pollen (e.g. Figs. 13.10 to 13.15 in West, 1968). On the basis of careful comparison of the *Betula* pollen in sample 8, it has been thought possible tentatively to distinguish between the tree-birch grains and those of the dwarf birch, *B. nana*; a total of 278 birch grains being compared with one another in that sample and with the relevant reference slides. According to Walker in Godwin (1956, p. 190), "the pores of *B. nana* grains seem to be less protuberant than those of tree-birch pollen" and "*B. nana* pollen has been found in all the British Late-glacial deposits in which it has been so far sought". By comparison with sample 8 and the reference material, pollen of *B. nana* type was also discerned in samples 4 to 10, sample 4 apparently having only *B. nana*. However, as there is a gradation in morphology and size from the tree-birch type to the *B. nana* type, I have considered it best to refer to the latter at Hengistbury as *B.* cf. *nana*, and to include it with the trees even though the real *B. nana* is a shrub. The other *Betula* pollen yielded by the Hengistbury samples presumably belongs to both *B. pendula* and *B. pubescens*.

A curious anomaly in the shrub pollen sequence is the absence of a sharp rise in *Corylus* in my Post Glacial Zone VI; this may mean that most of VI is absent and the pollen represented simply jumps from V to early VII. Of course, it must be stressed that I have constructed a "composite sequence" for both Figures 78 and 79 on the basis of deposit samples collected at separate points and/or columns at the site, their stratigraphic order being arranged according to the most reasonable interpretation of the actual layer sequence (see again Figs. 37, 39, 41 and 42 for sampling positions). This composite sequence allows for the most complete pollen profile possible, but it certainly does not guarantee an uninterrupted sequence. Thus there may well be quite a few minor gaps within the Post Glacial; however, the only major floral gap would appear to be that between what I have interpreted as Last Interglacial and Full Last Glacial in a portion of the sequence which is not composite, that is, sample 3 was collected directly below sample 4.

As may be seen on Figures 78 and 79, the main Later Upper Palaeolithic occurrence would appear to "date" from the transition between Late Last Glacial Zones I and II, whilst the main Mesolithic occurrence is from Post Glacial Zone V or possibly somewhat later. A noteworthy feature of late Zone I in sample 6 is the sequence's only high peak in *Juniperus* (see Fig. 79; *Juniperus* count was 125 grains, or about 260% of the total "tree" pollen in sample 6). Comparably high frequencies have already been observed elsewhere in late Zone I, for example, at Tadcaster in Yorkshire (Bartley, 1962, Fig. 3) and at Vance in Belgium (Hulshof, Jungerius and Riezebos, 1968, Fig. 2). The presence of the thermophilous genera, *Tilia, Alnus* and *Corylus*, in the Late Last Glacial at Hengistbury is presumably due either to derivation from the underlying Last Interglacial deposits or to downwashing from the overlying Post Glacial deposits, or most likely some of both. There certainly appears to have been some downwashing of humic matter, however fine, and considerable penetration by roots and rootlets into layer B2, as may be seen in the most unsatisfactory radiocarbon results from that layer (see Figs. 41 and 45, and Table 4). The apparently early rise of *Alnus* in the Post Glacial may also be due to some downwashing, although such would seem most unlikely, given the peaty structure of layer A1b. It may of course be that *Alnus* managed to colonize, however sparsely, some of the exposed bed of the English Channel during Zone IV, as it had already re-established itself in France during "Allerød", or Zone II, if not in fact slightly earlier (Leroi-Gourhan, 1964).

The above results of the pollen analyses, like those of the granulometric studies, are of great importance to any assessment of the chronology and ecology of Upper Palaeolithic settlement in Britain. However, it must be stressed that their interpretation or zonation is often very difficult, and the suggested origins and ages of the pollen and spores encountered are simply those considered by the present author to be the most reasonable; others would certainly be possible in some cases. Much more detailed examination of pollen morphology is possible but the aim of the present study is simply to determine the general nature of British Upper Palaeolithic floras.

In summary, it would appear from the pollen analyses of the author's samples that Earlier Upper Palaeolithic artifacts at Kent's Cavern, Badger Hole, Long Hole and possibly the Hyena Den, are associated with Middle to oncoming Full Last Glacial flora of generally Arctic type, whilst Later Upper Palaeolithic artifacts at Kent's Cavern, Gough's Cave, Sun Hole, Cathole, Dead Man's Cave, Mother Grundy's Parlour, Robin Hood's Cave and Hengistbury Head are associated with principally Late Last Glacial flora of generally Arctic to Boreal type. Such conclusions are in no way at variance with the broader picture presented in the section on chronology.

3. Faunal Evidence

The known fauna of the British Last Glacial is suggestive of a generally open terrain of tundra to steppe or cool parkland type. Remains include notably those of Mollusca (marine, fresh water and land shells), Coleoptera (beetles), Pisces (marine and fresh water fish), Amphibia (principally frogs and toads), Aves (birds) and Mammalia (principally placental land mammals). According to our present state of knowledge, the animal class most economically important to British Upper Palaeolithic hunter/gatherers would appear to be the Mammalia, particularly certain large land mammals such as *Ursus arctos* (brown bear), *Coelodonta antiquitatis* (woolly rhinoceros), *Equus* sp. (wild horse) and *Rangifer tarandus* (reindeer). *Mammuthus primigenius* (woolly mammoth), *Megaloceros giganteus* (giant deer) and many lesser forms would also be sought when either (a) they were available, or (b) the apparent principal preferences, *Equus* and *Rangifer*, were not available. As the land mammals do seem to be so important, I do not propose to deal in any great detail with the other animal classes which are represented in Last Glacial deposits, but simply to review very briefly the highlights of the evidence they offer.

The Mollusca can be useful as environmental indicators, and recent work has shown that when they are found stratified and in abundance, they may yield zonal patterns not unlike the pollen zones (e.g. Late Last Glacial molluscan zonation for south-eastern England by Kerney, 1963). Stratified samples of land Mollusca which were collected during excavations by the present author at Cathole in 1968 and Long Hole in 1969 (both Glamorganshire) have been studied by Dr. J.G. Evans of the Department of Archaeology at the University of Wales in Cardiff (results are reported in Appendix 2 and Table 27, as well as Fig. 80). Sadly, his results indicate that most, if not all, of the snails represented are Post Glacial in character, and in the case of Cathole in particular these are perhaps in part intrusive elements in the Late Last Glacial screes. They appear to be of little zonal value at these sites, although it seems to me that they may nonetheless be of some micro-ecological significance.

Mollusca have already been excavated from a number of Later Upper Palaeolithic sites including notably: Aveline's Hole, Somerset (Kennard and Woodward in Davies, 1923); Gough's Cave, Somerset (Donovan, 1955); Sun Hole, Somerset (Tratman, 1963); King Arthur's Cave, Herefordshire (Hewer, 1926); Ossum's Cave, Staffordshire (Bramwell, 1962); Mother Grundy's Parlour, Derbyshire (Jackson in Armstrong, 1925); and Pin Hole, Derbyshire (Jackson, 1967). From this list of sites, the Mollusca of particular interest include drilled shells of *Neritoides obtusatus* from Aveline's Hole which were presumably altered to become ornaments. Some of these Mollusca from Later Upper Palaeolithic sites may well have been gathered as a source of food: for example, the marine species *Mytilus edulis* (mussel) and *Ocenebra erinacea* (sting winkle) found at Gough's Cave and which were presumably originally brought from some ancient coast well off the present shoreline of south-western Britain. A possible Earlier Upper Palaeolithic example of shell ornaments is that found by Buckland in close association with the burial of the "Red Lady" at Paviland Cave (Glamorganshire): "Close to that part of the thigh-bone where the pocket is usually worn, I found laid together and surrounded also by ruddle about two handsfull of small shells of the *nerita littoralis* in a state of complete decay, and falling to dust on the slightest pressure" (Buckland, 1823, p. 87). Marine molluscs are also recorded for the various levels of the Upper Palaeolithic at Kent's Cavern (Devonshire) in Pengelly's Diary (Pengelly, 1865–80, but particularly 1865).

The Coleoptera are also useful as environmental indicators, and recent work has shown that they are quite capable of yielding very informative zonal patterns, again not unlike the pollen zones (e.g. Last Glacial coleopteran zonation for England and Wales by Coope, Morgan and Osborne, 1971; also see Figs. 59 and 83 in this present study). However, no beetle remains have as yet been found at any British Upper Palaeolithic site, even including Hengistbury Head (Hampshire) where they might have been preserved (M. Speight, personal communication following his visit to Site C2 during the present author's excavations).

Pisces, or fish, have been recovered from a few Last Glacial sites (e.g. *Perca fluviatilis* in Zone II at Grimston Hall, Yorkshire, Shotton and Williams, 1973) and one noteworthy Upper Palaeolithic site, namely Pin Hole (Derbyshire; White in Armstrong, 1928). The remains from Pin Hole include scales of the roach, *Rutilus rutilus*, from the "upper Mousterian level" (i.e. probably Earlier Upper Palaeolithic, see p. 47 on stratigraphy), and bones of the pike, *Esox lucius*, and lemon sole, *Microstomus microcephalus*, from the "Magdalenian level" (i.e. Later Upper Palaeolithic). The last named species is of immense interest as it offers yet another piece of evidence for Later Upper Palaeolithic contact with the sea, contact which must have involved crossing a greater distance than would be necessary today as many of the present sea beds off England would still have been exposed as dry land. Fishing was perhaps comparatively rare as a source of food, but it was certainly not unique to the British Upper Palaeolithic. Remains of *Esox lucius* are known from the Hamburgian at Meiendorf (Rust, 1937) and the Ahrensburgian at Stellmoor (Rust, 1943) in north-western Germany, as well as from a close association with a broken barbed "harpoon" from a Late Last Glacial clay at Abschruten (Gross, 1940) in northern Poland. *E. lucius* and various other fresh water fish, as well as salmon, are known from many of the south-western French Upper Palaeolithic cave-sites (Clark, 1948; Casteel, 1972), where they are sometimes even included in the art work (Leroi-Gourhan, 1971), e.g. the well known engraving of what appears to be a salmon on a Magdalenian bone "bâton" from La grotte d'Isturitz (Saint-Périer, 1936, pp. 124–25, Fig. 72.5). However, fishing was apparently much more common during the Mesolithic, as for instance is suggested by the widely distributed and often abundant remains of *E. lucius* at various Maglemosian sites from England to the Baltic (Clark, 1952, Fig. 20). Amongst the Nunamiut Eskimo today fish may constitute about 10 to 20% of their diet when reindeer are scarce, but they are not a preferred food source (Gubser, 1965, p. 252). But some other Eskimo groups, as well as many of the Athapaskan Indians, do take a considerable amount of fish (see Lee, 1968; Balikci, 1970; VanStone, 1974). No fish remains were found during the present author's own excavations at British Upper Palaeolithic sites, but in 1974 Mr. Simon Collcutt (personal communication) found what appear to be roach scales at about the Later Upper Palaeolithic level in Pin Hole (Derbyshire).

Remains of amphibians also appear to be comparatively rare in the Last Glacial and the Upper Palaeolithic in Britain. One species which is sometimes found in association with Later Upper Palaeolithic material is the common frog, *Rana temporaria* (e.g. Soldier's Hole, spit 9, Somerset, Jackson in Parry, 1931). However, although quite edible for man, the presence of this species in Last

Glacial cave-sites is probably most often the result of the activities of birds of prey.

Bird remains are, on the other hand, fairly common at some Last Glacial and/or Upper Palaeolithic cave-sites in Britain, and, particularly when abundant, they are useful as environmental indicators. The commonest birds found belong to the genus *Lagopus*, and consist of forms of grouse and ptarmigan. Ptarmigan, or *Lagopus mutus*, in Britain today are mainly concentrated in the Sub-Arctic zone of the higher Scottish mountains, whilst red grouse, or *L. scoticus*, and other Last Glacial forms are now found lower down in moorland to parkland areas more representative of Boreal conditions. Thus Bramwell (1960) is of the opinion that "Sub-Arctic to Boreal" rather than "Arctic" is the best brief description of Last Glacial bird communities. The presence of northern Last Glacial ice sheets would have excluded many tundra nesters from their modern breeding grounds, so that England and other regions south of the ice would have provided breeding grounds for many swans, ducks and waders. A high proportion of waterfowl in cave finds tends to confirm such a larger and more widespread population than that at present found in England. British bird remains of Last Glacial age are known chiefly from cave-sites in the following areas: south-eastern Devonshire (e.g. Kent's Cavern), the Mendip Hills (e.g. Aveline's Hole, Gough's Cave, Soldier's Hole and Sun Hole), the Gower Peninsula (e.g. Cathole), the Peak District (e.g. Dowel Cave and Ossum's Cave) and the Creswell Crags district (e.g. Langwith Cave, Robin Hood's Cave and Pin Hole) (Bramwell, 1960, and personal communications). Although there are quite frequent occurrences of Pleistocene avian remains in these areas, as well as certain isolated instances in other areas, the sites I have selected to form the basis of this section are deliberately chosen both as being prolific and as also being Upper Palaeolithic sites. Indeed, where there is any sort of stratigraphic record available, it is apparent that some of the "avifauna" was found in the same layer or layers as the Upper Palaeolithic material.

However, Bramwell (1960) feels certain that the bulk of the known Last Glacial bird remains formed the prey of predatory birds and mammals, which are often themselves well represented in the deposits, whilst only a few were brought into the caves by man, presumably for food and as a source of bone suitable for tool-making. The case which he cites as most likely to represent human activity is from the Earlier Upper Palaeolithic at Soldier's Hole (layer A2, or Parry's layer 4) where the strong metatarsal of a grey-lag goose, or *Anser* cf. *anser*, and the ulna of a whooper swan, or *Cygnus* cf. *cygnus*, had both been snapped across, presumably to extract the marrow (Bramwell, 1959b, p. 89). Another possibility, though one which Bramwell apparently ignores, for Upper Palaeolithic evidence on exploitation of birds is the unusual survival of a nearly complete egg-shell of a large species of duck from Armstrong's "Font-Robert level" (i.e. Earlier Upper Palaeolithic) inside Pin Hole (Armstrong, 1928, p. 27). This duck egg-shell was found in the proximity of a hearth, around which fragments of at least three other egg-shells were found, together with numerous split and charred bones of large mammals such as reindeer and various Earlier Upper Palaeolithic flint artifacts. During the course of excavations by the present author what appeared to be bird bones were found in Upper Palaeolithic layers at the following sites: Earlier Upper Palaeolithic layer A3a, Long Hole, Glamorganshire; Later Upper Palaeolithic layers B2–4 to B2–9, Sun Hole, Somerset; Later Upper Palaeolithic layers LOB to MSB, Cathole, Glamorganshire; Later Upper Palaeolithic layers LB and SB, Mother Grundy's Parlour, Derbyshire; and Later Upper Palaeolithic layers B/A to USB, Robin Hood's Cave, Derbyshire. As small mammal remains were found in all of these layers as well, it seems quite likely that they were mostly the prey of birds and that both the small mammal remains and the bird remains generally arrived at these sites naturally, their occurrence in Upper Palaeolithic layers being apparently coincidental. However, a few of the bird bones in the larger samples, i.e. Cathole and Robin Hood's Cave, appear to have been intentionally split and may therefore be the result of man's activities. Mr. D. Bramwell, who is certainly the foremost living British authority on Pleistocene birds, has kindly studied the bird remains which I have excavated from the above-mentioned sites, and his report is given in Appendix 5 and Tables 29 and 30. It should be noted that identifiable bird bones occurred only at Cathole and Robin Hood's Cave.

It is perhaps relevant that amongst the reindeer-hunting Nunamiut Eskimos today, birds are often an important subsidiary source of food and raw materials (Gubser, 1965, pp. 242–51). Two categories of birds are recognized; large birds such as ducks, geese and ptarmigan, and small birds such as snow bunting, sparrow, rosy finch, thrush, jay and woodpecker. The small birds are generally the less important group and are hunted by boys with slingshots, or catapults. The larger birds are normally caught with snares or arrows, and sometimes even with spears. Ducks and geese are hunted particularly when they are migrating in spring and autumn and are valued highly for their meat and fat; also their skins are sometimes used for insoles of boots and some clothing. Ptarmigan are hunted from autumn to spring and are known to have kept some Nunamiut alive when other hunting has been poor, even though ptarmigan have little fat.

Amongst the various Late Last Glacial reindeer-hunters of Europe, it would appear that certain birds were also sometimes important. For example, at some Magdalenian cave-sites in France, Switzerland and southern Germany forms of grouse and ptarmigan occur in large numbers which might suggest specialized fowling activities, e.g. at Petersfels in Germany those two bird forms account for most of the bird remains (Peters and Toepfer, 1932, p. 166). The interest shown by the Magdalenians in the Arctic willow-grouse, or *Lagopus albus*, is demonstrated by an engraving of one on a fragment of reindeer antler from Isturitz in France (Passemard, 1922, Fig. 38). In fact, birds are really quite frequent in Palaeolithic art (Leroi-Gourhan, 1968). However, it should be noted that the evidence for fowling in general is much greater for the Mesolithic than for the Upper Palaeolithic in Europe, including Britain (Clark, 1952). Nonetheless, considering the variety of British Last Glacial birds and their potential for exploitation by Upper Palaeolithic hunters when preferred mammals were scarce or not available, I think it is worth citing as a detailed example the full list of bird species reported by Bramwell (1959b, p. 90) from the Earlier and Later Upper Palaeolithic layers at Soldier's Hole (Somerset). This list is therefore given in Table 28. It is worth noting that grouse and ptarmigan are the most plentiful forms in both layers (Bramwell, 1959b, p. 88).

As there are only 7 species of birds in the Earlier Upper Palaeolithic layer at Soldier's Hole, none of them predatory, and all 7 classifiable as comparatively large birds, it seems to me reasonable to suppose that their presence could well be entirely the result of man's activities. Such a suggestion is reinforced by the fact, already mentioned above, that two bones from that layer (A2, or Parry's '4'), one of *Anser* cf. *anser* and one of *Cygnus* cf. *cygnus*, were apparently broken by man. The Later Upper Palaeolithic layer, on the other hand, has not only a big increase in the actual quantity of bird bones (Bramwell, 1959b), but a substantial increase in the number of species represented and the addition of one predator, *Asio flammeus* or the short-eared owl. Thus although the Later Upper Palaeolithic inhabitants may not have been responsible for any of the bird bones in their layer, the Earlier Upper Palaeolithic inhabitants would certainly seem to have been fowling for some of their food, and on the basis of the migratory species identified (see Table 28), one could even suggest that they were engaged in such hunting during either spring or autumn, or perhaps both.

From the data given above it would appear that there is fairly conclusive evidence for some Earlier Upper Palaeolithic bird-hunting at Soldier's Hole and at least egg-gathering at Pin Hole. As a result of my excavations at Cathole and Robin Hood's Cave, it certainly seems to me that there is now also evidence for some Later Upper Palaeolithic bird-hunting (see Appendix 5). In any case, Later Upper Palaeolithic hunters certainly seem to have employed some bird bones as tools, whether or not they hunted the birds whose bones they used, if we accept the evidence of the sharpened tool, "not well finished, but apparently much used", made from the ulna of a swan (*Cygnus* sp.) and found in spit 14 at Gough's Cave in Somerset (Gray in Parry, 1929, p. 115, plate 20.8).

The remains of small land mammals, or so-called mammalian "microfauna", occur in considerable abundance at many British Last Glacial and/or Upper Palaeolithic cave-sites, and when carefully excavated and studied they are very useful as environmental indicators. The commonest small mammals found are various forms of voles and lemmings, whilst the less common ones include various shrews, bats, the steppe pika and the varying or "Arctic" hare. Small carnivores such as polecats, weasels and otters also occur and might also be classed as "microfauna". Taken as a whole the environments which these small mammals could tolerate range from Arctic to Temperate, with the strongest emphasis in the Arctic to Boreal habitats. Most occurrences of their remains are probably due to predators other than man, but Upper Palaeolithic hunters may well be responsible for some of them, particularly those of the varying hare, *Lepus timidus*, and the small, fur-bearing carnivores. However, there is no British example which is comparable to the Magdalenian site of Petersfels in western Germany where *Lepus timidus* amounted to over 50% of the mammals, even greater than reindeer, and had clearly been caught in rather exceptional numbers (Peters, 1930; Peters and Toepfer, 1932). Amongst the Nunamiut today hares are not a significant food source, but they are valued for their soft and flexible fur (Gubser, 1965).

The remains of large land mammals, or so-called mammalian "megafauna", also occur in considerable abundance at many British Last Glacial and Upper Palaeolithic sites, and might be considered useful environmental indicators, although they would not seem to be so sensitive in this respect as are certain beetles and small mammals. However, as has already been stated, the large mammals would certainly appear to have provided the main source of Upper Palaeolithic food, as well as an important source of raw materials. They are therefore of paramount importance to this study. Unlike the Magdalenian, Hamburgian and Ahrensburgian hunters who often preyed almost exclusively upon reindeer, *Rangifer tarandus*, the most common prey of the British Upper Palaeolithic hunters, both Earlier and Later, would seem to have been frequently, though not always, wild horse. Both *E. germanicus* and *E. przewalskii* seem to occur, the latter smaller form being the more prevalent and the former only possibly associated with the Earlier Upper Palaeolithic. Zeuner (1963) suggests that the range of variation in Last Glacial *Equus* may have been due to different "ecotypes", rather than separate species, and he therefore recommends the use of the designation *E. caballus* with tundra, loess-steppe, grassland and forest ecotypes according to associated evidence. Although his suggestion may well be valid in part, in accordance with the recommendations made by Bohlken (1961) and followed by Kurtén (1968) and Mohr (1971), I prefer not to use those scientific names which have been based on domestic mammals. Hence *E. przewalskii* becomes a more satisfactory name than *E. caballus*, the domestic horse. Also, although there is doubtless some variation within *E. przewalskii*, there appears to have been a sudden change in size during the Last Glacial which Kurtén (1968) attributes to replacement by immigration during a shift from the extinct *E. germanicus* to the still living *E. przewalskii*.

Equus przewalskii survives today in a mountainous region called Tachīn-Shara-Nurū, or "Ridge of the Yellow Horses", in the Mongolian Gobi Desert. It lives in a principally steppe environment at an altitude of about 1,000 to 1,500 metres above sea level, and its range extends into desert conditions on the one hand and Alpine on the other, living mainly off various grasses (Isenbart, 1969; Mohr, 1971; Bökönyi, 1974; Benirschke, 1969; Kaszab, 1966). It spends its days grazing or migrating to new pastures, and it normally lives in herds of mares led by a stallion. Its tawny greyish-yellow to reddish-yellow colour, short erect mane and dark tail and legs compare well with many of the French Upper Palaeolithic cave paintings of horses, e.g. the right-hand wall of the Axial Gallery at Lascaux (Leroi-Gourhan, 1968, Fig. 73). Also, although it lacks any surviving pigment, the engraving of a horse's head on a rib segment which was found by Dawkins and Mello in an Upper Palaeolithic context at Robin Hood's Cave in Derbyshire (Dawkins, 1877, p. 592; Garrod, 1926, p. 129 and Fig. 31.5; also Fig. 105 in this present book) has a very similar mane and profile. Skeletally, the living *E. przewalskii* differs little from most of the horse remains from western European and British Last Glacial sites; those Last Glacial examples which do differ generally belong to the earlier part of the Last Glacial and to the apparently extinct so-called *E. germanicus* (Kurtén, 1968). Isenbart (1969, p. 42) has gone so far as to suggest that when hunting wild horse, Upper Palaeolithic man may have begun to specialize and follow one herd, depending on it as a main food source even though the herd remained free and wild. But its potential for great speed might have made such herds difficult to follow (Mohr, 1971). The apparent climatic and environmental adaptability of *E. przewalskii* would doubtless help to explain its abundance in what might initially seem a hostile setting, the rigours of Middle and Late Last Glacial Britain. In fact, it may well have found the steppe-tundra phases and regions of those times particularly well suited to its requirements and in response proliferated over much of southern Britain and the exposed areas of adjacent sea beds. If so, it would have been the most abundant herbivore available for man's exploitation.

The other most common prey of the British Upper Palaeolithic hunters, both Earlier and Later, would seem nevertheless to have been at times reindeer, *Rangifer tarandus*. In spite of great variation, all Pleistocene and living reindeer are considered as belonging to this single species, the present distribution of which is circumpolar (Kurtén, 1968; Burch, 1972). An important feature in the ecology of most wild reindeer or caribou is a seasonal migration between forest or low tundra, normally of the taiga type, in winter and high tundra in summer. When migrating in spring and autumn, they form very large herds, but particularly in winter they spread out to feed in small groups, often even above the timber-line as mountain winds tend to blow snow away from the requisite lichens and grasses (Gubser, 1965; Burch, 1972; also see Map 8B in this present book). They sometimes migrate as much as 300 kilometres, and their large migrating herds are susceptible to large scale exploitation (Gubser, 1965; Corbet, 1966), that is provided one manages to intercept them (Burch, 1972). In a normal year, the Nunamiut Eskimos at Anaktuvuk Pass kill and consume about 1,000 reindeer; this amount feeds about 96 humans and 200 dogs, and most of the killing is done by only about 20 major hunters (Gubser, 1965, p. 319). It should be noted that these reindeer form the main portion of the community's diet. Perhaps a similar pattern occurred from time to time in Upper Palaeolithic Britain, although the evidence for such is sadly still insufficient and/or unreliable; one can only say that reindeer remains are often "abundant" and sometimes even more "abundant" than those of horses, e.g. in the Earlier Upper Palaeolithic and Later Upper Palaeolithic at Soldier's Hole (Jackson in Parry, 1931) and in the Later Upper Palaeolithic at Pin Hole (Armstrong, 1931a). It may be that more horse was available in certain seasons (?summer) and more reindeer in others (?spring and autumn), so that the British Upper Palaeolithic hunters exploited both herbivores according to their availability, the horse perhaps being generally more abundant. There is also, of course, the likelihood that horse was rare or absent in some parts of Upper Palaeolithic Britain during the most severe stages of the Middle and Late Last Glacial, and almost certainly during the Full Last Glacial. Although the samples are small, such changes are perhaps apparent in the relative frequencies of horse and reindeer remains found by the present author during his own excavations of the Later Upper Palaeolithic layers at Robin Hood's Cave (Campbell, 1969; see pp. 125–31 below as well).

Coelodonta antiquitatis, the woolly rhinoceros, would also appear to have been fairly commonly exploited by British Upper Palaeolithic hunters, particularly during the Earlier Upper Palaeolithic when it may have been more abundant. When killed it would certainly have provided much more meat, fat, bones and skin than either a single horse or a single reindeer. Although it is now extinct, it is thought to have been less gregarious than horses and reindeer (Kurtén, 1968), and therefore perhaps harder to hunt. It is mostly thought of as an extreme tundra form, but unmistakable specimens of this species have also been discovered in deposits which pollen analysis indicates as having formed under dry, Temperate grassland conditions, e.g. at Cueva del Toll in Spain (Donner and Kurtén, 1958). Not only the head but also the neck was carried low in this species, showing extreme grass-eating specialization (Kurtén, 1968).

Ursus arctos, or the brown bear, occurs quite frequently at British Upper Palaeolithic cave-sites and may have been exploited more than has generally been assumed before. It would be useful not only for its fur, grease and bones, but being an omnivorous animal, despite its physical classification as a carnivore (Kurtén, 1968), it could well be useful for its meat. Like *Coelodonta antiquitatis*, however, it would also probably be more difficult to hunt than wild horse or reindeer. The frequent occurrences of *Crocuta crocuta*, the spotted or cave hyena, particularly in Earlier Upper Palaeolithic levels would, on the other hand, seem more likely due to an intercalation of *Crocuta* and Earlier Upper Palaeolithic cave-site occupations. This scavenger/predator was thus probably more often a competitor than a prey to man. Other large mammals which were sometimes hunted included notably: *Mammuthus primigenius*, the woolly mammoth, mainly or perhaps entirely in the Earlier Upper Palaeolithic; and *Cervus elaphus*, the red deer, *Megaloceros giganteus*, the giant deer, *Alces alces*, the elk or moose, and *Bison priscus*, the steppe wisent or bison, in the Earlier and the Later Upper Palaeolithic.

The above comments on the large mammals most likely to have been commonly exploited during the British Upper Palaeolithic are based on a careful analysis of the many faunal lists currently available for the British sites. Table 31 summarizes these faunal lists by naming most of the land mammals, both large and small, which are known or reasonably thought to occur in British Upper Palaeolithic deposits, with the exception of those mammals from poorly recorded excavations which are more likely to be Last Interglacial or Early Last Glacial at the latest, e.g. *Dicerorhinus hemitoechus* from Long Hole (Falconer, 1868; Garrod, 1926), and those which are most likely intrusive either from the Post Glacial, e.g. *Capreolus capreolus* and *Capra hircus* from Gough's Cave (Parry, 1929), or from very much older deposits, e.g. *Homotherium latidens* from Kent's Cavern (Campbell and Sampson, 1971). The mammals listed in Table 31 may all have been exploited by man in one way or another during the Upper Palaeolithic, but only in the case of those marked with an 'E' does definite evidence exist. This list is based on those in Garrod (1926) and Jackson (1962), which were checked against their references and on the results of the present author's own excavations and museum visits, as well as the information contained in Moore (1954) and Hallam, Edwards, Barnes and Stuart (1973). The list has also been checked against Kurtén (1968) and the scientific names acceptable to him are employed here. Those species marked with an 'I' are known to have reached Ireland during the Last Glacial (Savage, 1966); this information is well worth noting, since I feel there certainly remains a possibility, however remote, of eventual discovery of definite Upper Palaeolithic evidence in that country.

The Earlier Upper Palaeolithic sites which support the "EUP" column in Table 31 include the following cave-sites: Bench Fissure, Kent's Cavern (Devonshire); Badger Hole, Brean Down Shelter, Hyena Den, Soldier's Hole, Uphill Cave, Walton Cave (Somerset); Long Hole, Paviland Cave (Glamorganshire); King Arthur's Cave (Herefordshire); Cae Gwyn/Ffynnon Beuno Caves (Flintshire); Ash Tree Cave, Pin Hole, Robin Hood's Cave (Derbyshire) (Garrod, 1926 and references therein; Jackson, 1962 and references therein; present author's own excavations and museum visits). The Later Upper Palaeolithic sites which support the "LUP" column include the following cave-sites: Kent's Cavern, Tornewton Cave (Devonshire); Aveline's Hole, Bridged Pot Shelter, Gough's Cave, Soldier's Hole, Sun Hole (Somerset); Cathole, Paviland Cave (Glamorganshire); Hoyle's Mouth, Little Hoyle, Nana's Cave, Priory Farm Cave (Pembrokeshire); King Arthur's Cave (Herefordshire); Dead Man's Cave, Kinsey Cave, Victoria Cave (Yorkshire); Elder Bush Cave, Ossum's Cave, Thor's Fissure (Staffordshire); Church Hole, Churchdale Shelter, Dowel Cave, Fox Hole, Harborough Cave, Langwith Cave, Mother Grundy's Parlour, One Ash Shelter, Pin Hole, Robin Hood's Cave (Derbyshire); Yew Tree Shelter (Nottinghamshire) (Garrod, 1926 and references therein; Jackson, 1962 and references therein; White, 1970; present author's own excavations and museum visits), as well as the following open-sites: High Furlong (Lancashire) and Flixton 2 (Yorkshire).

There are thus 15 Earlier Upper Palaeolithic sites with fauna and 33 Later Upper Palaeolithic ones, and these totals represent the maximum possible counts of sites in the "EUP" and "LUP" columns, respectively. It must be stressed that the stratigraphic evidence at some of these sites, both Earlier and Later, is much less reliable than one would wish, because of the poor standards of some earlier excavations and/or natural disturbances. I have endeavoured to assess the proffered evidence critically in all cases.

Again, some of the mammal species themselves are quite difficult to identify, particularly the smaller kinds, and certain larger species such as *Bison priscus* are perhaps better considered generically only. However, even keeping these sources of error in mind, one may at least conclude from the general picture presented in Table 31, that the broad environmental/ecological implications of the occurrences of mammalian fauna are not at variance with those already presented from geomorphological, floral and other faunal lines of evidence, i.e. a range from Arctic to Temperate with an emphasis on Arctic to Boreal conditions, perhaps best described as Sub-Arctic/Boreal steppe-tundra. The obvious environmental differences implied between the two periods of Upper Palaeolithic occupation are also in accordance with the deductions made in previous sections of this chapter.

It is worth noting that simply on the basis of the number of sites at which they occur, the most commonly exploited mammals would again appear to include *Ursus arctos, Equus przewalskii* and *Rangifer tarandus*. (Indeed, the last named occurred at 30 out of 33 Later Upper Palaeolithic sites.) The next highest would appear to be *Bison priscus,* followed by *Canis lupus, Vulpes vulpes, Coelodonta antiquitatis, Cervus elaphus, Megaloceros giganteus* and *Lepus timidus*. Also on this basis, the Earlier Upper Palaeolithic in particular would appear to have seen comparatively common exploitation of *Mammuthus primigenius*. On the other hand, during the Later Upper Palaeolithic the rare herbivores, *Alces alces, Saiga tatarica* and *Capra ibex,* were doubtless exploited when available, as was the carnivore *Alopex lagopus*.

There now follows a brief review of the mammals identified by the present author from amongst the faunal remains found during his own excavations at British Upper Palaeolithic cave-sites. The discussion is arranged by site in the usual clock-wise geographic order. Complete lists, by site and layer, of all of the mammals identified and the antlers, teeth and/or bones on which the identifications are based, may be found in Tables 32 to 43. The mammals considered here in the text are generally only those actually found in Upper Palaeolithic layers.

As regards my identification of small mammals, it should be stressed that it is based almost entirely on teeth and/or jaws (see pp. 13–15 for discussion on identification of mammals). Their post-cranial bones do not normally permit accurate classification beyond a genus, whilst their more diagnostic teeth and jaws usually do. Hence there is naturally much disagreement amongst specialists on what species are actually represented by Last Glacial remains, and that in addition to the disputes over the specific classification of living insectivores, rodents and lagomorphs. I have chosen to follow Kurtén's comparatively comprehensive account of Pleistocene small mammals and their living representatives (Kurtén, 1968), in my choice of generic and specific names, though for brevity I have even shortened some of his groupings. These shortened groupings include particularly the voles, *Arvicola terrestris* and *Microtus agrestis*, which, if written out in full, would be better referred to as *Arvicola terrestris/amphibius* group and *Microtus agrestis/arvalis* group. My shortened names are merely intended as guide-lines to these specific/sub-specific groups, the habitats of the animals concerned not being greatly different within the two groups. Somewhat more difficult species to identify include the shrews, mice and lagomorphs, which when present I have sometimes only been able tentatively to assign to the species apparently represented, e.g. *Lepus* cf. *timidus*, the varying hare. My identification of both large and small mammals could not have been carried out without the use of the Oxford University Museum reference collection specimens and the kind assistance of Mr. P. Powell, Dr. J.C. Pernetta and the late Mr. D.F.W. Baden-Powell, to all of whom I again record my grateful thanks.

Badger Hole, 1968: Earlier Upper Palaeolithic Mammals

Table 32 gives the complete listing of undisturbed mammals which were found during the present author's excavations outside Badger Hole (Somerset) in 1968 (see Figs. 10 and 11 for excavation plan and sections). Remains were found at the interface of layers B/A3 and A2 and within the underlying layer A2, both positions being within the stratigraphic range of the neighbouring undisturbed unifacial leaf-point found earlier by Balch. The large mammals from interface B/A3–A2 include *Ursus* cf. *arctos, Coelodonta antiquitatis* and *Equus przewalskii*, all of which appear to have been exploited by the Earlier Upper Palaeolithic hunters at the site; the small mammals include *Microtus ratticeps* and *Dicrostonyx torquatus*, the latter a high tundra form according to Kurtén (1968). There are only remains sufficient to identify large mammals for layer A2 and these include *Crocuta crocuta, Felis* sp. and *Megaloceros giganteus* from near the base of the layer on bedrock, and *Lutra lutra, Vulpes/Alopex, Ursus* cf. *arctos, Equus przewalskii* and *Rangifer?* from the middle and upper portion of the layer. Those mammals from the middle and upper portion of layer A2 were most likely exploited by Earlier Upper Palaeolithic man as they were found within, or in close association with, a large cluster of burnt bone fragments (see Fig. 11). The scapula which represents *Lutra lutra* was one of the burnt remains, a most interesting direct suggestion of exploitation of the otter, presumably for fur and then for fuel. Those mammals from near the base of layer A2, on the other hand, would appear to represent an occupation of the site by the hyena, *Crocuta crocuta*, as not only his remains, but many unburnt and apparently gnawed bone fragments occurred there, although there is one enigma: the complete and possibly burnt metacarpal of *Megaloceros giganteus* (see Fig. 11), which one may only suggest as due to either human or hyena activity, its slight appearance of having been burnt being perhaps only due to mineralization (R. Burleigh, personal communication). Of course, it is just possible that a hyena brought a bone to the site which had been burnt elsewhere such as at the famous Wookey Hole Hyena Den, a site which was only briefly used during the Earlier Upper Palaeolithic (Tratman, Donovan and Campbell, 1971), and which is just below Badger Hole. It is worth noting that additional mammals found earlier at Badger Hole have included *Mammuthus primigenius* and *Bison priscus* (Wells Museum), and that these may also have been exploited by the Earlier Upper Palaeolithic inhabitants.

Sun Hole, 1968: Later Upper Palaeolithic Mammals

Table 33 gives the complete listing of undisturbed mammals which were found during the present author's excavations at Sun Hole (Somerset) in 1968 (see Figs. 12 to 14 for excavation plan and sections). The only identifiable remains were found in layer B2–7, from which the large mammals include *Vulpes vulpes, Ursus* cf. *arctos* and *Rangifer?*, the most "abundant" being *Ursus* with 6 bones and teeth out of the total 10 for the large mammals, though these 6 seem to represent a single individual of *Ursus*. The small mammals identified include *Arvicola terrestris, Microtus agrestis, M. ratticeps, Lemmus lemmus, Dicrostonyx torquatus* and *Lepus* cf. *timidus*, and this assemblage is probably indicative of a steppe-tundra ecosystem according to the environmental/ecological information in Kurtén (1968). The large mammals and *Lepus* were most likely all exploited by the Later Upper Palaeolithic inhabitants of the site, as their remains occurred well within the stratigraphic range of the diagnostic artifacts found earlier by Tratman (see Figs. 12 to 14), but the smallest mammals were probably the results of activities by other predators. Additional large mammals found earlier in the Later Upper Palaeolithic layers include *Felis silvestris, Canis lupus, Equus przewalskii* and specific identification of *Rangifer tarandus* (Tratman, 1955 and 1963). There is also a record of a mandibular fragment which has been attributed to the boar or wild hog, *Sus scrofa* (Tratman, 1955, p. 74), but as this species is normally interglacial or Post Glacial (Kurtén, 1968), its occurrence in a British Late Last Glacial context seems most unlikely; it is presumably either misidentified or derived. An unpublished record of *Crocuta crocuta*, on the other hand, might be acceptable, as this species is apparently tolerant of both interglacial and glacial, or near-glacial, conditions; however, the tooth on which its presence is based has the appearance of having been much rolled, so it may also be derived (Tratman, personal communication). Of course, this tooth may have been collected from somewhere else by the Later Upper Palaeolithic inhabitants simply as a curiosity. In short, the mammals exploited by them at Sun Hole for food and/or fur and other raw materials would appear to include *Ursus arctos, Equus przewalskii, Rangifer tarandus* and *Lepus timidus*, as well as possibly *Felis silvestris, Canis lupus* and *Vulpes vulpes.*

Cathole, 1968: Later Upper Palaeolithic Mammals and Small Mammal Faunal Analysis

Table 34 gives the complete listing of undisturbed mammals which were found during the present author's excavations outside Cathole (Glamorganshire) in 1968 (see Figs. 15 to 17 for excavation plan and sections). As abundant small mammal remains occurred in nearly all of the layers and included many identifiable specimens, it was decided to compute percentages from the minimum number of individuals represented for identified species by their respective layers and plot these in stratigraphic order on a "tree-pollen-type" diagram. The resulting picture of apparent changes in the local small mammal fauna is shown in Figure 81, where layer counts are given as well; the actual percentages are included in Table 35. Of course, it must be stressed that although five of the layer counts are reasonably large, the other five have less than 10 individuals each. The latter five must therefore only have their percentages accepted with caution, as larger samples might well alter their impression somewhat. Nevertheless, on the basis of the apparent appearance and disappearance of various species from layer to layer, I feel justified in suggesting the tentative zonation given on Figure 81. The information on the climatic implications of the various species is taken from Kurtén (1968), and it is based on the preferred ranges of the living species. The total spectra would certainly seem to lend further support to my interpretation, particularly as regards the abrupt faunal change from Arctic-Boreal types to Boreal-Temperate types between layers USB and C, a change which clearly represents the well-known transition from Late Last Glacial to Early Post Glacial. This small mammal analysis as a whole is not inconsistent with the results already presented on granulometric and pollen analyses from Cathole (see Figs. 64 and 73), again indicating that the Later Upper Palaeolithic evidence which I found during my own excavation belongs to the earlier part of the Late Last Glacial, whilst my Mesolithic evidence belongs to the earlier part of the Post Glacial. All of these results are however somewhat inconsistent with those from the apparently intrusive land snails (see Fig. 80 and Appendix 2).

The remains of large mammals associated with my Later Upper Palaeolithic artifacts include those of *Alopex lagopus* and *Ursus arctos* in layer LOB and *Vulpes vulpes* and *Rangifer?* in layer MSB, all of which may well have been exploited. The large mammal remains found by Wood in the 1860s which may also have been associated included the following according to Garrod (1926): *Crocuta crocuta, Felis silvestris, F. leo, Vulpes vulpes, Ursus* sp., *Mammuthus primigenius* ("very rare"), *Coelodonta antiquitatis, Equus przewalskii, Cervus elaphus, Megaloceros giganteus* and *Rangifer tarandus.* Of course, as Wood kept no stratigraphic record, these mammals could belong to various parts of the Last Glacial, and their presence inside the cave could easily have been partly due to *Crocuta*, with its well-known habit of dragging the bones of its prey to its lair. The large mammals found by McBurney in 1958 at his layer interface C/B or within his layer B and associated with his "Creswellian" artifacts (presumably the equivalent of my separate Later Upper Palaeolithic and Mesolithic assemblages) included the following (McBurney, 1959): *Meles meles, Vulpes* sp., *Alopex lagopus, Ursus* sp., *Ursus* cf. *arctos, Coelodonta?, Equus* sp., *Cervus elaphus, Capreolus capreolus, Rangifer tarandus, Ovis* sp. and *Bos* sp., as well as the small mammals *Lemmus lemmus* and *Lepus* cf. *timidus*. Quite frankly, McBurney's faunal list appears even more mixed than Wood's does, for I should have thought that *Capreolus, Ovis* and *Bos* were far more likely to come from well into the British Post Glacial, *Ovis* itself only occurring in Britain in its domesticated form (aside, of course, from the very much earlier occurrence of *Ovis savini* in the Cromer Forest Bed, Kurtén, 1968). In my excavations at Cathole, *Capreolus capreolus* occurred in layer D which is partly Mesolithic at the earliest, whilst *Ovis/Capra* and *Bos* sp. occurred in the overlying layers E and F, F being the modern topsoil.

On these bases the mammals contemporary with, and exploited by, the Later Upper Palaeolithic inhabitants of Cathole would seem most likely to have included the following: *Vulpes vulpes, Alopex lagopus, Ursus arctos, Coelodonta antiquitatis, Equus przewalskii, Cervus elaphus, Megaloceros giganteus, Rangifer tarandus* and *Lepus timidus*.

Table 36 provides an indication of the vertical distribution of all undisturbed teeth, bones and fragments thereof found during the present author's excavations outside the cave. It evidently reflects changes in the intensity of human and other predatorial activities, and suggests a wider vertical distribution of Later Upper Palaeolithic large mammal food débris than the vertical range of the artifacts found, as well as demonstrating the almost complete lack of large mammal remains in layer USB, before the accumulation of Mesolithic and later food débris in layers C and above. The occurrence of large mammal remains in layer A2 is presumably due to a predator such as *Crocuta*, although it is not inconceivable that Earlier Upper Palaeolithic man may have visited the site, as some of these bones appeared intentionally split but ungnawed.

Long Hole, 1969: Earlier Upper Palaeolithic Mammals

Table 37 gives the complete listing of undisturbed mammals which were found during the present author's excavations outside Long Hole (Glamorganshire) in 1969 (see Figs. 18 to 22 for excavation plan and sections). Identifiable remains associated with artifacts of Earlier Upper Palaeolithic aspect were only found in layer A3a and these include the following: *Vulpes/Alopex, Equus* cf. *przewalskii* and *Rangifer tarandus*, all three of which were presumably exploited. There is a more extensive list of mammals published by Falconer (1868), all of which were supposedly found in association with other artifacts from the site which I consider Earlier Upper Palaeolithic. These mammal remains were excavated by Wood in 1861 and include the following: *Palaeoloxodon antiquus, Dicerorhinus hemitoechus* and *Sus scrofa* which are more likely Last Interglacial; *Crocuta crocuta, Felis silvestris, F. Leo, Canis lupus, Vulpes vulpes, Ursus* cf. *arctos, Cervus elaphus, Megaloceros giganteus* and *Bison priscus* which may all be either Last Interglacial or Last Glacial; and *Martes* cf. *martes, Mustela putorius, Lutra lutra, Mammuthus primigenius, Coelodonta antiquitatis, Equus* cf. *germanicus/przewalskii, E.* cf. *hydruntinus, Rangifer tarandus* and *Lepus timidus* which are most likely Early to Middle Last Glacial as a group. Garrod herself suggested that "there had been some disturbance of the cave-earth, or that two levels were present but were not detected" (Garrod, 1926, p. 69), and on the basis of my own field and laboratory observations, I would claim that at least four separate faunal assemblages were present in the area I excavated, and were presumably also present in the area Wood excavated a century before me. My uppermost faunal assemblage was that which is described above from Earlier Upper Palaeolithic layer A3a, i.e. *Vulpes/Alopex, Equus* and *Rangifer*. In descending order the faunal evidence from my underlying layer A2c indicates an occupation by *Crocuta crocuta* with, as the débris of its scavenging, various steppe-tundra herbivores including the following: *Mammuthus primigenius, Coelodonta antiquitatis, Equus* sp., *Megaloceros giganteus, Rangifer tarandus* and *Lepus* sp. Then layer A2b with its scant evidence of possible Middle Palaeolithic activity (see p. 60) has a fauna including *Martes* cf. *martes*, the pine marten, and *Alces alces*, the true elk or moose, two species in remarkably close agreement with the Boreal coniferous forests indicated by the pollen from that layer (see Fig. 74); *A. alces* is represented by a freshly shed antler, which might suggest a winter occupation as its antlers are shed in January (Clark, 1954; Hall and Kelson, 1959). And finally at the bottom of the vertical distribution of faunal evidence, from layer A2a there is a single identification of *Bison/Bos* which would be quite at home in the Temperate woodland/steppe of the Late Last Interglacial as indicated by the pollen from this layer (see Fig. 74 again).

Thus on these bases the mammals contemporary with, and exploited by, the Earlier Upper Palaeolithic inhabitants of Long Hole would seem most likely to have included the following at most: *Lutra lutra, Canis lupus, Vulpes vulpes, Vulpes/Alopex, Ursus arctos, Mammuthus primigenius, Coelodonta antiquitatis, Equus* cf. *germanicus/przewalskii, Megaloceros giganteus, Rangifer tarandus, Bison priscus* and *Lepus timidus.* There is also the interesting possibility of the occurrence of *Equus hydruntinus*, the extinct European wild ass (or Falconer's "E. asinus", 1868; see Kurtén, 1968 as well), but if Falconer's identification were in fact correct, this would be the only example in Upper Palaeolithic Britain.

Mother Grundy's Parlour, 1969: Later Upper Palaeolithic Mammals and "Food and Fur Analysis"

Table 38 gives the complete listing of undisturbed mammals which were found during the present author's excavations outside Mother Grundy's Parlour (Derbyshire) in 1969 (see Figs. 23 and 24 for excavation plan and section). Identifiable remains associated with artifacts of Later Upper Palaeolithic aspect were found in layers LB and SB. The large mammals from layer LB include *Ursus arctos, Equus przewalskii* and *Megaloceros giganteus,* whilst those from layer SB include *E. przewalskii* and *Cervus elaphus*, both groups having presumably been exploited at their respective times. Identifiable small mammal remains were only found in layer LB and these include *Sorex araneus* and *Microtus gregalis*, both forms being not inconsistent with the Late Last Glacial pollen spectra of that layer (see Fig. 76). The occurrence of two molars of *Equus przewalskii* in the bottom of layer C is very likely to be due to derivation from the underlying layer SB, as C is definitely a Post Glacial formation and both layers are on a slope. However, it is worth noting that this vertical distribution of wild horse could just conceivably represent a rare British example of localized survival and adaptation from a steppe-tundra form to an Early Post Glacial woodland form, i.e. a transformation to Zeuner's forest horse ecotype which he terms "Equus caballus silvestris" (Zeuner, 1963), or assuming it was wild, *E. przewalskii* "silvestris". As McBurney found a similar vertical distribution of *Equus* remains during his 1959–60 excavations at the site and obtained radiocarbon age estimates suggestive of a Post Glacial age for them (see pp. 62–63, 104–05 and Fig. 76), he has tentatively come to just that conclusion, a possible case of an Early Post Glacial forest horse (C.B.M. McBurney, personal communication). Armstrong (1925) also found *Equus* in every one of his spits or "zones", but as is the case with my own results, his evidence for *Equus* was most abundant in the lower portion of his faunal sequence, i.e. in his "base and lower middle zones" (see Jackson in Armstrong, 1925, pp. 176–78), in what was almost certainly a Late Last Glacial context and about equivalent to my layers LB and SB. But despite these similar observations by three independent investigators, including myself, I am still not entirely convinced that the upper faunal evidence at the site offers an incontestable case for Post Glacial wild horse in Britain, as Mother Grundy's Parlour is one of the very few British sites where this has been seen; as regards the occurrence of three horse teeth at the prolific Early Mesolithic site of Thatcham in Berkshire (King in Wymer, 1962, pp. 355–61), John Wymer (personal communication) is now of the opinion that they may have been derived naturally from the underlying river gravel. But in the Peak District a steppe horse ecotype may have survived well into the Post Glacial (Bramwell, 1973), where open vegetation communities apparently continued until Zone VI (Tallis, 1964). Of course, if there were in fact Post Glacial wild horse in the vicinity of Mother Grundy's Parlour, that might help to explain an apparent survival into the local Mesolithic of certain typologically

Later Upper Palaeolithic tool-forms such as "Creswell points" and "shouldered points" (see pp. 171–75), if one accepts the view that there is a definite functional correlation between the pursuit of horse and the manufacture of those tool-forms.

The list of large mammal remains found by Armstrong and identified by Jackson (Jackson in Armstrong, 1925, pp. 176–78) is given in Table 39 according to the stratigraphic information recorded by Armstrong and Jackson. It should be noted that *Felis leo* is only represented by a broken vertebra and *Mammuthus primigenius* by two fragments of tusks, all of which come from the "base" spit (Jackson in Armstrong, 1925, p. 176), and may therefore possibly be older than the supposedly associated Later Upper Palaeolithic artifacts (see pp. 61–62). As has already been stated, a number of times above, the chronological and environmental evidence in Britain generally is against these two mammals having survived into the British Late Last Glacial. *Crocuta crocuta* and *Coelodonta antiquitatis*, on the other hand, occurred in both the "base" and "lower middle" spits and may well have survived into the local Late Last Glacial.

Table 40 summarizes the list of "food" and "fur" mammals which I found and is useful as a check on Armstrong's list in Table 39. From a comparison of these two lists and the available relevant stratigraphic information, it would seem that the large mammals most likely exploited by the Later Upper Palaeolithic inhabitants of Mother Grundy's Parlour included the following: *Ursus arctos, Coelodonta antiquitatis, Equus przewalskii, Cervus elaphus, Megaloceros giganteus, Rangifer tarandus* and *Bison priscus,* with the main emphasis on *E. przewalskii* and perhaps *B. priscus*. In my excavation *Bison/Bos* only occurred in Mesolithic layers C and D, and its remains are presumably attributable to the aurochs, *Bos primigenius*, rather than *Bison priscus* (see Kurtén, 1968, pp. 185–90).

It should be noted that the interface of layers C and SB yielded a hybrid artifact assemblage which, without being a mixture of two separate entities, seems to stand somewhere between what one considers formally "Upper Palaeolithic" and "Mesolithic" (see pp. 171–73), and as has already been stated (p. 123), the two molars of *Equus przewalskii* from layer C occurred near the bottom of that layer. On the basis of these observations and the information under "middle" spit on Table 39, it might be suggested that the large mammals exploited by the "transitional" Mesolithic/Later Upper Palaeolithic inhabitants of Mother Grundy's Parlour included the following: *Canis lupus, Ursus arctos, Equus przewalskii, Cervus elaphus, Bison priscus* and *Bos primigenius.*

Table 40 also presents estimates of the approximate weights of the various "food" mammals represented at Mother Grundy's Parlour, 1969. Both "dead weights" and "meat weights" are given, the latter being taken as about 50% of the former (see pp. 13–14 on methodology). Some interesting contrasts are apparent when one compares percentages based on weight with those based on minimum number of individuals. In the case of layer LB (L.U.P.) the total meat weight of

Megaloceros is almost twice that of *Equus* suggesting perhaps that the former was more important. But of course larger samples from any of these layers would probably alter the picture implied by Table 40 considerably, as may be seen by the data in Table 39. Nonetheless, it is worth noting that *Equus* represents over 80% of the meat in my sample from Later Upper Palaeolithic layer SB, whilst *Bison/Bos* represents about 67% and then 81% of the meat, respectively, in Mesolithic layers C and D. Further, the grand totals for meat per layer might suggest that, given some method of storage (e.g. freezing or drying) and given a daily meat consumption of about 1 kilogramme per person (see pp. 13–14), these variegated meat sources might have supported from about 10 to 20 people for about 1 to 3 months in each separate occupation of the site. Sadly, such a suggestion is almost entirely speculative.

Robin Hood's Cave, 1969: Later Upper Palaeolithic Mammals, Small Mammal Analysis and "Food and Fur Analysis"

Table 41 gives the complete listing of undisturbed mammals which were found during the present author's excavations outside Robin Hood's Cave (Derbyshire) in 1969 (see Figs. 27 to 31 for excavation plan and sections). As small mammal remains are abundant and include a fair number of identifiable teeth and jaws, as with those from Cathole (see Fig. 81), I have again felt justified in computing percentages and illustrating their vertical distribution in another "tree-pollen-type" diagram. The results of this small mammal analysis are given in Figure 82 and Table 42, but it must be stressed that the counts of minimum number of individuals are generally rather less than those at Cathole. Also no identifiable mammals, large or small, occurred in my layers C and D, two layers which had been almost entirely removed by earlier excavations, as indeed was a substantial portion of layer USB (see sections, Figs. 28 and 31). The small mammal remains were obtained by sieving both in the field and in the laboratory, their positions being recorded either according to 10 centimetre spits and the natural layers, or according to the levels of the deposit samples. For the purposes of Figure 82 and Table 42, as most of the small mammal remains occurred in or near grid square A1 where the deposit samples were collected, the various spits have been carefully correlated with the deposit sample levels, the sample numbers themselves being indicated on the diagram, as well as in the table. Despite the small counts, on the basis of observed "appearances" and "disappearances", the environmental preferences of the various small mammals represented seem to suggest two cold phases and an intervening milder phase, the most reasonable interpretation of which is Late Last Glacial Zones I, II and III, particularly when considered in conjunction with the Robin Hood's Cave 1969 pollen diagram (see Fig. 77 as well); the associated artifact assemblages would fit well with this reading. As indicated on Figure 82, there may also be a superficial indication of the fairly well-known sub-divisions of Zone I, the "microfauna" of layers B/A and lower LSB (samples 2 and 3, respectively) probably being either Zone Ia or Full Last Glacial, or both, as it consists almost entirely of remains of the Arctic lemming, *Dicrostonyx torquatus*. The habitat of the living *D. torquatus* is treeless tundra, whilst that of the Norway lemming, *Lemmus lemmus*, is today mountain birch woods and the zone immediately above the timber line (Kurtén, 1968). The climatic implications of the other small mammals listed on Figure 82 are also again extracted from Kurtén (1968). If my tentative zonation be correct, then the "L.U.P.1" artifact assemblage from layer B/A would be about Zone Ia in age, "L.U.P.2" from layer LSB would be Zone Ib, "L.U.P.3" from layer OB would be Zone II, and "Meso./L.U.P." from layer USB would be Zone III. Two of the small mammals which might have been exploited include the mustelid carnivore, *Martes martes* (see Table 41, layer OB), and the varying hare, *Lepus timidus* (see Fig. 82 and Tables 41 and 42). A particularly noteworthy occurrence is that of the steppe pika, *Ochotona pusilla*, near the bottom of layer OB. *O. pusilla* is a Boreal-Temperate grassland species which has only been previously recorded from southern English cave-sites (Jackson, 1967; Kurtén, 1968). However, its presence in the vicinity of Robin Hood's Cave would not be at all unlikely during the transition from Zone I to II.

The identified "food" and "fur" mammals found in 1969 are summarized in Table 43. Those found earlier by Dawkins and Mello in their "breccia and cave earth" in association with what I consider Middle Palaeolithic, Earlier Upper Palaeolithic and Later Upper Palaeolithic artifacts included the following (Dawkins, 1876 and 1877): *Crocuta crocuta* ("very abundant"), *Felis silvestris* ("very rare"), *F. leo* ("rare"), *Canis lupus* ("abundant"), *Vulpes vulpes* ("abundant"), *Ursus* sp. ("abundant"), *Mammuthus primigenius* ("abundant"), *Coelodonta antiquitatis* ("very abundant"), *Equus przewalskii* ("very abundant"), *Megaloceros giganteus* ("rather scarce"), *Rangifer tarandus* ("very abundant") and *Bison priscus* ("rather scarce"), as well as the small mammal, *Lepus timidus* ("abundant"). Of some relevance to these workers' faunal list is my discovery of a small flint handaxe and a deciduous molar of *M. primigenius*, both with traces of a similar red silt adhering, quite close together in their tip. Of course, some of the mammals in their list could perfectly well have been properly associated with any or all of their Palaeolithic assemblages, although some may also have been due to the activities of *Crocuta crocuta*.

It should be noted as regards the occurrence of *Crocuta crocuta* in layers A to USB that it is based entirely on teeth (see Table 41). Whilst the *Crocuta* teeth from layer A appear comparatively fresh, those from layers B/A to USB are distinctly more rolled and eroded. This difference in preservation is presumably due to the teeth in layer A being *in situ* whilst the others have been derived in one way or another, perhaps even through the curiosity collecting activities of Later Upper Palaeolithic man, although in the case of layer B/A perhaps partly by natural erosion of the upper portion of the underlying layer A (see pp. 66–67 on stratigraphy). For these reasons *Crocuta* is not listed under layers B/A to USB in

Table 43. The other faunal remains from layers A to USB all appear unabraded, although certainly they are usually broken, and were presumably left in primary position as their containing layers formed. However, for obvious reasons individuals represented by shed antlers are not included in Table 43 either.

As with Table 40 on Mother Grundy's Parlour, Table 43 presents estimates both of "dead weights" and of "meat weights" at Robin Hood's Cave, there again being a sufficient number of individuals represented to make these calculations worthwhile (see pp. 13–14 on method). In Middle Palaeolithic layer A *Cervus elaphus* accounts for 25% of the individual "food" and "fur" mammals, but *Mammuthus primigenius* and *Coelodonta antiquitatis* each account for 46%, or a combined total of 92%, of the meat weight. In Later Upper Palaeolithic layer B/A (L.U.P.1) *Coelodonta* alone amounts to over 75% of the meat weight, even though it is only represented by a single individual. Obviously, these weight percentage figures are affected by the great size of mammoth and woolly rhinoceros to begin with, so that, for example, it would require about 10 horses to equal 1 rhinoceros in total meat weight. But then the meat of a rhinoceros might proportionately include much more fat than that of a horse, although it is difficult to be certain what their actual food values were during the Last Glacial. Horse may at times have been less important in Last Glacial Palaeolithic economies, particularly when not readily available, but it, together with reindeer, probably accounted for a much greater proportion of the variegated "biomass" than did mammoth and woolly rhinoceros, though to be sure one would need to conduct some sort of analysis of the numerous remains of these animals found in river terraces and other locations attributed to the Last Glacial, a study which in itself might entail yet another full monograph.

In Later Upper Palaeolithic layer LSB (L.U.P.2) the meat percentage from *Coelodonta* falls to less than 60% whilst that of *Rangifer* rises to 23% and *Equus* to almost 18%. In terms of individuals, *Coelodonta* with 1 is 6%, *Equus* with 3 is 19%, *Rangifer* with 6 is almost 38% and the varying hare, *Lepus timidus* with 4 is 25%. *Alopex lagopus*, the Arctic fox, with 2 is almost 13% of the minimum number of individuals. *Alopex* and *Lepus* were probably important for their pelts/skins, and of the two at least *Lepus* was probably eaten from time to time as many of its bones are split and broken (not only in layer LSB but in all of the layers from A to USB, inclusively).

In layer OB (L.U.P.3) *Coelodonta* falls still further to about 46% of the meat weight, whilst *Ursus* appears with almost 14%, *Equus* stays about the same at 18%, and *Rangifer* falls to a mere 6% but is joined by *Cervus elaphus* with 6% and *Megaloceros* with 9% of the meat. By minimum number of individuals *Lepus timidus* is highest with 25%, followed closely by *Equus* at 20%. There is then a cluster lower down the percentage scale including *Ursus, Cervus elaphus* and *Rangifer*, each at 10% of the individuals represented. Simply by number of individuals horse is the most important large mammal here. But the hunting of brown bear is also probably of some significance. Aside from being a potential source of fur, the omnivorous habits of the latter may have made it rather good to eat. Amongst the Nunamiut Eskimo bear meat is not highly prized (Gubser, 1965), but amongst the Netsilik Eskimo it is (Balikci, 1970). Of course, the Ainu of northern Japan are well known for their bear hunts and festivals, the bear forming an important component of their cultural ecology (Watanabe, 1968). On the basis of the evidence for bear in layer OB at Robin Hood's Cave, as well as Bramwell's fascinating description of an apparently intentional burial of a bear skull in what might be a Later Upper Palaeolithic context at Fox Hole, Derbyshire (Bramwell, 1971 and 1973), I would suggest that brown bear sometimes played an important role in British Later Upper Palaeolithic society as well.

In "Mesolithic/Later Upper Palaeolithic" layer USB *Coelodonta* again rises to almost 69% of the meat weight and *Rangifer* rises to almost 14%, whilst *Ursus* falls slightly to about 10% and *Equus* falls drastically to only about 7%. According to minimum number of individuals, on the other hand, *Rangifer* rises even more dramatically to over 33%, whilst the percentages for *Canis lupus* and *Coelodonta* rise slightly to 11% each, *Ursus* stays about the same at 11%, *Equus* falls to 11% and finally *Lepus timidus* drops only slightly to 22%. Reindeer and perhaps woolly rhinoceros would appear to have been the most important large mammals, whilst the varying hare was perhaps the most important small mammal in the "Mesolithic/Later Upper Palaeolithic" ecosystem as represented by layer USB. Of course, such statements ought merely to be taken as suggestions, since so much of the evidence involves quite small samples. But nevertheless, even though they are small by continental European standards, the Robin Hood's Cave mammal samples are comparatively speaking amongst the largest reliable ones available thus far for Upper Palaeolithic Britain.

Early Mesolithic Britain is another story as Star Carr, for example, provides an eminently large sample of Early Post Glacial Zone IV mammals caught in the context of a lake-side winter camp. The "food" and "fur" mammals here include out of a total of 153 individuals represented 1.3% *Martes martes,* 0.7% *Meles meles*, 1.3% *Canis familiaris*, 1.3% *Vulpes vulpes,* 3.3% *Sus scrofa*, 52.3% *Cervus elaphus*, 21.6% *Capreolus capreolus*, 7.2% *Alces alces*, 5.9% *Bos primigenius*, 4.6% *Castor fiber* and 0.7% *Lepus europaeus*. These data are based on the faunal report by Fraser and King in Clark (1954, p. 91) and checked against Clark (1972). The red deer, *Cervus elaphus*, is clearly the most important mammal in this particular Earlier Mesolithic ecosystem, even though it would not have been quite as important in the natural ecosystem of the time (Fraser and King in Clark, 1954). Clark (1972) has worked out the approximate meat weights and *C. elaphus* is again the most important, although the diet on it was clearly supplemented with additions from *Sus, Capreolus, Alces* and *Bos*. The presence of just about the earliest definitely domesticated dog, *Canis familiaris* (Clark, 1972), is also quite significant, but whether it was raised for food or something else remains uncertain.

At Robin Hood's Cave the *Canis* represented seems still to belong to the wolf, *C. lupus*, even in layers OB and USB which belong, respectively and at least in part, to pollen Zones II and III. But this is not to say that some sort of "special relationship" was not already developing between wolf and man in the Later Upper Palaeolithic, a relationship which then evolved to the point where it became recognizable as *Canis familiaris* in Zone IV at Star Carr. Over on the Continent my colleague Michel Dewez at first thought he might possibly have an example of *C. familiaris* in Zone III at La grotte de Remouchamps (Belgium), but this has again turned out morphologically at least to be *C. lupus* (see Dewez *et al.*, 1974).

In the case of the deer family members represented at Star Carr Hallam, Edwards, Barnes and Stuart (1973) have even suggested that some of them, namely red deer and elk, may have been intentionally herded. Whether this actually happened and whether such practices also occurred earlier at Robin Hood's Cave in the case of horse, red deer, giant deer and reindeer, is very difficult to test. The animals represented are mostly fairly young adults, not juveniles, and mostly males, both at Star Carr and Robin Hood's Cave, and 1 or 2 of the horses in layer OB (Zone II) at the latter site were fairly old at death. But then whether the juveniles were "herded" elsewhere and elsewhen in each case or simply ignored altogether by the "herder/hunters" remains equally uncertain.

The grand totals for meat per layer in the Later Upper Palaeolithic at Robin Hood's Cave (see Table 43) might suggest that these were sufficient to support from about 10 to at most 30 people for about 1 to possibly 6 months in each successive occupation of the site. This assumption is partly based on what I consider to be the probability that each person ate about 1 kilogramme of meat per day (see pp. 13–14), as well as the probability that various means of meat storage (freezing, drying, etc.) were known and used when necessary. Such suggestions are unfortunately still rather speculative but nonetheless seem worth making until better data is available for the British Upper Palaeolithic.

The general sequence of large mammals at Robin Hood's Cave would seem to reflect at least in a broad manner the Last Glacial zonations already demonstrated from the pollen and small mammal studies (compare Table 43 with Figs. 77 and 82). That is, the large mammals found in layer A would be at home in the Early Last Glacial, those in layer B/A in Late Last Glacial Zone Ia/Full Last Glacial, those in layer LSB in Zone Ib, those in layer OB in Zones II/Ic, and finally those in layer USB in Zone III.

As regards seasons of occupation at the site these might be inferred from certain mammals and birds. Lacking newly born and juvenile mammal remains in any of the undisturbed samples it is not possible to examine that line of evidence, although their absence might suggest winter occupations, unless they were intentionally ignored during other seasons. But shed antlers offer another line of evidence: freshly shed antlers of *Cervus elaphus* in layers A and OB might imply, respectively, that the Middle Palaeolithic and Later Upper Palaeolithic (L.U.P.3) occupations occurred round about April; freshly shed female reindeer antlers, *Rangifer tarandus*, might indicate the calving season and Later Upper Palaeolithic occupations not later than June in layers B/A to USB; apparently unshed male reindeer antlers from layers B/A to USB might also suggest a similar season but these are mostly fragmentary and thus difficult to deduce anything from (males do not shed their antlers until after rutting in October or so). This data on red deer and reindeer antlers is drawn from Clark (1954 and 1972) and Hall and Kelson (1959), respectively. The reindeer remains seem to be those of the "barren ground" ecotype rather than the "forest" ecotype.

Seasons inferred from the bird remains at Robin Hood's Cave include the following: "?winter" in layers A and USB; "summer" in layers B/A to OB. The bird remains themselves are listed in Table 30 and further described in Appendix 5. The inference of "?winter" is drawn from the fact that only "residents" are represented in layers A and USB, whilst that of "summer" is based on the presence both of "residents" and of "summer tundra/taiga nesting" birds in layers B/A to OB. These inferences are not greatly at variance with the seasonal occupations inferred above from antlers. Where the birds and mammals suggest different seasons for the same layer, as in layer USB (birds: ?winter; reindeer: c. June), it must be remembered that it took far more than one season, or for that matter one year, for each layer to accumulate. Also, birds of prey may have left many of the bird remains during seasons when people were *not* at the site! But some of the bird remains, particularly those of mallard, goldeneye, grouse, ptarmigan and plover, were almost certainly snared or hunted by man, goldeneye duck having probably been taken in summer in layer OB, and plover in summer in layer B/A. Mallard, grouse and ptarmigan were probably residents in the region and therefore might have been taken at any time of year.

A noteworthy small mammal carnivore in layer OB is the pine marten, *Martes martes*. It occurred at about the level of deposit sample 6 (i.e., the middle of layer OB). *M. martes* usually keeps to pine forests according to Kurtén (1968), and so its presence even only as a single individual in mid-layer OB might imply that some stands of pine were growing at the same time not too far from Creswell Crags. According to the associated pollen, rodents and lagomorphs (see Figs. 77 and 82) mid-layer OB belongs to Zone II of the Late Last Glacial (i.e., to the so-called "Allerød Interstadial"), and therefore the right sort of habitat might have been available for *M. martes*. However, it is still debatable whether in fact pine grew in Britain during Zone II, even though birch trees certainly did (see West, 1968; Walker and West, 1970; Hallam, Edwards, Barnes and Stuart, 1973; Mitchell, Penny, Shotton and West, 1973). Of course, *M. martes* can range a bit beyond its normal habitat, though generally not very far. Its occurrence in this manner may also be seen in Zone IV at Star Carr (Clark, 1954).

A particularly noteworthy large mammal occurrence is that of the ibex, *Capra ibex*, in layer B/A. *C. ibex* is an extreme Alpine species which has been previously recorded from only a few British Last Glacial cave-sites, e.g. Victoria Cave in western Yorkshire (Kurtén, 1968), and its presence in layer B/A is quite consistent with the apparent Zone Ia/Full Last Glacial age of that layer. *C. ibex* is also known in the Belgian Upper Palaeolithic (Cordy, 1974).

On the basis of the above observations it would appear that the times and seasons of Later Upper Palaeolithic occupation, as well as the mammals contemporary with, and exploited by, the various Later Upper Palaeolithic inhabitants of Robin Hood's Cave included the following:

L.U.P.1 (Zone Ia summer)	*Alopex lagopus, Coelodonta antiquitatis, Equus przewalskii, Rangifer tarandus, Capra ibex* and *Lepus timidus;*
L.U.P.2 (Zone Ib summer)	*Alopex lagopus, Coelodonta antiquitatis, Equus przewalskii, Rangifer tarandus* and *Lepus timidus;*
L.U.P.3 (Zones Ic–II spring–summer)	*Martes martes, Canis lupus, Vulpes vulpes, Ursus arctos, Coelodonta antiquitatis, Equus przewalskii, Cervus elaphus, Megaloceros giganteus, Rangifer tarandus* and *Lepus timidus*;
Meso./L.U.P. (Zone III summer)	*Canis lupus, Ursus arctos, Coelodonta antiquitatis, Equus przewalskii, Rangifer tarandus* and *Lepus timidus.*

The above results of mammal identifications and analyses, like those of the granulometric and pollen studies, are of great importance to any assessment of the chronology and ecology of Upper Palaeolithic settlement in Britain. These mammal studies are also of primary importance to the interpretation of the hunting economies of the British Upper Palaeolithic. However, it must be stressed that their identification and interpretation is often very difficult, and the suggested ages and mammals selected for exploitation are simply those considered by the present author to be the most reasonable; others would certainly be possible in some cases. In addition to the palaeontological assistance from Mr. P. Powell, Dr. J.C. Pernetta and the late Mr. D.F.W. Baden-Powell, to whom I have already recorded my thanks, I am indeed thankful to Dr. J.W. Jackson, Dr. J. Jewel, Dr. K.P. Oakley and Dr. A.J. Sutcliffe for helpful advice on the interpretation of faunal lists which were published before my own work was begun. Dr. Jackson, who is now well over 90 years old, in fact worked under Sir William Boyd Dawkins at the Manchester University Museum before he himself became Curator, after which he regularly examined the faunal remains found by Armstrong, Parry and the University of Bristol Spelaeological Society, not to mention the samples excavated by many other workers.

Before leaving this discussion of Upper Palaeolithic faunal evidence, I think it is well worth adding comments on the Last Glacial mammalian remains discovered by those who excavated at various times at two further very important sites: Kent's Cavern and Soldier's Hole. At both of these sites the information available indicates Later Upper Palaeolithic artifacts overlying Earlier Upper Palaeolithic ones, as has already been demonstrated in the section on stratigraphy (see pp. 37–43). Similarly, there are separate faunal assemblages, whose stratigraphic relationship is known, associated with the artifacts.

Kent's Cavern, 1865–80: Earlier and Later Upper Palaeolithic Mammals

The relevant undisturbed mammal remains from Kent's Cavern (Devonshire) were found by Pengelly during his excavations of the "Cave Earth" (layers B/A2) particularly in the years 1865 to 1868 (see Figs. 4 to 6 for excavation plan and sections, and pp. 37–42 on stratigraphy). Those remains recorded by him from the Great Chamber are associated in part with Later Middle Palaeolithic ("Handaxe Mousterian") and Earlier Upper Palaeolithic artifacts, whilst those from the "Black Band" and its vicinity in the upper portion (layer B2) of the "Cave Earth" of the Vestibule are associated entirely with Later Upper Palaeolithic artifacts. Although a faunal list for the "Cave Earth" as a whole has been published (Pengelly, 1869 and 1884; Garrod, 1926), there has been no publication of the vertical and horizontal distribution of this evidence within the "Cave Earth", despite the fact that such information was duly recorded by Pengelly himself in his ever meticulous Diary (1865–80, pp. 1–600). Pengelly not only provides distributional data in his Diary, but actual counts of the number of mammal teeth found and the species or genus to which each tooth appears to belong. I have checked as many of his tooth identifications as I could in the Torquay Museum and in all cases I agreed with the identifications. I have therefore based Table 44 on his recorded tooth counts, although, as is my standard practice, I have altered the scientific names to those in current use (e.g. Kurtén, 1968).

It is certainly significant that both *Felis leo* and *Mammuthus* are apparently absent from the Later Upper Palaeolithic faunal list, whose Late Last Glacial age would certainly seem to be confirmed by my radiocarbon age estimates (see pp. 41–42). Also *Crocuta* and *Coelodonta* exhibit a marked decrease in the Later Upper Palaeolithic list in comparison with the earlier list from the Great Chamber, and, conversely, *Ursus, Equus* and *Megaloceros* exhibit an increase. These differences may indicate changes in the environment or changes in the preferences of the hunters, or both. However, the main similarity between the two lists is the high proportion of *Equus* teeth in both, a similarity which I would consider a reflection of the Earlier and Later Upper Palaeolithic hunters' common preference for, and dependence upon, wild horse as a major food source. A secondary, but important, food source for the Earlier Upper Palaeolithic was probably *Coelodonta*, whilst those for the Later Upper Palaeolithic perhaps included both *Ursus* and *Megaloceros*. The presence of teeth attributable to both *Equus* cf. *germanicus* and *E. przewalskii* in the list from the Great Chamber is of interest, as my radiocarbon age estimates (see pp. 40–41) for the Earlier Upper Palaeolithic of the Great Chamber and the adjacent Gallery indicate an age more or less compatible with the known age of the *germanicus/przewalskii* "shift" of Kurtén (1968). Also according to Pengelly's Diary (1865–80, pp. 1–236, 483–600) most of the horse teeth of *E. germanicus* form, as well as most of the *Crocuta* teeth, come from the 3rd and 4th foot spits in positions which are generally closer to the Middle Palaeolithic scatter of artifacts. However, as there is considerable overlap between the faunal evidence of the Middle Palaeolithic and that of the Earlier Upper Palaeolithic, due to Pengelly's use of 1 foot spits, I have not endeavoured to separate what must originally have been two or more quite separate faunal assemblages, the high proportion of *Crocuta* probably being accounted for by its own cave-dwelling activities. I have, however, deleted *Ursus spelaeus* from these Last Glacial lists as that species quite clearly only occurs *in situ* in the much earlier "breccia" layer, a layer which had already been largely or entirely eroded out of the Great Chamber and the Vestibule (see Figs. 4 to 6 and pp. 38–40; also Campbell and Sampson, 1971). Small mammals recorded from the "Cave Earth" but not specifically or generically cited in Pengelly's Diary probably include the following according to Evans (1872): *Gulo gulo, Castor fiber, Arvicola terrestris, Microtus agrestis, Ochotona pusilla* and *Lepus timidus*, all of which are "rare" or "scarce". Large mammals found by Ogilvie and Dowie in 1926–28 below Pengelly's work in the Vestibule partly in association with Middle Palaeolithic and Earlier Upper Palaeolithic artifacts (see Vestibule Section, Fig. 6) completely confirm Pengelly's observations in the Great Chamber. These large mammals include the following according to Dowie (1928): *Crocuta crocuta, Felis leo, Meles meles, Canis lupus, Vulpes/Alopex?, Ursus arctos, Mammuthus primigenius, Coelodonta antiquitatis, Equus* sp., *Cervus elaphus, Megaloceros giganteus, Rangifer tarandus* and *Bison/Bos; Crocuta* and *Equus* being "predominant" and *Coelodonta* and *Cervus elaphus* "very common".

Soldier's Hole, 1928–29: Earlier and Later Upper Palaeolithic Mammals

Table 45 gives the complete listing of undisturbed mammals which were found by Parry during his excavations inside Soldier's Hole (Somerset) in 1928–29 (see pp. 42–43 on stratigraphy). The remains were identified by J.W. Jackson (Jackson in Parry, 1931, pp. 58–62).

Although *Coelodonta* is not represented at this site, it is known from other Upper Palaeolithic sites in the Mendip region, e.g. Badger Hole (see p. 118). Also although *Equus* is not recorded from Later Upper Palaeolithic layer B at Soldier's Hole, it is the most abundant large mammal associated with the Later Upper Palaeolithic at the nearby site of Gough's Cave (Parry, 1929, p. 104). At Soldier's Hole, *Rangifer* is the most abundant mammal in both the Earlier and the Later Upper Palaeolithic layers. This preponderance of *Rangifer* might be due to the occupation of the site at a different season from many of the other Mendip sites, but I should think it at least as likely to be due to the site's position being nearer to the Mendip plateau where *Rangifer tarandus* might have been particularly abundant at certain times of the year, and thus worth exploiting. As at Kent's Cavern, an important aspect of the Soldier's Hole faunal evidence is the apparent absence of *Felis leo* and *Mammuthus* from any association with the Later Upper Palaeolithic. But the absence from the Later Upper Palaeolithic of some of the other large mammals associated with the Earlier Upper Palaeolithic is probably less significant, the site as a whole having yielded comparatively small artifact and faunal assemblages. The small mammals represented are amongst those one might normally expect in a mixed Arctic to Boreal steppe-tundra environment.

In summary, it would appear from the faunal assemblages excavated by the present author, as well as certain earlier workers, that the large mammals associated with the Earlier Upper Palaeolithic (at Kent's Cavern, Badger Hole, Soldier's Hole and Long Hole) included notably the following: *Crocuta crocuta, Felis leo, Canis lupus, Vulpes/Alopex, Ursus arctos, Mammuthus primigenius, Coelodonta antiquitatis, Equus germanicus?, E. przewalskii, Cervus elaphus, Megaloceros giganteus, Rangifer tarandus* and *Bison priscus*, with an apparent exploitational emphasis on *Equus* and *Rangifer*, and perhaps on *Coelodonta* as well. And it would appear that the large mammals associated with the Later Upper Palaeolithic (at Kent's Cavern, Soldier's Hole, Sun Hole, Cathole, Mother Grundy's Parlour and Robin Hood's Cave) included notably the following: *Crocuta crocuta, Canis lupus, Vulpes vulpes, Alopex lagopus, Ursus arctos, Coelodonta antiquitatis, Equus przewalskii, Cervus elaphus, Megaloceros giganteus, Rangifer tarandus, Capra ibex* and *Bison priscus*, and with an apparent exploitational emphasis on and possibly "special relationship" with *Equus* and *Rangifer*, as well as possibly with *Ursus* and *Coelodonta*.

Both large and small mammals and other fauna suggest ecological conditions mainly in the Arctic to Boreal systems during the Middle and Late Last Glacial. Such conclusions are in no way at variance with the broader picture presented in the sections on chronology, geomorphology and flora.

4. General Ecological Reconstruction

Figure 83 summarizes in a general way the basic climatic trends of the British Last Glacial, as well as the apparently world-wide changes in sea level during the Last Glacial. The names given for the various climatic phases immediately before, during and after the British Last Glacial are those which I find most practical to employ and which I strongly recommend in preference to "Ipswichian", "Weichselian"/"Devensian", "Flandrian" and similar names used by West (1968), Mitchell, Penny, Shotton and West (1973) and certain others, which I consider misleading and more value-laden. I certainly accept a pollen zonation of the Last

Interglacial, the Late Last Glacial and the Post Glacial, but for simplicity I have not included such zones on Figure 83. The temperature curve on Figure 83 is based on the estimated mean "thermal" requirements for beetles represented by remains from English Last Interglacial to Post Glacial deposits; although similar beetle remains are now being found in the other regions of the British Isles, they are generally not yet sufficient to construct temperature curves for those regions. In any case, England and Wales are the only two regions known definitely to have been exploited by man during the Upper Palaeolithic. As beetles are very sensitive climatic indicators, the July temperature curve presented for England may be taken as essentially correct, except of course where evidence is still sparse or lacking as indicated on Figure 83. The relevant July mean thermal estimates are drawn primarily from the work of G.R. Coope and his colleagues as given in the following references: Coope (1959, 1961, 1962, 1965a and b, 1967, 1968a and b 1969a, b and c, 1970, 1975 and personal communications); Coope, Morgan and Osborne (1971); Coope and Sands (1966); Coope, Shotton and Strachan (1961); Lambert, Pearson and Sparks (1962); Morgan (1969); Pearson (1962a and b and 1963); Penny, Coope and Catt (1969); and Shotton (1965). The inter-quartile ranges of the radiocarbon age estimates for these various beetle remains have already been presented together with other age estimates in Figures 58 and 59, and their actual readings may be found in Table 5. The striking thing about this temperature curve based on beetle requirements is that it agrees remarkably well with the palaeotemperature determinations by Dansgaard, Johnsen, Clausen and Langway (1971) from the Greenland Camp Century ice core and by Rona and Emiliani (1969) and Emiliani (1972) from deep-sea cores. These cores, both ice and deep-sea, also indicate that the coldest part of the Last Glacial occurred at about 17,000 years B.P. Segota (1967) in his reconstruction of palaeotemperatures for central Europe has suggested that the coldest part of the Last Glacial occurred at about 22,000 to 20,000 years B.P., whilst van der Hammen, Maarleveld, Vogel and Zagwijn (1967) have only been able to narrow this maximum cold down to about 26,000 to 13,000 years B.P. for the Netherlands. As regards the relationship of the British Upper Palaeolithic to the palaeotemperature curve in Figure 83, it has already been demonstrated on the basis of available radiocarbon age estimates (see Fig. 57) that the Earlier Upper Palaeolithic ranges from about 38,000 to 18,000 years B.P. at the latest, whilst the Later Upper Palaeolithic ranges from about 14,500 to 10,000 years B.P. or possibly somewhat later. Thus it would appear that Britain was not exploited by Upper Palaeolithic man during the High Arctic conditions of about 18,000 to 15,000 years B.P. (see Fig. 83), although, of course, he may have managed very sporadic summer visits which have yet to be detected. However, the apparent typological differences between the Earlier and Later Upper Palaeolithic are also in favour of such a gap in the occupation of Britain (see Chapters IV to VI), however brief, in terms of centuries or millennia, that gap may have been.

The estimates for changes in eustatic sea level given in Figure 83 are drawn primarily from Milliman and Emery (1968) and in part from the following references: Broecker, Thurber, Goddard, Ku, Mathews and Mesolella (1968); Donovan (1962); Jongsma (1970); Mercer (1970); Mörner (1969); Stearns and Thurber (1965) and Veenstra (1970). The available age estimates for these sea-level changes are based on both the radiocarbon and the thorium-230 methods of age determination. As may be seen on Figure 83, the time span relevant to the British Upper Palaeolithic had eustatic sea levels considerably lower than the present level, and the most obvious important effect would have been to expose large areas of the sea beds adjacent to southern Britain, allowing a dry-land connection with the European continent. There is also clearly close agreement between the maximum fall in land temperature and the maximum fall in sea level.

Maps 18A and B show the annual average of days with snow lying and the annual precipitation (measured as rainfall), respectively, in the modern British Isles. When considered together, these maps suggest a striking correlation between the modern maxima and the known ice fronts and assumed ice centres of the British Last Glacial (compare Maps 16 and 17 with 18A and B). In other words, the general pattern of precipitation and snow accumulation may have remained about the same over the past 30,000 years, even though the actual amount of snow and ice accumulation was considerably greater just before, during and after the Full Last Glacial. During the Middle and Late Last Glacial the southern British snow-line would certainly have been lower than at present, perhaps even as low as 300 metres O.D. (Seddon, 1957), but there may nonetheless have been comparatively little snow accumulation on the low lying lands of southern Britain and the adjacent, exposed sea beds. This would doubtless have affected the availability of flora and fauna for Upper Palaeolithic man even in winter in Middle and Late Last Glacial England and Wales.

Maps 19 to 23 are new attempts at the reconstruction of southern British Last Glacial environments, particularly the possible geographical distribution of vegetation zones. Only the Middle to Late Last Glacial phases are considered in this case, as being directly relevant to the British Upper Palaeolithic. It should be noted that Map 19 only presents the possible environments for the latter portion of the Middle Last Glacial, whilst Maps 20 and 21 present the earlier portions of the Full Last Glacial and Late Last Glacial Zone I, respectively. The presumed ice fronts shown on Map 19 are drawn on the premiss that ice was already advancing by the end of the Middle Last Glacial (see pp. 84–85) and that it may have reached as far south as the position of the known Scottish Readvance line of the end of the Full Last Glacial. The other glaciers (Maps 20, 21 and 23) are based on the known ice fronts as already presented in Maps 16 and 17, and discussed on pages 84–87. The presumed sea levels in Maps 19 to 23 are also based on data already presented on those pages and above in this present section, whilst the presumed lakes are partly my own suggestion and partly obtained from Stamp

(1946), West (1968), Sparks and West (1972) and their references. For the purposes of these maps, "highlands" are taken as those areas above 200 metres O.D., an altitude which may have seen snow for much of the year. The clock-face diagrams of relevant pollen spectra are based largely on my own pollen analyses and partly on selected published analyses. Precise percentage figures are not quoted: for sites analysed by myself, this information is supplied in the appropriate tables, while in most of the published accounts the information is not detailed enough for precise percentages to be calculated. The clock-face diagrams are centred as closely as possible on the positions of the sites on which they are based. All of these sites are listed with their references in Table 46 in my usual geographic order beginning with those in the south-west; the maps on which they are plotted are also indicated.

Map 19 is a proposed reconstruction of the ecological conditions for the Earlier Upper Palaeolithic occupation of southern Britain. The known distribution of Earlier Upper Palaeolithic sites does not extend further north than the suggested zone of lowland herb and moss tundra and only just borders on the barren highlands within that zone. The "land-bridge" to Europe would at no time in this phase have been wider than that shown, and at the start of the Earlier Upper Palaeolithic (c. 38,000 B.P.) it would probably have been considerably less (cf. sea-level curve on Fig. 83). There would probably have been no contemporary ice-free "land-bridge" to Ireland. During summers reindeer would have been particularly common in the highland and lowland herb-moss tundra, whilst horse and woolly rhinoceros would have been comparatively abundant in the lowland steppe-tundra of the south. During springs and autumns these large herbivores probably migrated across the "land-bridge" between Britain and the North-west European Plain, although some may have wintered in small, widely scattered groups on the wind-swept steppe-tundra of southern Britain where snow accumulation may have been kept to a minimum. Man presumably migrated out of Britain as well for the winters, unless a few small bands sheltered at such well situated and archaeologically prolific sites as Kent's Cavern (Devonshire) and Paviland Cave (Glamorganshire). During springs and autumns, open-air encampments may have been much more common in the obvious migration "funnel" of eastern England than has yet become apparent. However, with the continued advance of ice and fall of sea level, subsistence conditions became gradually poorer, so that by the time of the Full Last Glacial maximum ice advance (Map 20) man probably stopped visiting Britain altogether, except perhaps sporadically during summers when he may have found sufficient reindeer and migratory birds on which to subsist. Of course, some of the latest Earlier Upper Palaeolithic open-air encampments may have been under what is now the English Channel. At the start of the Full Last Glacial a few mammals probably migrated without man across the exposed bottom of St. George's Channel into southern Ireland, only to become generally extinct.

Maps 21 to 23 are proposed reconstructions of the ecological conditions for the Later Upper Palaeolithic occupation of southern Britain. Although the time span of the Later Upper Palaeolithic settlement is much briefer than that of the Earlier Upper Palaeolithic, the environmental changes of the Late Last Glacial occurred much more quickly, requiring faster responses and ecological adaptations by flora, fauna and man. Ice sheets were either at a halt stage or retreating into Scotland, and the sea was rapidly rising, so quickly in fact that it inundated parts of the present coasts of Scotland and northern Ireland during Zone II (see Map 22) before those regions had been able to recover sufficiently from their downwarping under the great weight of the Full Last Glacial ice caps. The known distribution of Later Upper Palaeolithic sites does not extend further north than the northernmost reaches of Lancashire and Yorkshire, so it may be that even in Zone II Scotland offered few inducements to settlement. At the very beginning of Zone I various mammals were probably able to migrate into Ireland (see Map 21) where *Megaloceros* and *Rangifer* became prolific during Zone II, but these do not seem to have been followed by man, who was perhaps too sparse on the ground even in England and Wales during the earlier portion of Zone I, and too late by the beginning of Zone II when the sea had reclaimed the Irish "land-bridge" (see Map 22). Throughout the Late Last Glacial there would have been a "land-bridge" between southern Britain and the European continent, but its size would have been progressively reduced. During the summers reindeer would have been abundant in the highland and lowland herb-moss tundra, whilst horse would have been even more abundant in the lowland steppe-tundra. These herbivores probably had seasonal migrations within Britain as well as perhaps to the European continent for winter. In fact, their wintering may well usually have taken place on the lowland steppe-tundra of southernmost Britain, open-air encampments such as Hengistbury Head (Hampshire) perhaps being employed by man to exploit them at that time of year, as well as when they were migrating in spring and autumn. Southern English and Welsh cave-sites may have also been more frequently used by Later Upper Palaeolithic man in the autumn-winter-spring portion of the year, whilst more northern English and Welsh cave and open-air sites, on the other hand, may have been more frequently used for spring-summer-autumn hunting and gathering. At times man may have migrated completely out of Britain for the winters, perhaps particularly in Zones I and III, but generally he may well have found it economically possible to stay for the whole year, although the total human population was probably never very great. But judging from the numbers of artifacts he produced, if that is anything to go by, the population was probably at least slightly larger than that of Earlier Upper Palaeolithic man in Britain, a difference which one might go so far as to attribute not only to natural environmental improvements, but to improved hunting techniques and social organization. This assumed increase of hunting expertise may even have encouraged a relationship between man and the horse and reindeer herds amounting almost to incipient domestication (cf. the Binfords, 1966a; Jarman, 1972); there is absolutely no conclusive evidence for such as yet, though I find suggestive the striking dependence on, or preference for, horse and reindeer as sources of food and certain raw materials. Other animals were certainly killed, but seldom in such numbers. During the amelioration of Zone II open forests or woodlands or birch and some pine, probably, managed to reach the easternmost portions of England, and certainly with them such Boreal herbivores as the true elk, *Alces alces*, e.g. at Neasham in Co. Durham (Blackburn, 1952), but strangely enough there is as yet scant evidence that Later Upper Palaeolithic man tried to avail himself of this potential source of much meat and bone, e.g. so far only the two barbed points associated with *A. alces* in Zone II at High Furlong, Lancashire (Hallam, Edwards, Barnes and Stuart, 1973).

Figures 84 and 85 are presented as possible models of British Upper Palaeolithic ecosytems. These models are unfortunately still largely intuitive, but future research ought to be able to analyse them and other possibilities much more thoroughly and hopefully sort out the probabilities from the mere possibilities.

Figure 84 shows the ecology of the Upper Palaeolithic in a "steady state", that is with a general "negative feedback" system which would result in stabilization of the ecological network between man and his environment at some "no-growth" level of so-called "dynamic equilibrium" (see pp. 15–16). This model might be suited in part to the circumstances of the long-lived but sparse Earlier Upper Palaeolithic tradition.

Figure 85 shows the British Upper Palaeolithic ecosystem with some degree of self-sustaining change, that is a "positive feedback" system resulting in more or less continual growth and/or decline, according to general environmental circumstances (see pp. 15–16). This model suggests that the Earlier Upper Palaeolithic people managed to re-adapt and eventually evolve into the more energetic Later Upper Palaeolithic, a view which I do not normally hold, but a real possibility nonetheless.

This chapter has set the complex stage on which Earlier and Later Upper Palaeolithic peoples acted out their lives in Britain. There now follows a discussion of the actual artifacts manufactured for that stage, beginning with the Earlier group.

CHAPTER IV

EARLIER UPPER PALAEOLITHIC CULTURAL MATERIAL: THE SO-CALLED "AURIGNACIAN" AND "PROTO-SOLUTREAN" INDUSTRIES OF BRITAIN

A. GENERAL CONSIDERATIONS

Chapters IV and V are meant to be read in conjunction with the Gazetteers of British sites and lists of artifacts. Gazetteer I lists both possible and definite Earlier Upper Palaeolithic cultural material, Gazetteer II lists both possible and definite Later Upper Palaeolithic material, and Gazetteer III lists some supposedly Upper Palaeolithic evidence. Each of these three Gazetteers is arranged in my normal clock-wise geographic order beginning with the south-west of England. Table 47 lists possible and definite British Upper Palaeolithic human skeletal remains according to the apparent number of individuals represented at each relevant site, and it is divided into two parts, Earlier and Later Upper Palaeolithic, respectively. Various supposedly Upper Palaeolithic human remains which have recently been shown to be of fully Post Glacial date are not considered in any way as they are dealt with in full in Oakley (1971).

Maps 24 and 33 show the distribution of all Earlier Upper Palaeolithic artifacts, both possible and definite, known to the author, Map 33 itself offering suggestions on possible patterns of land use. There is an obvious preponderance of finds from cave-sites, and this may well be due largely to a collecting bias in their favour. However, there is an apparent, dense clustering of finds around the Bristol Channel and in south-eastern Devonshire, and this may well be due to an actual preference for exploiting the presumed milder environments and in British terms "optimum" ecological conditions of the Middle to Full Last Glacial Bristol and English Channel plains (compare Maps 24 and 33 with Maps 19 and 20). Additionally, these prolific south-western sites are near the contact zone between lowlands and highlands and may therefore have been useful for exploiting the neighbouring higher grounds at certain times of the year. The lesser find-clusters in Flintshire and Derbyshire are presumably due to less frequent exploitation of those more northern areas, whilst the scatter of open-air finds across the south-east of England is perhaps due to the pursuit of migrating game such as wild horse and reindeer, as well as Earlier Upper Palaeolithic migrations to and from the continent.

In future, it seems most probable that more new open-sites than cave-sites will be found, but it also seems likely that the basic distribution pattern now available will not be drastically altered by the addition of such new open-sites, as they would probably only produce single or comparatively few finds, although more "base-camps" might be expected in certain areas (see Map 33).

The total number of extant artifacts from definite Earlier Upper Palaeolithic occurrences is 5,860, but it must be stressed that this total is heavily dominated by 4,464 stone waste products from Paviland Cave (E.U.P. site 15 in Gazetteer I). These may well contain an admixture of Later Upper Palaeolithic material, although there is good reason to suppose that most belong to the Earlier group. The total number of extant artifacts from "possible" Earlier sites is 47, whilst the total for known missing artifacts from definite sites is about 510. Thus the maximum figure for the grand total of all known British Earlier Upper Palaeolithic artifacts is less than 6,420.

Drawings of representative Earlier Upper Palaeolithic tool-forms from both definite and possible sites are presented in Figures 86 to 109. The principal results of metrical and statistical analyses of selected Earlier Upper Palaeolithic stone tool samples are given in section IV.J and in Figures 110 to 113. Sections IV.B to I, which now follow, are merely intended as a brief survey, region by region, of the cultural evidence presently available for Earlier Upper Palaeolithic settlement patterns in Britain. The reader is referred to Gazetteer I for more detailed information on the relevant sites, their bibliographic sources, their artifact typology and the raw materials employed, as well as to pages 20–23 and Table 1 for detailed descriptions of the artifact classification codes devised and used by the present writer.

B. DEVONSHIRE SITES

Map 25 shows the distribution of Earlier Upper Palaeolithic sites in Devonshire. Bench Fissure and Kent's Cavern are the only two definite sites, and of the two, the latter is by far the most important, not only because the material at Bench Fissure may have washed in from above (although it is not abraded), but because Kent's Cavern has provided one of the largest Earlier Upper Palaeolithic artifact assemblages in Britain. The other two Devonshire sites, Cow Cave and Tornewton Cave, are of uncertain affinity, although their stratigraphic evidence suggests that they might have some Earlier Upper Palaeolithic material.

A circle with a radius of 10 kilometres has been drawn round Kent's Cavern as that site was apparently an important "base-camp" for the Earlier Upper Palaeolithic exploitation of Devonshire, and the only one amongst the known Devonshire sites (see pp. 30–33 on so-called "site catchment analysis"). It is perhaps significant that the only other definite site, Bench Fissure, falls within that 10 kilometre radius. Kent's Cavern lies midway between the valleys of the Teign and the Dart and may have been particularly useful for exploiting both of these potential migration routes between the English Channel plain and the higher ground of Dartmoor to the west.

Representative tool-forms from the Earlier Upper Palaeolithic of Kent's Cavern are shown in Figures 86 and 87, and the single leaf-point from Bench Fissure is shown in Figure 97 (no.1). Of the 112 extant stone tools from Kent's Cavern (see E.U.P. site 3, Gazetteer I for detailed typology), 7% are burins, 41% are scrapers, 24% are saws/notches, 6% are of multiple classes (B/C, B/F, C/F and C/E), 10% are retouched flakes/blades and 10% are leaf-points. Scrapers are by far the most important class, and as they include such striking forms as nosed scrapers (CA5) and keeled scrapers (CA6), it is no wonder that they were attributed to the "Aurignacian" by Garrod (1926). Of the 11 leaf-points, 10 are unifacial (HE) and 1 is bifacial (HF1); these would have comprised the characteristic portion of Garrod's "Proto-Solutrean" group. As Pengelly's original field-notes (Pengelly, 1865–80, pp. 1–600) indicate a close, interlocking association between all of these various tool-forms, I think there is every reason to assume that they are all part of one group, combining the so-called "Aurignacian" and "Proto-Solutrean" traits in a manner which I prefer to term British Earlier Upper Palaeolithic on simply stratigraphic grounds (see Figs. 4 to 9, Kent's Cavern plans and sections). Of course, keeping in mind Pengelly's use of 1 foot (30 cm.) spits, his field-work, accurate though it was for its time, may well have obscured some original palimpsest (no more than 30–60 cm. thick) of closely related assemblages, resulting from more than one Earlier Upper Palaeolithic visit. An interesting find from Ogilvie's subsequent, less accurate excavations in the Vestibule of Kent's Cavern is a true "burin busqué" (BD4), a tool-form supposedly confined to France and there thought characteristic of "Aurignacian II" according to Sonneville-Bordes (1960) and her husband, Bordes (1968). This tool-form definitely exists in the British Earlier Upper Palaeolithic and is at its most frequent occurrence at Paviland Cave (see below and Gazetteer I, site 15). I have recently shown some of the actual "burins busqués" from Paviland Cave, as well as a cast of one from Ffynnon Beuno Cave, to Prof. Bordes, and although surprised, he has certainly agreed that they are true "burins busqués" (Bordes, personal communication). He also apparently had not seen the drawing in Garrod (1926, Fig. 24, no. 2) of the "burin busqué typique" from Ffynnon Beuno Cave. It would certainly seem that the presence of this tool-form alone gives a somewhat "Aurignacian" aspect to the Earlier Upper Palaeolithic of Britain. However, leaf-points do not occur in any quantity, if indeed at all, in the French Aurignacian, whilst they are a characteristic and diagnostic class in the British Earlier Upper Palaeolithic, clearly setting it apart from the typical Aurignacian of France. The classic Solutrean of France has various leaf-point forms (see Smith, 1966), but they are not precisely like the British leaf-points which, if anything, bear more resemblance to the few which have been found in Belgium in contexts which Eloy (1956) describes as "Proto-Solutrean elements in Aurignacian and Upper Perigordian of Font-Robert facies" and which Otte (1974b) separates into Belgian "Group B ?Aurignacian and Group A Upper Perigordian with Font-Robert points" (my translations), as well as perhaps to the somewhat earlier leaf-points of the Polish Jerzmanovician (see Chmielewski, 1961 and 1972) and the intervening German "Altmühlgruppe" (see Bohmers, 1951; Bosinksi, 1967; Toepfer, 1970).

C. MENDIP REGION SITES (SOMERSET)

Map 26 shows the distribution of Earlier Upper Palaeolithic sites in the Mendip region of Somerset. Badger Hole, the Hyena Den, Soldier's Hole and Uphill Cave are the only definite sites. Of the four, Badger Hole and Soldier's Hole are the most important, the former for its relatively large number of artifacts (54), and the latter for its fine bifacial leaf-points found stratigraphically below a characteristic Later Upper Palaeolithic backed tool assemblage. Representative tool-forms from these four definite sites are shown in Figures 88 to 93. A circle with a radius of 10 kilometres has been drawn from Badger Hole in Map 26 as that site might be interpreted as a "base-camp" for the Earlier Upper Palaeolithic exploitation of Somerset. It is ideally located for hunting on either the high ground of the Mendip Hills or the low, well-watered areas of the Somerset Levels. Also it may be significant that Soldier's Hole falls just within the 10 kilometre radius with only a few artifacts, whilst the Hyena Den is virtually at the circle's centre with nearly as many extant artifacts as Badger Hole. Of course, this is not to suggest that all of these sites may have been occupied at the same time, but rather that they may have been used in a more or less similar way from time to time so that we now see the greatest activity in Somerset as having taken place at or near Badger Hole.

The morphology of the Soldier's Hole leaf-points (Fig. 91, no. 1 and Fig. 92, no. 1) is remarkably similar to some of those from the Final German Mousterian "Altmühlgruppe" at Mauern II (see Bohmers, 1951; Bosinski, 1967), Figure 91 (no. 1) having a basal notch or "Kerbe" and both having generally denticulated and/or serrated edges and plano-convex sections (all the more the pity that I have thus far been unable to obtain any radiocarbon age estimates for this assemblage, as has already been stated). I nevertheless think Soldier's Hole is Earlier Upper Palaeolithic rather than Middle Palaeolithic, as a not dissimilar bifacial leaf-point (Fig. 86, no. 4) exists in the Earlier Upper Palaeolithic at Kent's Cavern and now has a radiocarbon age estimate of about 28,700 B.P. (see p. 41). In any case, various elements of "Middle Palaeolithic" aspect such as side scrapers (e.g. Fig. 89, no. 3–4) and what I term "heavy saws" (e.g. Fig. 88, no. 4 and Fig. 90, no. 3) are apparently present in the British Earlier Upper Palaeolithic and certainly present in the French Aurignacian; however, it should be noted that one of the British examples (Fig. 90, no. 3) is from the unstratified Hyena Den collections, a site which has also yielded Middle Palaeolithic handaxes (Tratman, Donovan and Campbell, 1971).

Uphill Cave may have originally yielded more artifacts than Badger Hole, but precisely how many have been lost through sale and re-sale or total destruction in the Second World War I have been unable to determine. The total number of extant artifacts is 30, and to this may be added 4 tools figured by Garrod (1926, Fig. 22) and reproduced here in Figure 93 (no. 2–5). Uphill Cave is ideally situated for exploiting the eastern reaches of what would have been the Bristol Channel plain (see Map 26), and it may well have served at various times as a "base-camp", perhaps with some subsidiary activity taking place near-by at what I have termed the Brean Down Shelter, as well as to the east at Picken's Hole.

D. SOUTH-WESTERN WELSH SITES (GLAMORGANSHIRE, CARMARTHENSHIRE AND PEMBROKESHIRE)

Map 27 shows the distribution of Earlier Upper Palaeolithic sites in south-western Wales. The only definite sites are Long Hole, Nottle Tor and Paviland Cave, the latter being by far the most important with the greatest number of artifacts produced by any Earlier Upper Palaeolithic site in Britain. The material from Long Hole is partly unstratified, whilst that from both Nottle Tor and Paviland Cave is completely unstratified, that is, except for the bone and ivory tools and ornaments from Paviland which are known to have been found in close association with the famous burial of the "Red Lady" in an ochreous deposit (see Buckland, 1823; also Sollas, 1913; Oakley, 1968). As many of the Earlier Upper Palaeolithic stone tools and waste products from Paviland have traces of red ochre adhering to them, whilst none of the tools of definite Later Upper Palaeolithic form do, there is good reason to assume some sort of relationship between the burial and the extensive Earlier Upper Palaeolithic stone artifact assemblage. Representative tool-forms from Paviland are shown in Figures 94 to 96, whilst those from Little Hoyle (or "Longbury Bank Cave", to use its other name), Long Hole and Nottle Tor are shown in Figure 97.

A 10 kilometre radius circle has been drawn around Paviland Cave as that site can surely be regarded as an Earlier Upper Palaeolithic "base-camp"; it was also the centre of the most intense Earlier Upper Palaeolithic activity known thus far in Wales, if not in the whole of Britain. It faces south and commands an excellent view all the way to Exmoor and Lundy over what would have been the Bristol Channel plain. It is also adjacent to a ravine which provides (or without the modern sea, would have provided) easy access to the plateau of the Gower Peninsula. With all of its useful topographic features, it might even have been occupied during the beginning of the Full Last Glacial according to the radiocarbon age estimate for the "Red Lady", c. 18,500 B.P., although I suspect this "date" is somewhat too young (see pp. 79–80). As has already been demonstrated, the maximum ice advance of the Full Last Glacial in southern Wales never actually covered the Gower Peninsula, having only reached the Swansea Bay to the east, as well as Cardiff.

It may be of some relevance that three ivory rings/bracelets nearly identical to that from the Paviland "Red Lady" burial but found at Magdalenahöhle in western Germany now have a radiocarbon age estimate of c. 25,540 B.P. (Weiss, 1974). However, the German bracelets are associated with an early backed blade rather than leaf-point industry.

Of the 554 Earlier Upper Palaeolithic stone tools from Paviland (see Gazetteer I, site 15), 24% are burins (of 131, 30 are "burins busqués"), 56% are scrapers (of 310, 94 are nosed scrapers), 4% are borers/awls, 7% are saws/notches, 3% are multiple classes (B/C, B/E, C/E and C/F), 4% are retouched flakes/blades and only 2% are leaf-points. Scrapers are clearly the most important tool class, followed by burins. Leaf-points are reduced to mere "fossiles directeurs" in terms of their percentage of the total stone tools, but their actual number is the same as at Kent's Cavern, 11 of which 1 is a broken bifacial leaf-point. For a comparison of the tool class frequencies, and in particular scraper type frequencies, from Kent's Cavern and Paviland Cave see Figure 110. There are interesting differences in the use of raw materials between these two sites; at Kent's Cavern 81% of the total 479 artifacts are in flint and 18% in Greensand chert, whilst at Paviland 61% of the total 5,040 artifacts are in flint and 9% in Greensand chert, 26% in Carboniferous chert and 3% in adinole (see pp. 27–29 and Map 3 on raw materials). Flint was probably obtained from quite a few secondary sources, but good quality flint may have been particularly readily available in the English Channel plain, to which Kent's Cavern was adjacent. Greensand chert may have been obtained from the Blackdown Hills of southern Somerset or neighbouring secondary sources, whilst Carboniferous chert and Cambrian adinole were doubtless obtained from various sources, perhaps both primary and secondary, in southern Wales. Ivory, teeth and bone only account for less than 1% of the total recognized raw materials in both assemblages, but more of these may have been used than I have allowed for and future studies might therefore include a detailed analysis of the available faunal remains from these sites. A higher angle of scraper retouch at Paviland than at Kent's Cavern (see Fig. 111) might be partly accounted for by the greater use of chert in the manufacture of scrapers at Paviland, although this would have to be tested by experimenting with knapping and using scrapers in both flint and chert.

The 5 undisturbed stone waste products excavated by the present author from his layers A3a and A3b at Long Hole are considered to be Earlier Upper Palaeolithic not only because they were found below what proved to be a Full Last Glacial layer A3c (see granulometric and pollen diagrams, Figs. 65 and 74), but because they include a snapped medial segment of a well-made prismatic blade, which suggests an industry well practised in blade manufacture, and snapped segments of 3 separate, thin flakes, 2 of which may have resulted from the manufacture of a leaf-point as they have maximum thickness/breadth ratios of 0.121

in the one case and 0.095 in the other, and are therefore similar to the French so-called "éclats solutréens" of Smith (1966). It is also worth recalling that a fragment of a leaf-point was found at this site by Wood in 1861, although its stratigraphic position was poorly recorded, as was that of his other artifacts including a keeled scraper.

E. SEVERN BASIN SITES (GLOUCESTERSHIRE AND HEREFORDSHIRE)

Map 28 shows the distribution of Earlier Upper Palaeolithic sites in the southern Severn basin. The only definite sites are the Forty Acres gravel pit and King Arthur's Cave, neither of which yielded very many artifacts. The finds from Forty Acres Pit were stratified in a closely compacted bed of sand (layer 3, c. 45–60 cm. thick) above earlier Palaeolithic handaxes and other artifacts in layer 1 (with an intervening sterile layer 2) and below Mesolithic and later artifacts in surface layer 4 (Clifford, Garrod and Gracie, 1954). Although the relevant Forty Acres assemblage does not include any typical leaf-points, it does possess a possible tanged point with some flat, invasive retouch near its distal end on both faces (Fig. 98, no. 1), in addition to a well-made side scraper and a flat, oval-sectioned ivory point (Fig. 98, no. 2–3). Sadly enough, the ivory point has lost its base, so one does not know whether it was originally of the split-based variety, or some other form. However, judging from the well-defined occurrence of these artifacts, it seems not unlikely that they present the remnants of a small open-air encampment just east of what is now Gloucester City.

Most of the Earlier Upper Palaeolithic finds from King Arthur's Cave were destroyed in the Second World War, but luckily a proximal segment of a leaf-point has survived (Fig. 97, no. 2). The break on this specimen is original and not of wartime origin, and the piece was originally found in close association with a hearth in layer A3c (see pp. 44–45 on King Arthur's Cave stratigraphy). The single blade from the Eastington gravel pit is of uncertain affinity, but is is noteworthy that it has been retouched at a low angle along both edges and around the proximal end of its dorsal face (see Burkitt, 1938, Fig. 3), in a manner approaching but not quite attaining the status of an HE2 unifacial leaf-point. I have therefore classified it as an HB1 "retouched blade". King Arthur's Cave might have served from time to time as a "base-camp", whilst Forty Acres may have been a "transit-camp" for the exploitation of the migrating large mammals which may have been funnelled through the Severn basin; however, the evidence available does not offer sufficient support for such suggestions.

F. NORTHERN WELSH SITES (FLINTSHIRE)

Map 29 shows the distribution of Earlier Upper Palaeolithic sites in north-eastern Wales. The only two known sites are Cae Gwyn Cave and Ffynnon Beuno Cave, both of which have stratified, definite material, but neither of which have produced very much of it. However, these two sites are adjacent to one another and together were very probably the focus of more activity than is at present apparent. The representative tool-forms from both sites are presented in Figure 99, and the reader is referred to pages 45–46 for their important stratigraphic positions and possible radiocarbon age; as was shown on those pages, these sites could not have been occupied during, or for that matter immediately after, the Full Last Glacial maximum ice advance (see also small insert of southern Britain on Map 29; Flintshire was well behind the maximum ice front as indicated by the dotted line). Before the Full Last Glacial maximum ice advance covered these sites, they would have been ideally situated for exploiting not only the Vale of Clwyd, but both the southern reaches of an Irish Sea plain and the adjacent highlands of north-eastern Wales. Figure 99 (no. 3) is a typical "burin busqué" in the manner of the classic French Aurignacian ones; it is even better made than the more numerous examples from Paviland Cave (compare Fig. 96, no. 1–3). Figure 99 (no. 2), however, is quite "unAurignacian" in the French sense as it is the sort of leaf-point which Prof. Bordes would term a "Jerzmanovice point" (Bordes, 1968 and personal communication). The apparent association of the two tool-forms at Ffynnon Beuno Cave reinforces my view of the "mixed" nature of the British Earlier Upper Palaeolithic, a view which discounts the *a priori* attempts of previous workers such as Garrod (1926) to divide the relevant material into separate "cultures" such as "Aurignacian" and "Proto-Solutrean" on purely typological grounds. Away from this confusion comes McBurney's concept of a "hybrid Altmühl-Aurignacian complex" with an overall distribution including both Central Europe and Wales (McBurney, 1965 and personal communications), but even this approach is still somewhat *a priori*, although it may well be partly supported in Britain by the stratified evidence already presented from Kent's Cavern.

G. CRESWELL AND PEAK DISTRICT SITES (DERBYSHIRE)

Map 30 shows the distribution of Earlier Upper Palaeolithic sites in Derbyshire. The only definite sites are Pin Hole and Robin Hood's Cave at Creswell Crags, some representative tool-forms from both of which are depicted in Figures 100 to 105. At Pin Hole, although the exact nature of the containing deposits is uncertain, the recorded spits from a horizontal datum are reasonably reliable, even though there are various discrepancies and apparent typological overlaps (see pp. 47–48 on Pin Hole stratigraphy). Figures 100 to 102 are arranged from greatest depth to least depth and from furthest inside Pin Hole to nearest its entrance; there thus appears to be a slope from the entrance downwards towards the "Inner Chamber" (see Kitching's plan, 1963, Fig. 1) where the single unifacial leaf-point (Fig. 100, no. 1) was found. The site has produced at least 6 apparently Earlier Upper Palaeolithic side scrapers, 4 of which are shown in Figures 101 (no. 1–3) and 102 (no. 1), the one in Figure 102 being a double side scraper. These were all found below definite Later Upper Palaeolithic backed tool-forms and above the Earlier Upper Palaeolithic backed tools shown in Figure 100 (no. 2–3) and the apparently Earlier Upper Palaeolithic decorated ivory and bone tools in Figure 102 (no. 3–5). The latter examples of British Upper Palaeolithic art work are somewhat enigmatic to say the least, as they are what many would expect from the Later Upper Palaeolithic rather than the Earlier by comparison with continental Upper Palaeolithic material. Bevel-based points occur in the French industries from Aurignacian to Magdalenian (see Sonneville-Bordes, 1960; Smith, 1966), but it is in the Magdalenian that the double-bevelled types occur most frequently and within that complex the best comparisons may certainly be found for my Figure 102 (no. 4), e.g. at Isturitz (Saint-Périer, 1936, Fig. 22, no. 13). However, examples from the French Solutrean are not totally dissimilar, e.g. at Isturitz again (Saint-Périer, 1952, Fig. 7, no. 4 and 11, Fig. 8, no. 8). The fish-like motif on the Pin Hole bevel-based point could occur at almost any level of the European Upper Palaeolithic, but in Britain the fact that this point has been made in ivory rather than bone or antler suggests that it may indeed be the handiwork of Earlier Upper Palaeolithic hunters, as the above outlined stratigraphic evidence similarly suggests; as has already been shown in the faunal section of this book, there is little, if any, evidence to support the view that *Mammuthus* may have survived into contemporaneity with the Later Upper Palaeolithic, so the use of ivory in the Later Upper Palaeolithic is quite unlikely.

Thus, although Pin Hole has yielded a much smaller Earlier Upper Palaeolithic assemblage (only 72 artifacts at most) than either Kent's Cavern or Paviland Cave, it nonetheless provides a highly varied array of tool-forms. The large tanged point (Fig. 100, no. 3) is the only complete specimen from a British Earlier Upper Palaeolithic cave-site so far, but the tangs of what were probably once tanged points are known from Kent's Cavern (Fig. 87, no. 1) and Paviland Cave (Fig. 95, no. 5). The double-side scraper (Fig. 102. no. 1) is almost identical to an isolated scraper of uncertain affinity from Ravenscliffe Cave, except that the latter lacks cortex on the dorsal face and has been thermally fractured on its ventral face (Bramwell, 1973). The possibly Earlier Upper Palaeolithic material from Ash Tree Cave is less convincing but includes a "chunky" short end scraper comparable to some from Kent's Cavern and Paviland Cave.

The Earlier Upper Palaeolithic material from Robin Hood's Cave now available is entirely from Dawkins and Mello's Breccia/"Cave Earth" Layers (layers B3/2; see pp. 64–69 on Robin Hood's Cave stratigraphy) where their records provide no evidence whatsoever of any internal sub-division of the site's Upper Palaeolithic series. As many of the artifacts of Earlier Upper Palaeolithic type have a buff-red to red silt adhering to them, whilst those of Later Upper Palaeolithic type have only buff silt, it seems reasonable to assume that the Earlier series came principally from the equivalent of my later B/A whilst the Later series came from higher up in the equivalent of my layers LSB to USB. The Earlier Upper Palaeolithic tool-forms shown in Figures 103 to 105 include several characteristic pieces, with the exception of yet another example of poorly stratified, enigmatic British Upper Palaeolithic art (Fig. 105, no. 6). This problem piece has what appears to be red silt adhering to its underside, although some of its redness might be due to an original application of red ochre which has since generally eroded away. The piece is a large herbivore's rib segment with a well-executed engraving of the forequarters and head of a wild horse with a short erect mane and profile not at all unlike the living *Equus przewalskii* (see pp. 113–14 on this species; as well as Mohr, 1971; Bökönyi, 1974). In my opinion, this decorated rib segment is a genuine piece of Palaeolithic art which may be either Earlier or Later Upper Palaeolithic; as it has a matrix similar to that surviving on the definite Earlier Upper Palaeolithic tool-forms, it is here tentatively included with their assemblage.

Both Pin Hole and Robin Hood's Cave have over 50 surviving artifacts and both may have served from time to time as "base-camps" for exploiting the surrounding well-watered plains (see Map 30). As their region was not covered by ice during the Full Last Glacial, they may even have been used for shelter during that severe climatic phase, although it seems more likely that they would have been used primarily during the Middle Last Glacial. Brief summer forays into the Peak District may even have taken place and may thus far be represented by the scanty evidence from Ravenscliffe Cave.

H. EAST ANGLIAN (NORFOLK, SUFFOLK AND NORTHERN ESSEX) AND HERTFORDSHIRE SITES

Map 31 shows the distribution of Earlier Upper Palaeolithic sites in the East Anglian region. The only definite sites in my opinion are at the gravel pits known as Bramford Road Pit and White Colne Pit I, and of these two the latter is a single find-spot, although better stratified than the former. Some representative tool-forms from Bramford Road are shown in Figures 106 and 107, and the bifacial leaf-point from White Colne is shown in Figure 108 (no. 1). Examples of less certain Earlier Upper Palaeolithic leaf-points from Rikof's Pit and Charsfield are given in Figures 98 (no. 4) and 108 (no. 2), respectively. The specimen from White Colne was found at a depth of 8 feet (c. 245 cm.) just above an occurrence of remains of *Mammuthus, Equus, Capra ibex* and a large bovid (Layard, 1927). The interest of the supposedly "Levallois" artifact from Rikof's Pit is that it was reputedly found about 5 feet (c. 150 cm) above the so-called "Lea Valley Arctic Bed" in the immediately overlying gravels (Warren, 1938); if this be the case then it might belong to the Earlier Upper Palaeolithic, as the "Arctic Bed" itself has recently been shown to be Middle Last Glacial and perhaps about 28,000 to 21,530 years B.P. in radiocarbon age (Godwin and Willis, 1960; West, 1968;

Shotton and Williams, 1971). Although I see nothing of a characteristic "Levallois style" in Figure 98 (no. 4), it is perhaps relevant that the "Levallois technique" was used from time to time in the continental Upper Palaeolithic, e.g. Proto-Solutrean and Solutrean in France (cf. Sonneville-Bordes, 1960; Smith, 1966).

Bramford Road is the only British open-air site to have yielded unifacial leaf-points morphologically identical to many of those known from the cave-sites (see Fig. 107, no. 1–2, and compare with those already reviewed from the cave-sites in Figs. 86, 88–90, 93, 95, 97, 99–100, 103–104). It has also yielded a fairly convincing, large tanged "point" (Fig. 107, no. 3), more or less comparable to that from Pin Hole (Fig. 100, no. 3), as well as a number of crude bifacial leaf-points, e.g. Figure 106. Although the total extant collection of pieces of Earlier Upper Palaeolithic aspect from Bramford Road amounts to only 21 artifacts, they were obtained by dredging from well below the water table in the gravel pit and may therefore be only a small part of what might originally have been a substantially larger assemblage, perhaps even one resulting from the débris of an important open-air "transit-site" or "base-camp". According to remains in the Ipswich Museum, the fauna from Bramford Road includes *Mammuthus, Coelodonta, Equus* and *Rangifer*, but sadly enough it cannot be reliably established whether these had really been associated with the artifacts in question. Finally, on the basis of the evidence outlined above, I think there can be but little question that the relevant tool-forms from Bramford Road and White Colne are definitely Earlier Upper Palaeolithic rather than the result of the manufacture of Neolithic, Beaker, etc. leaf-points.

I. SOUTHERN ENGLISH SITES (SURREY, WILTSHIRE AND HAMPSHIRE)

Map 32 shows the distribution of Earlier Upper Palaeolithic sites in southern England; although the Somerset cave-sites have already been reviewed, they are shown again on this map to complete the picture. No definite open-air Earlier Upper Palaeolithic finds have yet been made in southern England, but the three possible sites indicated on Map 32 might help to point the way to where better stratified material could eventually be found. The bifacial leaf-points from Fir Hill and Cameron Road are shown in Figure 109. I suspect that both of these may be either Later Middle Palaeolithic or Earlier Upper Palaeolithic, rather than anything of post-Palaeolithic age. Figure 109 (no. 2) almost verges on being a thin handaxe, but I think it is better described as a thick leaf-point or "Blattspitze", not very unlike some of those from the "Altmühlgruppe" at Mauern II (see Bohmers, 1951; Bosinski, 1967).

J. REPRESENTATIVE TOOL-FORMS AND METRICAL AND STATISTICAL ANALYSES OF SELECTED EARLIER UPPER PALAEOLITHIC MATERIAL

Figures 86 to 109 show examples of Earlier Upper Palaeolithic tool-forms, both definite and possible ones. Most have already been mentioned in the immediately preceding brief discussion by region and site, but there now follows a typological key to the specimens illustrated (see pp. 20–23 and Table 1 for detailed descriptions of the classes and types recognized by the author and Gazetteer I for present whereabouts):

Fig. 86 KENT'S CAVERN (Devonshire), 1865–68, layer B/A2 (c. A2). 1–4. Leaf-points HE1, HE3, HE1 and HF1. All in flint. Estimated radiocarbon age average c. 30,720 B.P.

Fig. 87 KENT'S CAVERN (Devonshire), 1865–68, layer B/A2 (c. A2). 1. Backed tool AF?; 2. burin BA4; 3–6. scrapers CA2, CA5, CF and CA6; 7. multiple class GC (CA2/FA). All in flint except 5 in Greensand chert. Estimated radiocarbon age average c. 30,720 B.P.

Fig. 88 BADGER HOLE (Somerset), 1938–53, layer A2?. 1–2. Leaf-points HE3 and HE1; 3. scraper CF; 4. saw FA. All in flint. Estimated radiocarbon age greater than 18,000 B.P.

Fig. 89 BADGER HOLE (Somerset), 1938–53, disturbed. 1–2. Leaf-points HE3 and HE1; 3–4. scrapers, both CA3; 5. retouched blade HB/HD; 6. saw FB. All in flint.

Fig. 90 HYENA DEN (Somerset), 1859–74, layer A2b?. 1. Leaf-point HE3; 2. scraper CA6; 3. saw FA; 4. leaf-point HF1? (or Middle Palaeolithic handaxe?). 1 and 4 in flint, 2–3 in Greensand chert.

Fig. 91 SOLDIER'S HOLE (Somerset), 1928–29, layer A2. 1. Leaf-point HF1; 2. awl IIA; 3. saw FB. 1 in flint, 2 in ivory, 3 in Greensand chert.

Fig. 92 SOLDIER'S HOLE (Somerset), 1928–29, layer A2. 1–2. Leaf-points HF1 and HF2. Both in flint.

Fig. 93 UPHILL CAVE (Somerset), 1826–1926, layer unknown. 1–3. Leaf-points HE3 and 2 HE1; 4. multiple class GF (BA/HE3); 5. leaf-point HE1; 6. burin BC2; 7. scraper CA5. All in flint. 2–5 redrawn after Garrod (1926, Fig. 22, no. 1 and 3–5).

Fig. 94 PAVILAND CAVE (Glamorganshire), 1823–1911, layer unknown but c. level of "Red Lady" burial. 1. Leaf-point HE2; 2–4. burins BA1, BA2 and BA4; 5–6. scrapers CA2 and CF; 7. multiple class GB (BD1/CA2); 8. retouched flake HC2 ("leaf-shaped"); 9. "spatula" IIG. 1–4 and 6–8 in flint, 5 in adinole, 9 in bone. Estimated radiocarbon age c. 18,460 B.P. (too young?).

Fig. 95 PAVILAND CAVE (Glamorganshire), 1912, layer unknown but c. level of "Red Lady" burial. 1–4. Leaf-points HE3, HE1, HE1 and HF2; 5. backed tool AF?. 1 in Carboniferous chert, 2 in adinole, 3–5 in flint.

Estimated radiocarbon age c. 18,460 B.P. (too young?).

Fig. 96 PAVILAND CAVE (Glamorganshire), 1912, layer unknown but c. level of "Red Lady" burial. 1–3. Burins, all BD4 (no. 3 a "busqué" on a "pebble tool"); 4–7. scrapers CA3, CA5, CA5 ("inverse") and CA6. 1–2 and 4 in flint, 3 and 5 in Carboniferous chert, 6–7 in adinole. Estimated radiocarbon age c. 18,460 B.P. (too young?).

Fig. 97 SOUTH-WESTERN BRITISH CAVES. 1–2. Leaf-points HE3 and HE1; 3. scraper CA3; 4. awl IIA (or stout IIB?); 5. multiple class GB (BA1/CB); 6. notched blade FD; 7. leaf-point HE3. 1–2 in flint, 3 in adinole, 4 in bone, 5–6 in Carboniferous chert, 7 in very cherty flint.

Fig. 98 CENTRAL ENGLISH OPEN-AIR FINDS. 1. Backed tool/leaf-point? AF/HE3?; 2. scraper CA3; 3. carved-base point? IIC?; 4. leaf-point HE3. 1–2 and 4 in flint, 3 in ivory. 4 redrawn after Warren (1938, Fig. 1, no. 3).

Fig. 99 CAE GWYN AND FFYNNON BEUNO CAVES (Flintshire), 1884–87, layer 2/1, below outwash and boulder clay. 1. Scraper CB; 2. leaf-point HE3; 3. burin BD4. All in flint. Estimated radiocarbon age c. 18,000 B.P. (too young?).

Fig. 100 PIN HOLE (Derbyshire), 1924–38, layer 2. 1. Leaf-point HE3; 2–3. backed tools AB1 and AF1. All in flint.

Fig. 101 PIN HOLE (Derbyshire), 1924–38, layer 2. 1–3. Scrapers, all CA3. All in flint.

Fig. 102 PIN HOLE (Derbyshire), 1924–38, layer 2. 1. Scraper CA3 (double side scraper); 2. awl EB2; 3. rod segment IIDA (with fish motif); 4. carved-base point IICB (with fish motif?); 5. decorated rib segment IIDB (with chevron or fish-net motif?). 1–2 in flint, 3–4 in ivory, 5 in bone.

Fig. 103 ROBIN HOOD'S CAVE (Derbyshire), 1874–76, layer B/A?. 1. Leaf-point HE3; 2. multiple class GF (BB4/HE1); 3–6. Leaf-points HE3, HE1, HF2 and HE2. All in flint. All redrawn after Garrod (1926, Fig. 30, no. 1–6).

Fig. 104 ROBIN HOOD'S CAVE (Derbyshire), 1874–76, layer B/A?. 1–4. Leaf-points HE1, HE3, HE3 and HF2. All in flint.

Fig. 105 ROBIN HOOD'S CAVE (Derbyshire), 1874–76, layer B/A?. 1. Multiple class GF (BA5/HE2); 2. burin BE5; 3. multiple class GC (CB/FB); 4. scraper CA2; 5. saw FA; 6. decorated rib segment IIDB (with horse head motif). 1–5 in flint, 6 in bone.

Fig. 106 BRAMFORD ROAD (Suffolk), c. 1936, layer unknown. Leaf-point HF1 in flint.

Fig. 107 BRAMFORD ROAD (Suffolk), c. 1936, layer unknown. 1–2. Leaf-points HE3 and HE1; 3. backed tool/leaf-point? AF/HE2?. All in flint.

Fig. 108 EAST ANGLIAN OPEN-AIR FINDS. 1–2. Leaf-points, both HF1 and both in flint.

Fig. 109 SOUTHERN ENGLISH OPEN-AIR FINDS. 1–2. Leaf-points, both HF1 and both in Portland chert.

Kent's Cavern and Paviland Cave are the only two Earlier Upper Palaeolithic sites to have yielded sufficiently large stone tool samples to permit any sort of simple typological analysis or comparative classification on the basis of percentages of the total stone tools in each of the two samples. The percentages for just the stone tool class headings are plotted for comparison in the upper half of Figure 110; the reader is referred to Gazetteer I, sites 3 and 15 for the actual counts and percentage calculations for both these class headings and individual tool-types. As may be seen in Figure 110 (upper half), there is a reasonable broad agreement between the two sites in that they both have a large proportion of their tools given over to the scraper category. However, there are also apparent differences, Paviland Cave having a higher frequency of burins, and Kent's Cavern having a higher frequency of saws/notches and a slightly greater proportion of retouched flakes/blades and leaf-points. The lower half of Figure 110 provides a comparison of percentages for scraper types as that tool class appears the most important; in this case the percentages are based on the total number of scrapers in each of the two samples. There are again apparent and interesting differences, Paviland Cave having higher frequencies of scrapers worked all around (CA1) and nosed scrapers (CA5), and Kent's Cavern having somewhat higher frequencies of side scrapers (CA3), blade end scrapers (CB) and broken end scrapers (CE), although as regards similarities, both sites appear to possess about the same frequency for short end scrapers (CA2). I think the similarities between the two assemblages and their characteristics, as a whole, are certainly sufficient to support the claim that both are attributable to the British Earlier Upper Palaeolithic at large, although it must be remembered that the material from Paviland Cave is virtually unstratified and may therefore be in part contaminated by other Palaeolithic, Mesolithic or even later artifacts, even though its nature seems predominantly Earlier Upper Palaeolithic.

Figure 111 presents the results of a metrical analysis of burin facet breadths at Paviland Cave and angles of scraper retouch at both Paviland Cave and Kent's Cavern (see pp. 23–24 and Fig. 3 for technique and location of these measurements). I think these particular features of burins and scrapers are important to measure: (a) because they are at the assumed business or working end of such implements; and (b) because I suspect there are profound differences in these attributes not only within the British Upper Palaeolithic but within the greater European Upper Palaeolithic at large; work on these and related attributes is currently underway on more reliable samples at the Abri Pataud in France, for example (Movius, David, Bricker and Clay, 1968; Movius and Brooks, 1971; Movius, 1975). As the total number of burins from the Kent's Cavern Earlier Upper Palaeolithic assemblage is less than 10, it was not considered worthwhile to plot their burin facet breadths on Figure 111. Table 48 gives the information on which Figure 111 is based, as well as the measurements for the Kent's Cavern burins. It is thought significant that the mean breadth of the Kent's Cavern burin facets is very close to that of the Paviland ones. Also both the mean (5.2 mm.) and the mode (4 mm.) at Paviland are greater than what one normally encounters in Later Upper Palaeolithic burins (e.g. Gough's Cave mean 2.2 mm. and mode 2 mm.; see Chapter V, section M, Fig. 170 and Table 50); of course, the secondary peak at 2 mm. at Paviland may therefore be due to the uncontrollable inclusion of a few possibly Later Upper Palaeolithic burins, but the effect of such a potential source of error would seem at least to be at a minimum. The more abundant scrapers from Paviland Cave and Kent's Cavern afford both striking differences in their modal patterns and a pronounced overlap in their total ranges (see Fig. 111). The distribution of angles of scraper retouch at Kent's Cavern is apparently bimodal, the peak at 65° presumably being caused by the stronger occurrence of low-angle side scrapers at that site comparable to those already described from the Pin Hole Earlier Upper Palaeolithic. The Kent's Cavern peak at 80°, together with the rise at Paviland at that point, is probably due, on the other hand, to the occurrence of fairly steep short end scrapers at both sites, whilst the great, unimodal Paviland peak at 90° is perhaps due not only to a frequent occurrence of very steep round scrapers (? equivalent to the "grattoir circulaire" of the French Aurignacian; see Sonneville-Bordes, 1960) and nosed scrapers, but to a more common use of chert in the manufacture of scrapers than at Kent's Cavern. It is noteworthy that there is a difference of about 11° between the means for the two samples, and that these means fall on either side of the mean (72.8°) and mode (75°) for the prolific Later Upper Palaeolithic scraper sample at Gough's Cave (see Chapter V, section M, Fig. 171 and Table 51).

Figure 112 shows an individual plotting based on two metrical analyses of nearly all available Earlier Upper Palaeolithic unifacial leaf-points, the exceptions being doubtful specimens and/or open-air single finds. The attributes considered in this graph are the ratio of maximum thickness over maximum breadth against the average angle of edge retouch on each specimen (see pp. 23–24 and Fig. 3 for technique and location of these measurements). The information on which the graph is based is also given in Table 49. A regression line has been calculated for all of these unifacial leaf-points as shown on Figure 112, and the mean and standard deviation for all of them has been calculated as well. The latter is indicated by the heavy cross near the centre of the graph. The scatter of individual specimens is obviously wide, but it may be significant that there are two apparent clusterings of these, one above and one below the mean for all of their T/B ratios. As an experiment, I have therefore divided the diagram in half and calculated the "secondary" means and standard deviations for the two clusters: these are indicated by the two light crosses on Figure 112. It may be significant that the mean of the lower cluster falls virtually on the calculated linear regression, perhaps suggesting it is a more reliable cluster than the upper one. However, it should be noted that the unifacial leaf-points from sites with more than one specimen occur in both "clusters", although as regards the two largest series, most of those from Paviland Cave are confined to the "upper cluster", whilst most of those from Kent's Cavern are confined to the "lower cluster". Thus it may be that one is here confronted with suggestive, but insufficient, evidence for at least two sub-groups within the Earlier Upper Palaeolithic unifacial leaf-point complex, and in the case of the more prolific sites, perhaps at least two occupational phases.

The possibility of two sub-groups within the Earlier Upper Palaeolithic is perhaps brought out more clearly in Figure 113. Here the means and standards deviations for the unifacial leaf-point samples from the five sites with more than one specimen are projected on to the same sort of graph as in Figure 112, and these are again based on the data in Table 49.

This time one sees what may be a northern/western sub-group including Robin Hood's Cave and Paviland Cave as against a southern/eastern sub-group including Badger Hole, Kent's Cavern and Bramford Road. But it must be stressed that the sample representing Robin Hood's Cave is incomplete, because the unifacial leaf-points from that site which are housed in the British Museum have been temporarily misplaced, and it is impossible to obtain even rough estimates of thickness and angle of retouch from the drawings in Garrod (1926). In any case, there appears to be a clear separation between the Kent's Cavern etc. sub-group and Paviland Cave, regardless of where a complete Robin Hood's Cave sample might fall. These two sub-groups might not only reflect regional differences, but differences in time. However, as has been stated many times the only truly reliable finite age estimates thus far are those for Kent's Cavern, the average of which is c. 30,720 B.P., although even these might reflect separate occupations (one at c. 38,270 B.P. and a later averaged at c. 28,200 B.P., see Table 4 and Fig. 57). If it could be properly established (instead of merely seeming) that most of the tool-forms from Paviland Cave were in fact more in the manner and frequency of those from the classic French Aurignacian than those from most other British Earlier Upper Palaeolithic sites, then the Paviland material might as a working hypothesis be supposed to "date" principally from the time range of Aurignacian 0 to II, c. 34,000 to 32,000 B.P. at Abri Pataud (Vogel and Waterbolk, 1967) (compare my tool-form drawings and Gazetteer I with the drawings and tables for French Aurignacian tool-forms in Sonneville-Bordes, 1960; as well as in Movius, 1975). However, remembering the unstratified nature of the Paviland collections, such suggestions rest on very loose footings.

These speculations should not be thought any more reasonable merely because their starting point is a small number of metrical analyses. My period of research has left me disappointed in the British Earlier Upper Palaeolithic material as a suitable subject for elaborate statistical treatment, which I had hoped to employ when I began. The time required for the other lines of research (field-work, and faunal, pollen and granulometric sample processing especially) might well, as things turned out, have left me unable to proceed far in that direction in any case, but I nevertheless record my opinion that the existing samples are almost all highly unsuitable for statistical treatment. The preceding pages are best considered as the setting-down of a few lines of thought which others may care to follow, and which I might myself take as a starting point for future work which would include the analysis and comparison of large valid samples taken from prolific sites in continental Europe (e.g. see Otte, 1974b on Belgian Earlier Upper Palaeolithic studies).

K. CONCLUSIONS

As noted above, I consider available British Earlier Upper Palaeolithic material unsuitable for elaborate statistical treatment. But as regards so-called "type-fossils", one can say that the Earlier material is characterized by generally robust burins including the "burin busqué" (BD4) form, various stout scrapers including the nosed (CA5) form, various unifacial leaf-points (HE) and rare bifacial leaf-points (HF). In the two largest assemblages, Kent's Cavern and Paviland Cave, scrapers are the predominant tool class. Although the burins and scrapers are generally "Aurignacian" (cf. Sonneville-Bordes, 1960) in character, the leaf-points range from being "Altmühlgruppe" (cf. Bohmers, 1951) and "Jerzmanovician" (cf. Chmielewski, 1961) in character to a few being actually closely comparable with the classic French Proto-Solutrean (cf. Smith, 1966) unifacial leaf-points which, unlike the British Earlier Upper Palaeolithic, are mostly of the HE2 form (e.g. at Paviland Cave, see Fig. 94, no. 1). In conclusion I still think it is best for the moment to refer to the relevant British material as Earlier Upper Palaeolithic rather than by some value-laden phrase such as "Aurignacian/Proto-Solutrean" or "Altmühl-Aurignacian". But even so, it may be significant that the average of the four most reliable radiocarbon age estimates thus far for the British Earlier Upper Palaeolithic, those which I obtained from Kent's Cavern, is c. 30,720 years B.P., and that that age is later than the beginning of the continental European Aurignacian sequence, and earlier than the beginning of the classic French Proto-Solutrean/Solutrean sequence (though it should be stressed again that the Kent's Cavern E.U.P. really has two "dating" clusters, one at c. 38,270 and a second at c. 28,200 B.P. comprising the other three estimates). It is now to be hoped that future field-work and laboratory analyses will yield sufficient new information both in Britain and in continental Europe to clarify accurately the relationship of the British Earlier Upper Palaeolithic to the continental sequence. Happily, work in the critical region of Belgium, western Germany and northern France is already underway (M. Otte, J. Hahn and F. Bordes, personal communications; also see Appendix 3 by Monsieur Otte in this present book, as well as the following references: Dauchot-Dehon and Heylen, 1975; Otte, 1969–70, 1974a and b; Heinzelin, 1973; all on Belgium; Hahn, 1973 and 1974; Hahn, Müller-Beck and

Taute, 1973; Bosinski, 1972; all on western Germany; Agache, 1971; Schmider, 1971; both on northern France), but many further efforts are needed to clear what is largely still a foggy, tradition-bound archaeological picture (e.g. see Binford, 1962, 1964, 1965, 1968 and 1972 on alternate anthropological/archaeological approaches).

CHAPTER V

LATER UPPER PALAEOLITHIC CULTURAL MATERIAL: THE SO-CALLED "CRESWELLIAN" AND "CHEDDARIAN" INDUSTRIES OF BRITAIN

A. GENERAL CONSIDERATIONS

Chapters IV and V, as was stated at the beginning of Chapter IV, are meant to be read in conjunction with the Gazetteers of British sites and lists of artifacts (see also p. 20 above for further details). Gazetteer I lists both possible and definite Earlier Upper Palaeolithic cultural material, Gazetteer II lists both possible and definite Later Upper Palaeolithic material, and Gazetteer III lists some supposedly Upper Palaeolithic evidence. Each of these three Gazetteers is arranged in my normal clock-wise geographic order beginning with the south-west of England. Table 47 lists possible and definite British Upper Palaeolithic human skeletal remains according to the apparent number of individuals represented at each relevant site, and it is divided into two parts, Earlier and Later Upper Palaeolithic, respectively. Various supposedly Upper Palaeolithic human remains which have recently been shown to be of fully Post Glacial date are not considered in any way as they are dealt with in full in Oakley (1971). But a new addition to the definite Later Upper Palaeolithic human remains is an ulna of a young adult at Ossum's Cave (Staffordshire); this was originally misidentified as reindeer (D. Bramwell, personal communication). It is worth noting here, however, that selected British Later Upper Palaeolithic human crania from Flint Jack's Cave, Gough's Cave and Langwith Cave have recently been subjected to rigorous multivariate analyses as part of a wider study of European hominids (see Stringer, 1974).

Maps 34 and 46 show the distribution of all Later Upper Palaeolithic artifacts, both possible and definite, known to the author, Map 46 itself offering a number of suggestions on possible patterns of land use. As with the Earlier Upper Palaeolithic (see Maps 24 and 33), there is again a preponderance of finds from cave-sites, and this may equally be due to a collecting bias in their favour. However, with the Later Upper Palaeolithic, sites extend further to the north, presumably under the influence of generally milder Late Last Glacial conditions. Also there is an actual increase in both number of sites and number of finds, and the open-air sites even include a few prolific occurrences. The distribution of the sites in general appears to be split into two main groups, a northern one with outliers and a southern, or south-western, one with outliers. The apparent separation of these two groups might be accounted for in part by a lack of suitable intervening cave-sites, but the lack of even possibly Later Upper Palaeolithic open-sites in the intervening region is in my opinion harder to dismiss so easily.

It seems to me that these two distribution groupings might well be at least in part due to different seasonal activities during the Later Upper Palaeolithic and/or to an intervening socio-ecological "buffer zone" (see Map 46; cf. Hickerson, 1965 and 1970). If such an interpretation(s) be correct, then the south-western clusters might represent a region where most of the sites are winter encampments, whilst the northern occurrences may mostly be summer hunting stations. Alternately, with an intermediate "buffer zone" acting as a food reservoir, seasonal scheduling of hunting (as well as fishing and gathering?) may be more on an east to west basis in the north and south. A good continental parallel would perhaps be the Hamburgian and Ahrensburgian summer reindeer hunting camps of north-western Germany (see Rust, 1937, 1943 and 1958), whose inhabitants presumably made seasonal movements to either near-by Dutch and Belgian sites or to southern regions to winter perhaps at such "peripheral Magdalenian" cave-sites as Petersfels, which was certainly not occupied later than May (see Peters, 1930; Peters and Toepfer, 1932). The scatters of sites around the clusters in the two British groups, on the other hand, might be due to spring and autumn hunting of the main migratory food sources, wild horse and reindeer, as indeed might the smaller, isolated clusters of open-air sites in East Anglia. The latter, i.e. East Anglia, may also represent connections with the European continent over a southern North Sea plain. Of course, during the more severe phases of the Late Last Glacial (i.e. Zone Ia and Zone III), the south-western region may itself have been the usual northern limit even of summer hunting activity, and Britain may have been completely abandoned during winters (compare Maps 34 and 46 with Maps 21 to 23), although I think such long annual migrations out of Britain quite unlikely on purely ethnographic grounds (i.e. distances of 100 to 200 km. are seldom exceeded by any hunter/gatherer, or for that matter herder, group; e.g. see Leeds, 1965; Lee and DeVore, 1968; Lee, 1969; Paine, 1972; Spooner, 1973; Peterson, 1975).

Other explanations could be offered to account for this discontinuous distribution of British Later Upper Palaeolithic finds with the clear "south-western" and "northern" concentrations. For example, it might be a case of different access routes bringing different hunting groups to separate areas of Britain from separate areas of origin on the European continent. However, if the reader will patiently study the detailed evidence set out in this chapter, in Chapter III, in Gazetteer II and in the relevant maps, figures and tables, having regard to the differences in the character, location and size of the settlements, and the variations in the tool-forms (or, rather, to the less perishable, tenuous remains of such information that survive today in the British archaeological record and are all we have to guide such judgements), then I think he will probably come as I have to the conclusion that the "local seasonal occupation" and/or the "buffer zone" interpretation is the best one. Correlations with known continental Late Last Glacial Palaeolithic centres are all very well, if one has the detailed evidence for them, but in the British Late Last Glacial situation we may well care first to think of a small population mobile and intelligent enough to go, like such ethnographically observed hunter/gatherers as the Nunamiut Eskimos of the Brooks Range (Gubser, 1965; Burch, 1972) or the !Kung Bushman of the Kalahari (Lee, 1969 and 1972), where the food was most plentiful at different times of the year; again provided the distances were not too great and/or other territories were not transgressed. This is after all why the study of the contemporary Quaternary animals and plants far beyond mere identification or gross climatic "value" has become so important to the archaeology (or rather anthropological ecology) of the Palaeolithic and Mesolithic. Hence, indeed, the title and whole approach of this book.

In addition to the possibilities that different regions were occupied at different seasons, that these regions were grouped and the groups separated by an intentional "buffer zone" or food reservoir during the British Later Upper Palaeolithic, there seems clear evidence that all three available topographic ecological zones were exploited: the Lowland Zone, the Highland Zone, and the Highland/Lowland Contact Zone, with an emphasis on the last named, according to present site distribution (again see Maps 34 and 46). It seems unlikely that this distribution pattern will be greatly changed in future, although the discovery of more new open-sites than cave-sites is now to be expected; but these will probably generally yield only a few finds, although some further "base-camps" might be expected (see Map 46), perhaps particularly outside the assumed "buffer zone".

The total number of extant artifacts from definite Later Upper Palaeolithic occurrences is 11,962, whilst the total number from "possible" occurrences is 4,260. The total for artifacts known to be missing from definite occurrences is about 4,470. Thus the maximum figure for the grand total of all known British Later Upper Palaeolithic artifacts is 20,692, or over three times the grand total for the Earlier Upper Palaeolithic (6,420). The ratio between Later and Earlier Upper Palaeolithic human skeletal remains is even more striking; the total number of individuals represented by the Later Upper Palaeolithic remains is about 105, whilst that of the Earlier Upper Palaeolithic is only about 7 (see Table 47). If the remains tentatively ascribed to the Earlier Upper Palaeolithic are accepted as belonging to that phase, there seems no reason to suppose any different treatment of the dead in the two Upper Palaeolithic stages. But there would certainly seem to be substantial evidence for a marked population increase during the Later Upper Palaeolithic in Britain, although the total population attained (perhaps 500 to 5000 people) was certainly never as great as that of Post Glacial Prehistoric

Britain (cf. Brothwell, 1972).

Drawings of representative Later Upper Palaeolithic tool-forms from both definite and possible sites are presented in Figures 114 to 166 and described in section V.M. at the end of this chapter. The principal results of metrical and statistical analyses of selected Later Upper Palaeolithic stone tool samples are also given in section V.M. and in Figures 167 to 174. Sections V.B. to L, which now follow, are merely intended as a brief survey, region by region, of the cultural evidence presently available for Later Upper Palaeolithic settlement patterns in Britain. The reader is referred to Gazetteer II for more detailed information on the relevant sites, their bibliographic sources, their artifact typology and the raw materials employed, as well as to pages 20–23 and Table 1 for detailed descriptions of the artifact classification codes devised and used by the present writer. It should be noted that the artifacts found by the present author during his excavations at the Later Upper Palaeolithic sites of Cathole (Glamorganshire), Mother Grundy's Parlour (Derbyshire), Robin Hood's Cave (Derbyshire) and Hengistbury Head (Hampshire) are included in Gazetteer II.

B. DEVONSHIRE SITES

Map 35 shows the distribution of Later Upper Palaeolithic sites in Devonshire. The only definite sites are Kent's Cavern, Three Holes Cave and Tornewton Cave, the latter two being two of the three "Torbryan Caves" shown on the map. Of these sites, Kent's Cavern is certainly the most important as it originally yielded over 500 Later Upper Palaeolithic artifacts, although more than half of these artifacts are now missing. A circle with a radius of 10 kilometres is shown around Kent's Cavern in order to indicate its idealized 2 hour economic territorial limit as an assumed "base-camp" (see pp. 30–33 on so-called "site catchment analysis"). It is perhaps significant that the only other definite sites, Three Holes Cave and Tornewton Cave, are just beyond the edge of that limit. Kent's Cavern would have been useful as a base for exploiting not only the exposed English Channel plain to the east, but the potential mammal migration routes along the valleys of the Teign and the Dart to the west. The Torbryan Caves may then have served as subsidiary sites for exploiting approaches to the higher grounds of the Dartmoor still further to the west. The great palimpsest of hearths known as the "Black Band" in the Vestibule of Kent's Cavern certainly suggests a fairly long series of Later Upper Palaeolithic visits to Devonshire (see Figs. 4, 6 and 9 for relevant Kent's Cavern plans and section).

Representative tool-forms from the Later Upper Palaeolithic of Kent's Cavern are shown in Figures 114 and 115, and the single backed tool from Tornewton Cave is shown in Figure 129 (no. 8). Of the extant stone tools from Kent's Cavern (see L.U.P. site 1, Gazetteer II), backed tools comprise 30% of them and are the most important class, including some of the forms which typify the Later Upper

Palaeolithic, e.g. the so-called "Creswell points" (Fig. 114, no. 1, 4–6) and to a lesser extent, the "Cheddar points" (Fig. 114, no. 7). (See pp. 18–21 and Table 1 on these terms; "Creswell point" and "Cheddar point" are names coined by Bohmers, 1956, for what would be my forms AC1–5 and AC6–10, respectively; whether they were "points" is a subject for debate in my opinion, and I think what future research should include is a "micro-wear analysis" of selected samples; meanwhile, I employ these terms simply for convenience, although as is shown in section V.M. and Fig. 169, the evidence available does not support Bohmer's claim that "Cheddar points" are more common in Somerset and "Creswell points" more common in Derbyshire.) But the most striking feature of the industry at Kent's Cavern is the occurrence of three antler "harpoons", one biserially barbed and exceptionally well-made (Fig. 114, no. 2) and two uniserially barbed (Fig. 115, no. 4 and 5). There is also a finely worked bone eyed-needle (Fig. 115, no. 3), which incidentally suggests a more sophisticated technique of clothing manufacture than that which was presumably carried out with heavy bone awls during the Earlier Upper Palaeolithic. Both the barbed points and the needle are quite comparable with similar finds from many continental Magdalenian sites (see especially Sonneville-Bordes, 1960; Müller-Karpe, 1966), but this is not to say that the Kent's Cavern specimens are in fact formally "Magdalenian", although Garrod (1926) considered them so. I think rather that they are just as likely an indigenous, but albeit rare, element of the British Later Upper Palaeolithic, regardless of whatever "ethnic" relationship that complex may have originally had with the various Late Last Glacial "cultures" of north-western continental Europe. However, the Hamburgian and Ahrensburgian harpoons are in fact generally less similar to the British ones (see Rust, 1937, 1943, 1958 and 1962), although one does not yet know what the Tjongerian bone and antler work may have been like as none has yet survived with its assemblages. But it is with the Tjongerian and related "Federmesser-Gruppen" that one finds what appear to be some of the best parallels for the British Later Upper Palaeolithic stone tools, including a frequent occurrence of "Creswell points" (see Schwabedissen, 1954; Bohmers, 1956 and 1963; Paddayya, 1973; Noten, 1975), although that and many of the other tool-forms also occur in the Magdalenian and Hamburgian to one lesser extent or another.

C. MENDIP REGION SITES (SOMERSET)

Map 36 shows the distribution of Later Upper Palaeolithic sites in the Mendip region of Somerset. The definite sites include Aveline's Hole, Banwell and Hutton Caves, Callow Hill, Flint Jack's Cave, Gough's Cave, Soldier's Hole and Sun Hole. Of these, Aveline's Hole and Gough's Cave are the two most important sites, the former having originally yielded over 300 artifacts and the latter over 7,000, making Gough's Cave the most productive Later Upper Palaeolithic site thus far in all of Britain. Both of these sites probably served as "base-camps" for the exploitation of the Mendips (see Map 36), and it is indeed remarkable how well their 10 kilometre radii encompass every one of the numerous other sites in Somerset, with the exception of one which lies just beyond that distance and may have been on the trail to southern Wales or perhaps a look-out post for checking reindeer and horse migrations on the Bristol Channel plain. Of the two supposed "base-camps", Gough's Cave was employed the longest, judging from its number of artifacts and the thickness of their occurrence (c. 3 m. as against c. 1 m. at Aveline's Hole); however, it is not certain whether they were ever used at the same time, so that one may really be viewing a palimpsest of exploitational patterns.

Representative tool-forms from the Mendip sites are shown in Figures 116 to 128, 129 (no. 1–7) and 130 (no. 3–6). Although most of the material from Aveline's Hole was destroyed by bombing during the Second World War, a fairly good series of casts of the biserial "harpoon" have survived, and it is from the one now housed at the University of Bristol Spelaeological Society Museum that I have made by drawing (Fig. 117, no. 15). I have added a dashed outline at the base of the harpoon to show what the shape of the base of the original specimen was like; this is taken from Davies (1921, Fig. 10, no. 1). The Aveline's Hole harpoon is certainly well-made, but even so it is not as neatly and precisely designed and cut as the Kent's Cavern biserially barbed point (Fig. 114, no. 2), which although smaller has its barbs set well back from the tip or distal end, presumably to allow for a greater penetration in horse, reindeer, or whatever game it was intended for use against. However, the Aveline's Hole "harpoon" has the addition of decoration in the form of incised grooves on its barbs and along its back; these may be merely conventional marks, or on the other hand they might be "blood-letting" grooves. Whilst the Kent's Cavern biserial "harpoon" appears most comparable with those of the classic Magdalenian VI in France (cf. Sonneville-Bordes, 1960), the Aveline's Hole example appears more closely similar to a perhaps slightly later "harpoon" found at Goyet Cave in Belgium (see Dupont, 1873, Fig. 16). Indeed, the Kent's Cavern specimen was closely associated with a bone sample from which I have since obtained a radiocarbon age estimate of about 12,180 years B.P. (see p. 42), whereas I would myself expect the Aveline's Hole one to be associated with a younger age estimate (a "dating" of c. 9,114 years B.P. is available from Aveline's Hole, but the association is uncertain; see Table 4, "?Meso./L.U.P." section). In fact, the stone tools from Aveline's Hole also suggest this as they have a more Mesolithic aspect than those at many other Later Upper Palaeolithic sites, so that one may not be incorrect in interpreting the Aveline's Hole assemblage as perhaps transitional from Later Upper Palaeolithic to Mesolithic (see Figs. 116 and 117, and compare with Gough's Cave Figs. 121 to 126, as well as with Kent's Cavern Figs. 114 and 115).

It is Gough's Cave which provides us with our most comprehensive series of Later Upper Palaeolithic stone tools, 799 extant ones to be exact (see L.U.P.

site 15, Gazetteer II). Of these, 34% are backed tools, 14% are burins, 10% are scrapers, 2.5% are "Zinken", 5% are borers/awls, 3.5% are saws/notches, 7% are of multiple classes (GA, GB, GC and GE) and 24% are retouched flakes/blades. The most important tool classes are backed tools and retouched flakes/blades, with the backed tools slightly stronger numerically. As this is the largest Later Upper Palaeolithic tool sample in Britain, it is doubtless very significant that still no leaf-points of any form whatsoever are present in it. Of 4,525 extant artifacts from Gough's Cave, 99.2% are in flint and the next highest percentage is for artifacts in bone or antler, 0.4%. The lesser, but nevertheless interesting, raw materials include 0.2% Greensand chert, 0.1% Portland chert, 0.04% Carboniferous chert, 0.04% Carboniferous limestone and 0.02% Baltic-type amber. The cherts suggest collecting activities over a fairly wide range in the south and south-west of England, perhaps from both primary and secondary sources (see section on raw materials, pp. 27–29 and Map 3). Although the single, large chunk of amber is of ultimately Baltic origin, the Gough's Cave inhabitants would probably have obtained it from some secondary source such as boulder clay (see Beck, 1965 on the analysis of this amber).

Representative tool-forms from Gough's Cave are shown in Figures 119 to 126. Two noteworthy bone and antler tools are two classic "bâtons-de-commandement", the antler one being reproduced in Figure 120. These are the only two "bâtons" known thus far from the British Upper Palaeolithic, but they are not an uncommon ornament/tool in many continental European Upper Palaeolithic industries (see Müller-Karpe, 1966) and are thought a typical feature of the classic Magdalenian (see Sonneville-Bordes, 1960). As the oblique perforation in the Gough's Cave "bâtons" is quite worn in both of them, particularly near the edges of the bevelled part, it seems to me not unlikely that these "ornaments" might have actually served as straighteners for the wood or bone shafts of projectiles, more or less in the manner of the morphologically comparable Eskimo shaft-straighteners (e.g. Boas, 1901, Fig. 117), an explanation which has indeed often been suggested for this artifact type as a general interpretation, although it may also have served as a holder or link for a lasso for reindeer, more or less in the manner employed by the Lapps (M. Dewez and D.A. Sturdy, personal communications; also see Manker, 1954, Fig. 14).

Another interesting example of Later Upper Palaeolithic bone-work from Gough's Cave is a large decorated rib segment, which I have reproduced in my Figure 119 (no. 9). In their report on this specimen, Hawkes, Tratman and Powers (1970) suggest that it may have originally been intended as more than a mere "tally", perhaps serving as a ruler for spacing barbs or bindings on "harpoons", or as a sort of frame for spacing in netting or weaving, or even as a kind of calculator, the groups of lines on the edges each totalling more than five. As regards their suggestion of its being a ruler, I have noticed that the groups of lines on the edges are in fact of about the same length as the barbs on the Kent's Cavern biserial "harpoon" (Fig. 114, no. 2), but of course this may simply be coincidence. The awl made from a tibia of *Lepus timidus* and depicted in Figure 119 (no. 8) also has "tally-marks" incised on its three edges. On the other hand, whether these sort of incised lines represent "lunar calendars" remains an interesting but uncertain alternate hypothesis (cf. Marshack, 1971, 1972a, 1972b, 1972c and 1975). Further, the possibility that they were also "gaming" pieces cannot be excluded (cf. Dewez, 1974).

With no less than 274 extant backed tools, Gough's Cave offers one of the most comprehensive British Later Upper Palaeolithic series of backed forms. Of this total number, 33.6% or 92 are in forms AC1–5, the so-called "Creswell points" (Fig. 119, no. 2–3; Fig. 121, no. 8–10; Fig. 125, no. 10–11; Fig. 126 no. 10–11), whereas only 10 or 3.7% are in forms AC6–10, trapeziform backed blades, the equivalent of Bohmers' (1956) "Cheddar points" (Fig. 121, no. 11; Fig. 125, no. 12–13). These and the other backed tool frequencies from Gough's Cave are shown in Figure 169. Bohmers' published sample of backed tools from the site totals only 88 (Bohmers, 1956, plate 1), so it is no wonder that he obtained considerably different frequencies, which he thought, together with certain other characteristics of the portion of the assemblage which he studied, were sufficient to warrant introducing a completely new "cultural" name, his "Cheddarian" of Somerset, which he regarded as a distinct entity on the basis of comparison with a similarly incomplete sample which he selected from Mother Grundy's Parlour and considered "Creswellian" (borrowing Garrod's term, 1926). I myself find no evidence in my wider survey of the British Later Upper Palaeolithic assemblages to support his concept of a "Cheddarian" (see section V.M. at the end of this chapter for further discussion), but sadly enough the term came quickly into use in Somerset, particularly in the *Proceedings of the University of Bristol Spelaeological Society* after 1956.

Most of the AC1–5 backed tools at Gough's Cave are of the AC2 form, i.e. 63 out of 92; this is the most common type or sub-type of what have variously been called typical sub-triangular backed blades or "Creswell points" (see Fig. 119, no. 2–3 and Fig. 121, no. 8 and 10). In addition to convex backed blades (AB1) and straight backed blades (AB2), other interesting backed forms include a series of shouldered points all in the AD class; these comprise 6.9% of the total 274 backed tools. Examples of these shouldered points may be seen in Figures 119 (no. 4), 122 (no. 1–7) and 126 (no. 1); the last-cited example is a rather good one of a "trapeziform" shouldered point. These various shouldered point forms would be quite at home in the Hamburgian (see Rust, 1962; Müller-Karpe, 1966) and some could even occur in the Magdalenian (see Sonneville-Bordes, 1960), but for the whole of north-western Europe, their highest frequencies would certainly seem to be in the Hamburgian (see Bohmers, 1956, 1960 and 1963). In small frequencies they appear to be an integral part of the British Later Upper Palaeolithic. "Microlithic" forms also occur in the Later Upper Palaeolithic at Gough's

Cave, but very infrequently, e.g. the rhomboid (AC22c) in Figure 121 (no. 12), and in the Somerset Later Upper Palaeolithic as a whole, I think "microliths" are only really characteristic of the Aveline's Hole assemblage (see Figs. 116–117), which I suspect is generally younger than many of the other assemblages. In the upper spits of the Gough's Cave sequence, however, "microlithic" obliquely blunted points (AB6/12) were originally found in greater numbers than now survive (see Davies in Parry, 1929, pp. 106–11); of the two extant ones I include an example in Figure 126 (no. 9). This may represent a mixture with Mesolithic material, but I am inclined to think it indicative of an actual transition stage between what we may term the "classic" tool-kits of the British Later Upper Palaeolithic and Earlier Mesolithic, scanty though the evidence may yet be. The radiocarbon age of "Cheddar Man", whose precise stratigraphic relationship to the artifacts we do not really know, is c. 9,080 B.P. (see Table 4) which may be suggestive of a "Mesolithic date" for at least part of the Gough's Cave series. But I would have thought that most of the Gough's Cave artifacts, particularly those from below about spit 12, were at least a millennium, and perhaps twice or four times as much, older, by comparison with the radiocarbon dated artifacts from Kent's Cavern, c. 14,275 and c. 12,180 (see Table 4); as well as the age estimate for a site directly opposite to Gough's Cave in Cheddar Gorge, Sun Hole c. 12,380 B.P. (see Table 4). In this light it is of some relevance that Sun Hole (see pp. 51–55 on stratigraphy) has Later Upper Palaeolithic stone tool-forms (Fig. 128) that are quite similar to those at Gough's Cave, though much smaller in actual number and often in size as well. "Microliths" certainly occur in the French Upper Palaeolithic and particularly in the Magdalenian (see Sonneville-Bordes, 1960) and they also occur in the German Ahrensburgian (see Rust, 1962; Müller-Karpe, 1966); it is reasonable enough to expect them in small numbers in the British Later Upper Palaeolithic, with increasing frequency towards its end.

Whatever may be the inadequacies of the archaeological work at Gough's Cave, and the unsatisfactory stratigraphic basis for parts of the assemblage, this must remain one of the most important Later Upper Palaeolithic industries in Britain, and one which will be vital to any future project of statistical analysis of Upper Palaeolithic assemblages including those of Britain. For this present study, I have merely made a few simple comparative analyses, which will be found in section V.M. of this chapter.

D. SOUTHERN WELSH SITES (GLAMORGANSHIRE AND PEMBROKESHIRE)

Maps 37 and 38 show the distribution of Later Upper Palaeolithic sites in southern Wales. In Glamorganshire (Map 37) the only two sites are Cathole and Paviland Cave, both of which are definite ones and both of which may have served at one time or another as "base-camps", Cathole perhaps for more inland exploitation and Paviland for concentration on the resources of what would have been a great Bristol Channel plain. These two sites are nearly within 10 kilometres of each other and may have even been employed simultaneously from time to time, as together they would have given the hunters an increased advantage over the surrounding territory, but this would certainly be very difficult to prove. Representative tool-forms from Cathole are shown in Figures 131 and 133 (no. 1–5) and some from Paviland are shown in Figure 133 (no. 11–13). Figure 131 of Cathole artifacts includes both Later Upper Palaeolithic and Mesolithic ones for direct comparison (see pp. 55–58 on Cathole stratigraphy). Those found in 1968 were excavated by the present author, whilst the earlier series was excavated by C.B.M. McBurney and has been arranged according to his original field-notes (by his kind permission). The pattern one sees here is Later Upper Palaeolithic in layers B, LOB and MSB, and Mesolithic at interface C/B or C/USB and interface D/C. Figure 131 (no. 7) is of particular interest as it is a good example of a large Later Upper Palaeolithic "micro-burin" (NB); the Mesolithic ones from the site are much smaller (Fig. 131, no. 13 and 23). The material from the 1860 excavations at Cathole includes two striking but unstratified additions to the Later Upper Palaeolithic assemblage(s), a shouldered point (AE) and a tanged point (AF) as depicted in Figure 133 (no. 2–3). Originally, both Cathole and Paviland may have had more Later Upper Palaeolithic artifacts than now survive.

In Pembrokeshire the definite Later Upper Palaeolithic sites include at least five of the seven shown on Map 38, the most productive site and presumed "base-camp" being Hoyle's Mouth, whose 10 kilometre territory encompasses all but two of the other sites. Representative tool-forms from Hoyle's Mouth are shown in Figure 132, whilst some from Nana's Cave are reproduced in Figure 133 (no. 6–10) and from Priory Farm Cave in Figure 133 (no. 14–15). At Hoyle's Mouth in a total of 60 extant stone tools, 41.7% are backed tools which include various normal Later Upper Palaeolithic forms (Fig. 132, no. 1–9). Raw materials for all 146 extant artifacts at Hoyle's Mouth include 83.6% flint, 15.8% adinole and 0.7% Carboniferous chert, the flint and adinole being apparently the most important. Adinole can be obtained locally in south-western Wales and was probably a useful substitute for presumably scarce good quality flint (see section on raw materials, pp. 27–29 and Map 3).

E. HEREFORDSHIRE AND BRECKNOCKSHIRE SITES

Map 39 shows the distribution of Later Upper Palaeolithic sites in Herefordshire and Brecknockshire. The only three sites are Ogof-y-Darren Ciliau, Arrow Court and King Arthur's Cave, the last two of which are definite ones and the last-named of which was probably a repetitively employed "base-camp". Whilst King Arthur's Cave would have been very useful for exploiting any migrating game in the Wye and Monnow valleys, as well as for slightly longer trips to the Severn basin and the upper reaches of the Bristol Channel plain, Arrow Court and Ogof-y-Darren Ciliau may have served as "transit-sites" on the way to what may have been good summer hunting grounds in the Welsh Highlands. Representative tool-forms from King Arthur's Cave are shown in Figure 134 and the two backed tools from Arrow Court are shown in Figure 130 (no. 1–2). King Arthur's Cave may have originally yielded just over 200 artifacts (see pp. 43–45 on stratigraphy), but most of these were destroyed during the Second World War.

F. NORTHERN WELSH SITES (CAERNARVONSHIRE AND DENBIGHSHIRE)

Map 40 shows the distribution of Later Upper Palaeolithic sites around the eastern reaches of what would have been an Irish Sea plain. The only definite site in northern Wales is Plas-yn-Cefn Cave which has yielded a mere 4 artifacts. The single backed tool, a "Creswell point" (AC1), from Plas-yn-Cefn is reproduced in Figure 155 (no. 1). Although the evidence is scanty, it certainly seems definite that Later Upper Palaeolithic hunters reached northern Wales at least for brief visits, and now perhaps one might even expect the eventual discovery of a more productive site in that region, as surely during summers it would have been quite pleasant and abundantly supplied with food sources, particularly during Late Last Glacial Zone II. Of course, many of the local camps may now be submerged, but Kendrick's Cave may just possibly have served as a burial ground (see Gazetteer II, site 35; also Sieveking, 1971) with "art objects", or "offerings".

G. NORTH-WESTERN ENGLISH SITES (NORTHERN LANCASHIRE AND WESTERNMOST YORKSHIRE)

As stated above in section V.F, Map 40 shows the distribution of Later Upper Palaeolithic sites around the eastern reaches of what would have been an Irish Sea plain. The only definite sites in Lancashire and westernmost Yorkshire are High Furlong, Kirkhead Cave and Kinsey Cave, but Jubilee Cave and Victoria Cave may also have some Later Upper Palaeolithic material included in what otherwise appears to be a mixed collection of later age. High Furlong is really the death site of a male elk (*Alces alces*) which had been hunted with barbed points, as well as possibly with stone-tipped arrows and a stone axe, during Late Last Glacial Zone II (see Fig. 135 and Gazetteer II, site 38). Where the hunters camped remains uncertain, but it was presumably close at hand.

Representative tool-forms from Kirkhead Cave are shown in Figure 136, whilst those from Kinsey and Victoria Caves are shown in Figure 137. The bevelled-base point (Fig. 137, no. 5) and three others not figured are, I suspect, genuine Upper Palaeolithic at Victoria Cave, partly because they are in reindeer antler and partly because there is some reason to suppose that they came from the glacial outwash, layer 2 (Breuil, 1922a; Garrod, 1926), but the other Victoria artifacts are from a context higher up in the sequence which may really be Mesolithic, e.g. the obliquely blunted point (AB6/12, Fig. 137, no. 7) and the robust biserial "harpoon" (IIFB2, Fig. 137, no. 9) in red deer antler (not reindeer as originally assumed by Breuil and Garrod; its structure is much too cancellous to be reindeer antler). This "harpoon" is somewhat of a problem piece, for it was originally ascribed to the "Azilian" (Breuil, 1922a) with whose "harpoons" it may to some extent be compared but with which an affinity seems most unlikely (cf. Sonneville-Bordes, 1960). Morphologically it might be British Later Upper Palaeolithic, but it is most unlike the seven definite barbed points thus far available (Kent's Cavern, Fig. 114, no. 2 and Fig. 115, no. 4–5; Aveline's Hole, Fig. 117, no. 15; High Furlong, Fig. 135, no. 1–2; Sproughton, Fig. 156, no. 4), and I am therefore inclined tentatively to compare it rather with the so-called "Obanian harpoons" of the Scottish Mesolithic (see Lacaille, 1954; Mellars, 1970; Rich, 1974), with which I think it has several important similarities. It might also be noted that the Zone II High Furlong barbed points (Fig. 135, no. 1–2) appear quite similar to some of the Zone IV "Maglemosian" ones at Star Carr (cf. Clark, 1954).

But in any case, as regards the sites themselves, Jubilee, Kinsey and Victoria caves would have been very useful for summer exploitation of the western Yorkshire Pennines when reindeer should have been quite abundant there, whilst Kirkhead Cave would have been ideal for summer exploitation of the southern Lake District and the adjacent parts of the Irish Sea plain, where both horse and reindeer should have been abundant at that time of year. The High Furlong elk was apparently pursued in winter, perhaps initially with a view to keeping it alive as "meat on the hoof", as it were (see Hallam, Edwards, Barnes and Stuart, 1973).

H. NORTH-EASTERN ENGLISH SITES (LOWLAND YORKSHIRE AND LINCOLNSHIRE)

Map 41 shows the distribution of Later Upper Palaeolithic sites in the Yorkshire, Lincolnshire and Derbyshire regions, those from the Derbyshire region being discussed in the next two sections. The only two definite open-air find-spots in lowland Yorkshire and Lincolnshire are Flixton Site 2 and Messingham. Flixton 2 has yielded a "microlithic shouldered point" (AC22a/AE4) and an unretouched blade (KB) in close association with remains of wild horse in a stratified context which has been assigned to the interface between Zones III and II of the Late Last Glacial (Moore, 1954). There is also a radiocarbon age estimate for the deposit just above the position of the artifacts and horse remains, c. 10,400 B.P. (see Table 4). Flixton 2, for all of its stratigraphic usefulness, probably does not represent even as much as a "transit-site" in terms of its economic importance; it presumably is just the result of the pursuit of the horses represented into a Late Last Glacial muddy bog, although the unretouched blade might suggest there had been some attempt made at butchering the prey before it became completely inundated. The fact that a "microlith" was employed in the hunt is of immense interest as it suggests the manufacture of flint-barbed bone or wood projectiles, perhaps in the manner suggested by some of the finds at the near-by Early Mesolithic site of Star Carr (see Clark, 1954). This "microlith" is reproduced here in Figure 141 (no. 5).

Some of the less convincing open-air finds from Yorkshire and Lincolnshire are shown in Figures 138 to 141 (except of course the Flixton 2 specimen in 141, no. 5). Those shown in Figures 138 to 140 are from Brigham Hill and are representative of the tool-forms in a large assemblage (total: 5,145 artifacts) which has been ascribed to the "Creswellian" by Manby (1966). I have handled all of the material and measured most of the pieces from Brigham, and I have been only able to draw the conclusion that almost the entire, if indeed not the entire, assemblage is Mesolithic and more comparable with Star Carr (Clark, 1954), Flixton Site 1 (Moore, 1950) and other Earlier Mesolithic assemblages in England, than with definite Later Upper Palaeolithic material. Of 69 backed tools at Brigham, 30 are of the obliquely blunted form (AB6/12) (Fig. 138, no. 2–4), which when occurring in high frequencies is certainly more characteristic of a Mesolithic assemblage than an Upper Palaeolithic assemblage. Also there are various robust core-tools which appear much more likely to be Mesolithic than Later Upper Palaeolithic, e.g. the "core-burin" (IF1) and "core-burin/scraper" (IF1/2) in Figure 140 (no. 1–2), while there are 3 clear examples of axe-sharpening or "tranchet" flakes (NC), 2 of which are shown in Figure 140 (no. 3–4) and 1 of which has been adapted as a burin (no. 3). There are also at least 19 "microburins" (NB) present, which when frequent by comparison with the backed tools are more typical of the Mesolithic. In fact, the only 3 pieces which I could rather unwillingly accept as Later Upper Palaeolithic, as this is a surface collection, are the "penknife point" (AC23) and the 2 "shouldered points" (AD7 and AE6) shown in Figure 138 (no. 10–12), but these "shouldered points" do not seem of typical Later Upper Palaeolithic forms (AD1–5 and AE1–5), and "penknife points" occur widely in the Later Upper Palaeolithic and the Mesolithic (see Gazetteer II; and Clark, 1932) and may therefore be either/or in an uncertain context.

Edlington Wood is in my opinion only a possibility for an open-air "base-camp" of Later Upper Palaeolithic age, as I have seen only one diagnostically Later Upper Palaeolithic tool-form from it, namely a "Creswell point". However, it has thus far yielded over 100 stone artifacts which my colleague P.A. Mellars (the director of the more recent excavations) interprets as quite likely "Creswellian". A selection of tool-forms found in 1958 by an earlier excavator, M.J. Dolby, is shown in Figure 141 (no. 1–4). However, as the site was certainly a focus of some sort of intense activity which may have been Later Upper Palaeolithic, I have drawn a 10 kilometre circle round it on Maps 41 and 42 to indicate its probable hunting territory. It would definitely have been advantageously situated for exploiting the surrounding plains and the migration of any reindeer or horse up the valleys of the Don and Dearne. But alas we must await future discoveries at the site before we may categorically say it is Later Upper Palaeolithic.

I. CRESWELL REGION SITES (SOUTHERNMOST YORKSHIRE, EASTERN DERBYSHIRE AND NORTHERN NOTTINGHAMSHIRE)

Map 42 shows the distribution of Later Upper Palaeolithic sites in the Creswell Crags region. The definite sites include Church Hole (Nottinghamshire), Dead Man's Cave (Yorkshire), Langwith Cave, Mother Grundy's Parlour, Pin Hole and Robin Hood's Cave (all Derbyshire). Selected tool-forms from these sites are shown in Figures 142 to 154, and those particularly representative of the tool-forms found by the present author during his excavations at Mother Grundy's Parlour and Robin Hood's Cave are reproduced in Figures 150 and 152–53 (see pp. 60–69 on the stratigraphy of these two sites as well as the relevant plans and sections, Figs. 23–36). Both Mother Grundy's Parlour and Robin Hood's Cave probably served as "base-camps" from time to time, perhaps particularly during the springs/summers of the Late Last Glacial according to the inferences on seasons drawn from the faunal evidence at Robin Hood's Cave (see pp. 129–31). As regards their close proximity to each other, we do not really know enough of the size and habits of the Later Upper Palaeolithic hunting groups to say whether both sites would have been occupied at the same time or only one. It is probably significant that all of the lesser cave-sites of the Creswell region, with both definite and possible Later Upper Palaeolithic evidence, fall at or within the 10 kilometre or 2 hour territorial limit of Mother Grundy's Parlour and Robin Hood's Cave, or, we should perhaps say, of Creswell Crags itself (see Map 42). Also there is no overlap with the assumed hunting territory of the possible open-air "base-camp" at Edlington Wood (see Maps 41 and 42). The caves at Creswell Crags and those of its region are ideally situated for exploiting the surrounding lowland plains as well as the neighbouring potential routes of migration for horse and reindeer going from spring and early summer feeding on the plains to middle and late summer feeding in the highlands of the Peak District to the west. In this respect it is noteworthy that "several young antlers" of reindeer were found in the basal portion of the section excavated at Mother Grundy's Parlour in 1924 (Jackson in Armstrong, 1925, p. 177), again presumably suggestive of late spring or early summer occupations.

The total number of artifacts from Mother Grundy's Parlour which I consider to be definite Later Upper Palaeolithic is 378; of these, 337 were obtained by Armstrong in 1924 from his base and lower spits (or "base and lower middle zones"), 8 were obtained by McBurney in 1959–60 from his layer B, and 33 were obtained by me from my layers LB and SB, so in other words simply in terms of the artifacts themselves, Armstrong's sample is the most important, although material found by McBurney and myself is of better stratigraphic reliability. A selection of representative tool-forms from Armstrong's base and lower spits is shown in Figures 144 to 147. Although there are many similarities between the assemblage and other British Later Upper Palaeolithic series from the Creswell region and further afield, there are also a number of striking differences which I think are a result of the Mother Grundy's Parlour material being fairly late in the British Later Upper Palaeolithic sequence, later than most of what one might term the "classic Creswellian" material from Robin Hood's Cave, Gough's Cave, Kent's Cavern, etc. One of these differences is the far greater proportion of "penknife points" (AC23, Fig. 144, no. 1–3, Fig. 146, no. 10–14) at Mother Grundy's Parlour than at any other prolific British Later Upper Palaeolithic site; of 47 backed tool-forms, 10 or 21.3% are "penknife points" (see Fig. 169). This feature alone might bring Mother Grundy's Parlour closer to the Dutch Tjongerian where "penknife points" would be counted together with other pointed convex backed blade forms as so-called "Tjonger points" (see Bohmers, 1956, 1960 and 1963). However, the occurrence of a number of shouldered points (AD, Fig. 144, no. 4–6) amongst the Mother Grundy's backed forms keeps the assemblage from becoming completely "Tjongerian", as shouldered points supposedly never occur in the "classic Tjongerian". "Creswell points" (AC1–2, Fig. 146, no. 7–9) only account for 8.5% of the backed forms at Mother Grundy's Parlour, or 4 out of 47. Scrapers also present a number of contrasts with most other British Later Upper Palaeolithic scraper samples, as they include scrapers worked all the way, or nearly all the way, around (CA1, Fig. 144, no. 15, Fig. 147, no. 1–3) and a side-and-end scraper (CA4, Fig. 147, no. 8). Also, out of the total of 19 scrapers, only 2 are end-of-blade scrapers (CB–D, Fig. 144, no. 16), whereas at Gough's Cave, out of 79 scrapers, 27 are end-of-blade scrapers (CB–D) and at that site there are no round scrapers or side scrapers of any form. In terms of awls, Mother Grundy's Parlour possesses the highest frequency of any British Later Upper Palaeolithic assemblage, 23.9% or 28 awls out of a total of 117 stone tools. All of these awls are of the stout EB2 form (Fig. 145, no. 1–9 and 11, Fig. 147, no. 10–17) and as a group they are very similar to those excavated from the Earlier Mesolithic settlement at Star Carr (see Clark, 1954).

With all of the above-cited differences between the industry of Mother Grundy's Parlour and that of most other British Later Upper Palaeolithic sites, it is perhaps small wonder that Bohmers (1956) felt inclined on the basis of his "statistical evidence" from only Mother Grundy's Parlour and Gough's Cave to attribute the latter to a separate "culture", his "Cheddarian", having assumed that Mother Grundy's Parlour was representative of the "true Creswellian". However, I think on the basis of my much wider survey of British material that if one were to continue the use of the term "Creswellian", the separation would be the other way around. In other words, Mother Grundy's Parlour would be "atypical

Creswellian" whilst Gough's Cave, Robin Hood's Cave, Kent's Cavern, etc. would be "typical Creswellian". As I have already said, I am inclined to attribute the distinctive features at Mother Grundy's Parlour to the industry's occupying a late or even final position within the British Later Upper Palaeolithic sequence that is beginning to emerge; one might say that the Later Upper Palaeolithic at Mother Grundy's Parlour stands on the threshold of the British Mesolithic.

Indeed, the technological and typological character of the material from Armstrong's middle spit (or "middle zone"), McBurney's interface C/B and my interface C/B, appears even closer than that just discussed to the general run of the Earlier Mesolithic in the northern Midlands and northern England (see Figs. 148 and 150). For convenience, this "transitional" assemblage from C/B may be referred to as "Mesolithic/Later Upper Palaeolithic" or simply "Meso./L.U.P."; on the basis of the stratigraphic evidence available, as well as the characteristics of the tool-forms themselves, I am prepared to suggest that it actually represents an intermediate phase between "typical" Later Upper Palaeolithic and "typical" Earlier Mesolithic in Britain. "Penknife points" (AC23, Fig. 148, no. 9–13, Fig. 150, no. 5), as in the lower assemblage at Mother Grundy's Parlour, again hold the highest percentage amongst the backed forms, 24.1% or 7 out of 29. However, it is in the obliquely blunted points (AB6/12, Fig. 148, no. 2) that there is the most striking increase, with 17.2% or 5 out of 29, while the lower assemblage had only 8.5% or 4 out of 47. This evident increase in the importance of obliquely blunted points could well be interpreted as indicative of a "trend" towards the Mesolithic (in which period they become extremely common), particularly when considered together with the stubby scrapers (Fig. 148, no. 21–22), another "Mesolithic" type. But the presence of as many as 20.7% or 6 "Creswell points" (AC1–5, Fig. 148, no. 3–6) and 2 trapeziform "Cheddar points" (AC7, Fig. 148, no. 7–8) would certainly suggest that the industry was still within the "influence" of the Later Upper Palaeolithic; it is perfectly true that these backed forms do occur in small numbers in British Mesolithic assemblages, including the Star Carr material (where I observed them in the collections of both the British Museum and the Cambridge University Museum of Archaeology and Ethnology), but their percentage value in a Mesolithic assemblage is greatly reduced. For comparison, fully Mesolithic tool-forms from Armstrong's upper spit (or "upper middle zone") at Mother Grundy's Parlour and my layers C and D and interface C/D, are depicted in Figures 149 and 150 (no. 8–19). It will be recalled that these layers date from Post Glacial Zones VI and VII.

Some representative tool-forms from Pin Hole are shown in Figures 143 (no. 5) and 151. Figure 151 (no. 1–6) are from Armstrong's 1 foot to 4 foot, 6 inch spits nearer the entrance, whilst Figures 143 (no. 5) and 151 (no. 7–9) are from his 0 to 3 foot spits further inside the cave (see Gazetteer II, site 69). I think Figure 151 (no. 1–6) indicative of a small, typical Later Upper Palaeolithic assemblage, and Figure 151 (no. 7–9) more suggestive of a "Mesolithic/Later

Upper Palaeolithic" assemblage, although here one has even less to examine than at Mother Grundy's Parlour, and such judgements are therefore rather arbitrary. Nevertheless, one may again receive the impression, if no more, of a Later Upper Palaeolithic occurrence followed by an occurrence more transitional to the Mesolithic. At Pin Hole the assumption of a transitional phase is somewhat strengthened by the fact that its small quantity of waste materials includes a good example of an axe-sharpening or "tranchet" flake (NC), as well as a small "micro-burin" (NB). A noteworthy bone artifact which may belong to the "Mesolithic/Later Upper Palaeolithic" at Pin Hole is that shown in Figure 143 (no. 5); this is the only example in British Upper Palaeolithic "art" of a human figure or outline known thus far.

As a result of my excavations at Robin Hood's Cave in 1969, I obtained at least 4 stratigraphically separate Later Upper Palaeolithic assemblages, which I labelled for convenience: "L.U.P.1" (layer B/A), "L.U.P.2" (layer LSB), "L.U.P.3" (layer OB) and "Meso./L.U.P." (layer USB). Representative tool-forms from these assemblages are shown in Figures 152 and 153. Although the actual numbers are comparatively small from assemblage to assemblage (see Gazetteer II, site 70), I think their tool-forms may nevertheless reflect certain changes within the British Later Upper Palaeolithic. The lowest assemblage, "L.U.P.1" from layer B/A, has "Creswell points" (AC1–2, Fig. 152, no. 3–4) associated with a robust, convex backed blade (AB5, Fig. 152, no. 2), whilst the next assemblage proceeding upwards, "L.U.P.2" from layer LSB, has more "Creswell points" (AC2, Fig. 152, no. 7, Fig. 153, no. 1) but this time associated with a shouldered point (AD1, Fig. 152, no. 9). Then still proceeding upwards, "L.U.P.3" from layer OB has no "Creswell points" but has shouldered points (AD1 and AD5, Fig. 153, no. 8–9), and finally the uppermost assemblage, what I have termed "Meso./L.U.P.", has the return of a "Creswell point" (AC2, Fig. 153, no. 15) associated with the initial appearance of a "penknife point" (AC23, Fig. 153, no. 16) as well as an apparent scraper/"micro-burin" (CE/NB, Fig. 153, no. 14). If these differences are truly reflective of changes within the British Later Upper Palaeolithic, then one might adopt as an hypothesis, hoping eventually to confirm or reject it with stronger evidence, the following sequence:

(a) an initial phase with frequent "Creswell points";
(b) a second phase with frequent "Creswell points" and some shouldered points;
(c) a third phase with either no or few "Creswell points" but frequent shouldered points;

and (d) a final phase with either a return or a rise in "Creswell points" associated with the beginning of frequent "penknife points" and a general transformation *in situ* from Later Upper Palaeolithic towards what we now term Mesolithic.

Whether such a sequence would be valid more than strictly locally, does not appear; it was however suggested above that this site was a "base-camp". Amongst the other tool-forms at Robin Hood's Cave, three noteworthy ones from "L.U.P.3" of layer OB include 2 finely made bone awls (IIA, Fig. 153, no. 13) and a burin in quartzite rather than flint (BB2, Fig. 153, no. 10).

Whatever may be the inadequacies of the various collections from the Later Upper Palaeolithic sites at Creswell Crags (Church Hole, Mother Grundy's Parlour, Pin Hole, and Robin Hood's Cave), there is no doubt that considered together they must remain one of the most important series in Britain, second only to Kent's Cavern and Gough's Cave. Future work should certainly include sophisticated analyses of the Creswell material, particularly that at Mother Grundy's Parlour and Robin Hood's Cave. In this present study I have only carried out a few simple comparative analyses, essentially pilot studies, of the tool-forms from Mother Grundy's Parlour and Robin Hood's Cave; these will be found in section V.M. of this chapter.

J. PEAK DISTRICT SITES (WESTERN DERBYSHIRE AND NORTHERN STAFFORDSHIRE)

Map 43 shows the distribution of Later Upper Palaeolithic sites in the Peak District. None of these sites has yielded very much and the only definite sites appear to be Dowel Cave and Fox Hole on the Derbyshire side, and Elder Bush Cave, Ossum's Cave and Thor's Fissure on the Staffordshire side. Some representative tool-forms from the Peak District caves are shown in Figure 155 (no. 2–10). As may be seen in Figure 155, there is considerable variation in the backed tool-forms of the Peak District, and this, taken together with the scantiness of the evidence in general, might be suggestive of mere sporadic visits to the area by various Later Upper Palaeolithic hunting bands during the Late Last Glacial. Most of the Peak District artifacts are associated with either wild horse or reindeer, or both, and these were probably the main animal food sources their makers sought during the brief summer months of the area. Particularly noteworthy "cult" practices include the apparently intentional burials of a bear skull at Fox Hole and the thoracic region of a young reindeer at Elder Bush Cave (Bramwell, 1964, 1971 and 1973).

K. EAST ANGLIAN SITES (NORFOLK AND SUFFOLK)

Map 44 shows the distribution of Later Upper Palaeolithic sites in East Anglia. All of these sites are represented by surface or near-surface collections, and the only material which appears to include definite Later Upper Palaeolithic tool-forms is in my opinion that from London Bottom near Icklingham and Sproughton near Ipswich. The Sproughton series is still being studied, having just been excavated in 1972 and 1974 (see Gazetteer II, site 81). It may belong somewhere between Zones II and IV, but the stratigraphy is far from clear just yet. For the moment I am pleased to include my own drawing of the barbed point (Fig. 156, no. 4) which was found separately in 1974.

The London Bottom series came from a gravel and includes at least 8 large, convex backed blades (AB1), 2 somewhat smaller, straight backed blades (AB2) and 2 large, typical "Cheddar points" of the usual trapeziform outline (AC7) (see illustrations in Sturge, 1914, especially plate 46). These backed forms are closely associated with various burins (BA1 and BB2), blade end scrapers (CB) and at least 2 typical "Zinken" (DA), the latter being a rare tool-form in the British Later Upper Palaeolithic, but an important feature nonetheless, as it occurs as a characteristic element in certain continental Later Upper Palaeolithic industries, probably the best examples being found in the Hamburgian (see Rust, 1937, 1943 and 1962). But sadly enough this material from London Bottom is mixed with Mesolithic and later artifacts, so it is difficult to determine what else from the site might originally have been Later Upper Palaeolithic. Another mixed East Anglian assemblage which may include some Later Upper Palaeolithic backed forms is that from Cranwich, Norfolk. Personally, I am less certain about this material, but Wymer (1971 and personal communications) thinks those pieces which I here illustrate in Figure 156 (no. 1–3) are definite Later Upper Palaeolithic. My own opinion is that both the shouldered point (Fig. 156, no. 2) and the tanged point (Fig. 156, no. 3) from Cranwich are quite atypical forms for the British Later Upper Palaeolithic, even when compared with the shouldered and tanged point series from Hengistbury Head (Hampshire), which is discussed below in the next section, nor do the Cranwich points greatly resemble the shouldered and tanged point forms of the various "Stielspitzen-Gruppen" of northern Europe (see Taute, 1968; Becker, 1971). The shouldered point (AD1, Fig. 156, no. 5) from Wangford Warren in Suffolk is more the sort one should expect in the British Later Upper Palaeolithic, but this particular example is apparently from a Mesolithic assemblage; more precisely, it belongs to a large collection, which contains many typically Mesolithic obliquely blunted points (AB6/12), e.g. Figure 156 (no. 6), and much other certainly Mesolithic material.

Of all of the East Anglian Later Upper Palaeolithic sites now known or claimed, perhaps only London Bottom and Sproughton may have originally served as open-air "base-camps", but even the evidence surviving from these sites is at present too scanty to support such a suggestion, although the sheer abundance of artifacts at Sproughton might go a long way in that direction. In the case of London Bottom it is striking how most of the East Anglian sites on Map 44, whether possible or definite, cluster within 30 kilometres, more or less, of one another in its region, an area which might have intersected a main migration route for wild horse and reindeer herds possibly proceeding seasonally to Europe and back; as well as perhaps a potential route for similar human movement. In future, perhaps new finds of more satisfactory open-air sites will be made within this particular area, as well as near Sproughton. Finally, the East Anglian material found thus far falls mostly it would seem in my "Mesolithic/Later Upper Palaeolithic" category (i.e., both vague and late), the Sproughton barbed point itself being rather similar to some of those in the "Maglemosian" (see Fig. 156, no. 4; cf. Clark, 1954; Radley, 1969; Wymer, Jacobi and Rose, 1975), though that at Sproughton is quite robust and could belong to Zone III.

L. SOUTHERN ENGLISH SITES (KENT, SUSSEX, BERKSHIRE, SURREY, HAMPSHIRE AND DORSET)

Map 45 shows the distribution of Later Upper Palaeolithic open-air sites in southern England, with the exception of three dubious cases in Dorset at Pilsdon Pen and on the Isle of Portland which lie beyond the range of this map: Pilsdon Pen Hillfort, Portland Bill Site 1 and Verne Ditch (see Gazetteer II, sites 97 to 99; also Fig. 130, no. 7–11). In my opinion, the definite southern English open-air sites include only Oare (Kent), Newhouse Farm (Sussex), Crown Acres (Berkshire), Hengistbury Head and Long Island (both Hampshire). The material from Oare, the exact find circumstances of which are unknown, includes two shouldered points, one of which is a trapeziform shouldered point (AD2, Fig. 166, no. 2) almost identical to those at Gough's Cave (Fig. 122, no. 5–6, Fig. 126, no. 1), Soldier's Hole (Fig. 127, no. 6) and Sun Hole (Fig. 128, no. 5). These trapeziform shouldered points occur on the European continent in both the Hamburgian and "Stielspitzen-Gruppen" (see Rust, 1962; Müller-Karpe, 1966; Taute, 1968) and, less frequently, in some of the industries of the "Federmesser-Gruppen" (see Schwabedissen, 1954), particularly the more "Creswellian" and "Cheddarian" ones of Belgium and Holland, as Bohmers (1956) would describe them. The tanged point (AF1) from Newhouse Farm, which has just recently come to my attention, seems quite similar to many of those in Taute's (1968) "Stielspitzen-Gruppen". The shouldered point from Long Island is shown in Figure 166 (AE4, no. 5), together with a burin (BA3) and a multiple class scraper/saw (GC – CB/FA) from the same find-spot. This shouldered point is comparable to some from Hengistbury Head and to a broken one which may be Later Upper Palaeolithic from Rush Corner in Bournemouth (Fig. 166, no. 4); it is a type which occurs in both the German Hamburgian (see Rust, 1962; Müller-Karpe, 1966) and the French Magdalenian (see Sonneville-Bordes, 1960).

Both of the other two definite southern English Later Upper Palaeolithic open-air sites, Crown Acres and Hengistbury Head, would appear to have served as important "base-camps", presumably for the exploitation of migrating wild horse and reindeer in spring and autumn, although it is certainly not inconceivable that some wintering may have taken place at these sites as well, perhaps particularly Hengistbury which has more artifacts than Crown Acres and which commands an excellent view of what would have been the adjacent English Channel plain (see Map 45). The "site exploitation territory" of Hengistbury, i.e. the area within a 2 hour walking distance or about 10 kilometres, also includes the confluence of the Christchurch rivers Stour and Avon, as well as an area of the adjacent Hampshire plains. Also both Crown Acres and Hengistbury are within easy range of abundant, good quality flint, in both local gravels and chalk outcrops. In fact the quantity and quality of the flint available would certainly seem to be reflected in the size of the cores abandoned at the two sites: the mean and standard deviation for maximum core length at Crown Acres is 90.8 ± 34.7 mm. (total: 18 cores) and that from my excavation at Hengistbury Site C2 (B layers) is 90.9 ± 28.7 mm. (total: 14 cores). For comparison, the only other samples of British Later Upper Palaeolithic cores with a reasonable number are Gough's Cave and Sproughton. Gough's Cave is somewhat further from sources of good quality flint, although not as far as many other Later Upper Palaeolithic cave-sites: the mean and standard deviation for maximum core length at Gough's Cave (spits 25–12) is 65.8 ± 10.4 mm. (total: 5 cores). Sproughton is near good sources of East Anglian flint: the mean and standard deviation for maximum core length is 120.5 ± 58.6 mm. (total: 31 cores). This data is plotted in Figure 174, together with that on mean breadths.

In the light of the closely similar Crown Acres and Hengistbury Site C2 (B layers) means (see Fig. 174), it is of some relevance that the Mesolithic cores from my excavations at Hengistbury Site C2 exhibit a marked decrease in size and hence, presumably, a less strong preference for large nodules or cobbles of flint, or at least a habit of working them down much more. The mean and standard deviation for Mesolithic maximum core length at Hengistbury Site C2 (A layers) is 58.6 ± 9.3 mm. (total: 12 cores). Also that for the cores from the supposedly Later Upper Palaeolithic assemblage excavated by Mace (1959) from the adjacent Hengistbury Site C1 is 61.4 ± 12.2 mm. (total: 30 cores), which really agrees better with my Mesolithic figure than with my Later Upper Palaeolithic one from Hengistbury Site C2 (see Fig. 174). In fact, not only on this ground, but on several other lines of suggestive evidence to be discussed below and in the next section on statistical analyses, I strongly suspect that most of Mace's supposed Later Upper Palaeolithic assemblage is really Mesolithic.

The total number of artifacts thus far from the Later Upper Palaeolithic occurrence at Crown Acres is 332, of which only 4 are recognizable tool-forms. However, more material might be obtained in future by excavating, as what has been obtained thus far has only been brought up by deep ploughing (J. Wymer, personal communications). All of the raw material in the present collection is flint. The convex backed blade (AB1) from Crown Acres is shown in Figure 166 (no. 1).

As implied above and already stated in the relevant section on stratigraphy (see pp. 69–76, Map 15 and Figs. 37–55), the Later Upper Palaeolithic material at

Hengistbury Head, whether definite or possible, is divided between at least four occurrences, some separate, some partially overlapping: Sites A, B, C1 and C2. The material from Site A was brought up by deep ploughing and although mixed with what are obviously later flint artifacts and pot sherds, it includes a small number of diagnostic Later Upper Palaeolithic backed tool-forms, four of which are reproduced here in Figure 157 (no. 1–4). Figure 157 (no. 1) is a "Creswell point" (AC2), (no. 2) is a shouldered point (AD1), and (no. 3–4) are tanged points (AF2). Tanged points are normally rare in the British Later Upper Palaeolithic, but Figure 157 (no. 3) is comparable with one from Cathole (Fig. 133, no. 3). However, it should be noted that tanged points also occur in the Earlier Mesolithic at Star Carr (see Clark, 1954), although here they are again a rare element. The material from Site B at Hengistbury was excavated by Calkin (see Mace, 1959, Fig. 11), but is really less convincing than that from Sites A and C, and if it is a single entity at all, it is probably Mesolithic or later. The material from "Site C" is divided between two "sub-sites", C1 and C2. A few diagnostic backed tool-forms have been gathered from the vicinity of "Site C" by various enthusiastic local collectors, and two of these, both "Creswell points" (AC1–2), are shown in Figure 157 (no. 5–6). Material from Site C1 itself was excavated by Mace and interpreted by her as entirely "Late Upper Palaeolithic" (see Mace, 1959); however, from my detailed study of her finds and comparisons with definite Later Upper Palaeolithic and definite Mesolithic samples from other sites in Britain, I have been forced to conclude that most of her assemblage is Mesolithic on the basis of its typology and morphology. Also, in contrast to the stratified finds from my own excavation, the general lack of patination amongst Mace's material is disconcerting. In the absence of any precise stratigraphic record, these factors are all one has to go on; Mace's standard of excavation was crude to say the least. There are in fact only 9 backed tools from Mace's excavation at Site C1 which I consider to be definite Later Upper Palaeolithic; 5 of these are illustrated in Figure 158 (no. 1–5). Most or all of the other 3,048 artifacts from Site C1 (see Gazetteer II, site 92) I regard as much more likely to be Mesolithic, or in some cases even later. Examples of the tool-forms from this ?Mesolithic assemblage are shown in Figures 158 (no. 6–11), 159 and 160. Of 108 backed tools, 86 are straight backed blades (AB2/8, Fig. 158, no. 7–8), of which most are of "microlithic" proportions (AB8). There are also 3 obliquely blunted points (AB6/12, Fig. 158, no. 9) and 5 "penknife points" of an unusual form—straight backed (AC23?, Fig. 158, no. 10–11). The burins are generally robust and quite unlike Later Upper Palaeolithic ones; the mean and standard deviation of their burin facet breadth is 6.2 ± 3.9 mm. (see Figs. 159 and 170), whereas that at Gough's Cave (spits 25–12) is 2.2 ± 1.0 mm. (total: 92 burins; see Fig. 170). Scrapers are also unlike most Later Upper Palaeolithic ones (see Fig. 160, no. 1–6), and there is even a possible small core-axe (Fig. 160, no. 9). The reader should carefully compare these illustrations with those of material from the best of the Later Upper Palaeolithic sites elsewhere in Britain.

Material from Site C2 at Hengistbury Head was excavated by the present author (see Figs. 37–55 for excavation and distribution plans and sections), and it includes two separate assemblages, a lower one from the B layers with definite Later Upper Palaeolithic tool-forms, and an upper one from the A layers with definite Mesolithic tool-forms. There are also traces of still later activity. The raw material for both assemblages is flint, but that from the Later Upper Palaeolithic is generally patinated, whereas that from the Mesolithic is generally unpatinated to only slightly patinated (cf. the comment on Mace's artifacts above). Representative tool-forms from the Later Upper Palaeolithic at Site C2 are shown in Figures 161 to 163 (B layers), and for comparison, those from the Mesolithic at Site C2 are shown in Figures 163 to 165 (A layers). Aside from the usual common factors, both assemblages possessing backed tool, burin and scraper classes, it may be seen at a glance from these illustrations (Figs. 161–65) that there are obvious differences in typology, morphology and dimensions between the Later Upper Palaeolithic tool-forms (B layers) and the Mesolithic tool-forms (A layers). For one thing, the Later Upper Palaeolithic has generally larger backed tools including notably 4 "Creswell points" (3 AC2 and 1 AC4, Fig. 161, no. 5–8), 1 of which is exceptionally large even for the Later Upper Palaeolithic (Fig. 161, no. 6, which appears to be an AC2 modified as an awl, EB2; the robustness of some of these Later Upper Palaeolithic backed tools is presumably due in part to the quality of the raw material available and selected for their manufacture; the same factor might also hold true for some of the Hengistbury Mesolithic material), 1 tanged point (AF3, Fig. 161, no. 9) and 2 shouldered points (AE1, Fig. 162, no. 1). The Mesolithic at Site C2, on the other hand, has backed tools which are predominantly "microlithic" to "medium-sized" (see Fig. 163, no. 5–10, Fig. 164, no. 1–5). Also the burins of the Later Upper Palaeolithic assemblage are generally less robust in that they have thinner burin facets than the Mesolithic burins (compare Fig. 162, no. 2–4 and Fig. 163, no. 4 with Fig. 164, no. 6–10). In fact, Figure 164 (no. 10) is quite like one of the burins from Mace's Site C1 assemblage (Fig. 159, no. 6), which, as already stated, I consider predominantly Mesolithic. There are also differences in the scrapers, the Mesolithic ones having the addition of round scrapers (CA1, Fig. 164, no. 11) and side-and-end scrapers (CA4, 165, no. 1). The reader is referred to Gazetteer II, sites 93 and 94, for typological details of the author's assemblages from Hengistbury Head Site C2. It is worth noting here, however, that the total number of artifacts in the Later Upper Palaeolithic assemblage is 1,163, of which 33 are recognizable tool-forms.

Thus together with Kent's Cavern, Gough's Cave, Mother Grundy's Parlour and Robin Hood's Cave (as well as possibly Sproughton and Crown Acres), Hengistbury Head would appear to be one of the more important British Later Upper Palaeolithic sites, particularly Hengistbury Head Site C2, the one excavated by the present author, since there the stratigraphy is known. Future excavations at Hengistbury Head would certainly be worthwhile, as it can be expected that they would yield still more Later Upper Palaeolithic artifacts, the concentration of Later Upper Palaeolithic activities at the eastern end of the headland having apparently been quite intense. In fact, it is to be hoped that more definite tool-forms will be found, so that elaborate statistical analyses may eventually be carried out on material from the Hengistbury sites. Even tools made in bone and wood might possibly be preserved in the small bog to the north of Site C2.

M. REPRESENTATIVE TOOL-FORMS AND METRICAL AND STATISTICAL ANALYSES OF SELECTED LATER UPPER PALAEOLITHIC MATERIAL

Figures 114 to 166 show examples of Later Upper Palaeolithic tool-forms, both definite and possible ones, as well as certain Mesolithic tool-forms from particularly Cathole, Brigham Hill, Mother Grundy's Parlour and Hengistbury Head. Most of these drawings have already been mentioned in the immediately preceding brief discussion by region and site, but there now follows a typological key to the specimens illustrated (see pp. 20–23 and Table 1 for detailed descriptions of the classes and types recognized by the author, and Gazetteer II for present whereabouts):

Fig. 114 KENT'S CAVERN (Devonshire), 1866–67, layer B/A2 (c. B2). 1. Backed tool AC2; 2. "harpoon" IIFB1; 3–7. backed tools AB6, AC1, AC2, AC4 and AC7; 8. burin BA1. All in flint except no. 2 in antler. Estimated radiocarbon ages c. 14,275 and c. 12,180 B.P. (the latter for the "harpoon").

Fig. 115 KENT'S CAVERN (Devonshire), 1866–67, layer B/A2 (c. B2). 1–2. Scrapers CD and CE; 3. needle IIB; 4–5. "harpoons", both IIFA1. 1 in flint, 2 in Greensand chert, 3 in bone, 4–5 in antler. 3–5 redrawn after Garrod (1926). Estimated radiocarbon age c. 14,275 B.P.

Fig. 116 AVELINE'S HOLE (Somerset), 1919–25, layer "red cave-earth". 1–14. Backed tools AB1, 2 AB2, 3 AC2, AC23, AE4, AB6, AA1, AB1, AB2, AC7, AB2(/NB); 15. scraper CE; 16–17. awls EB2; 18. rubbed end piece MC. All in flint. All redrawn after Davies (1921–23), except 2, 9 and 14. Estimated radiocarbon age c. 9,114 B.P. (but association uncertain).

Fig. 117 AVELINE'S HOLE (Somerset), 1919–25, layer "red cave-earth". 1–11. Backed tools 4 AB1, 2 AB2, AB6, AC6, 2 AC23, AC22a; 12. awl EB2; 13. multiple class GB (BB2/EB2); 14. saw FB; 15. "harpoon" IIFB2; 16–17. backed tools AC22a and AB8/NB; 18. scraper CE; 19. burin BB2. All in flint except no. 15 in antler. All redrawn after Davies (1921–25), except no. 15 from a cast and 11 and 16 from surviving specimens. Estimated radiocarbon age c. 9,114 B.P. (but association uncertain).

Fig. 118 FLINT JACK'S CAVE (Somerset), c. 1893, layer unknown. 1–4. Backed tools 3 AC2, AC10; 5–6. scrapers CB and CE; 7. saw FA; 8–10. multiple classes GA (AC2/BA1), GA (AC2/BB2) and GB (BD1/CB); 11. rubbed end piece MC. All in flint except no. 7 in Greensand chert.

Fig. 119 GOUGH'S CAVE (Somerset), c. 1892–1958, layers A2–B?. 1–4. Backed tools AA2, 2 AC2, AD1; 5. burins BA1 and BA2; 7. scraper CF; 8. awl IIA; 9. decorated rib segment IIDB. 1 and 3–7 in flint, 2 in Greensand chert, 8–9 in bone (8 in tibia of *Lepus timidus*). Estimated radiocarbon age c. 9,080 B.P. (but association uncertain).

Fig. 120 GOUGH'S CAVE (Somerset), 1927, layer A2. "Bâton" IIEB in antler.

Fig. 121 GOUGH'S CAVE (Somerset), 1927–31, layer B. 1–14. Backed tools AA1, 5 AB1, AB2, 3 AC2, AC7, AC22c, 2 AC23. All in flint. Estimated radiocarbon age c. 9,080 B.P. (but association uncertain).

Fig. 122 GOUGH'S CAVE (Somerset), 1927–31, layer B. 1–7. Backed tools 4 AD1, 2 AD2, AD5; 8–12. burins BA2, BA4, 2 BB2, BC1. All in flint. Estimated radiocarbon age c. 9,080 B.P. (but association uncertain).

Fig. 123 GOUGH'S CAVE (Somerset), 1927–31, layer B. 1–5. Burins BC3, BD2, BB4, BE2, BE5; 6–10. scrapers CA2, 2 CB, CC, CE. All in flint. Estimated radiocarbon age c. 9,080 B.P. (but association uncertain).

Fig. 124 GOUGH'S CAVE (Somerset), 1927–31, layer B. 1. Scraper CB; 2–3. "Zinken" DA; 4. borer EA; 5. "Zinken" DC; 6–8. saws FA, FB, FC; 9. notch FD; 10. retouched blade HC1; 11. multiple class GA (AB4/BA1). All in flint. Estimated radiocarbon age c. 9,080 B.P. (but association uncertain).

Fig. 125 GOUGH'S CAVE (Somerset), 1927–31, layer B. 1–5. Multiple classes GA (AB2/BA2), GA (AF/BA2), GB (BA1/CB), GC (CB/EA), GC (CE/FA); 6–13. backed tools 2 AA1, AA2, AB1, AC5, AC1, AC6, AC7. All in flint. Estimated radiocarbon age c. 9,080 B.P. (but association uncertain).

Fig. 126 GOUGH'S CAVE (Somerset), 1927–31, layers B and E/B. 1. Backed tool AD2; 2–4. burins BA2, BE5, BE3 ("chamfered"); 5. scraper CB; 6. borer EA1; 7. awl EB2; 8. notch FD; 9–12. backed tools AB6, 2 AC1, AC21. All in flint. Estimated radiocarbon age c. 9,080 B.P. (but association uncertain).

Fig. 127 SOLDIER'S HOLE (Somerset), 1928–29, layer B. 1–6. Backed tools AC1, AC3, AC6, AC10 (/BB2), AD1, AD2; 7. burin BA2; 8. scraper CB; 9. borer EA3. All in flint.

Fig. 128 SUN HOLE (Somerset), 1927–53, layers B2. 1–5. Backed tools AA1, AB2, AB6, AC7, AD2; 6. burin BB2; 7. scraper CE; 8. awl EB2; 9. rubbed end piece MB. All in flint. Estimated radiocarbon age c. 12,378 B.P.

Fig. 129 SOUTH-WESTERN ENGLISH CAVES. 1–4. Backed tools AC10, AA1, AB2, AC2; 5–6. scrapers, both CE; 7–8. backed tools AB3 and AC2. All in flint except no. 2 in Greensand chert.

Fig. 130 SOUTH-WESTERN ENGLISH OPEN-AIR FINDS. 1–4. Backed tools AC23, AD1, 2 AC23; 5. awl EB2; 6–11. backed tools AC23, AE6?, AF1?, AF4, AF2, AF4. 1–6 in flint, 7–11 in Portland chert. 6 redrawn after ApSimon (1957), 7–11 redrawn after Palmer (1970).

Fig. 131 CATHOLE (Glamorganshire), 1958 and 1968, layers B to interface D/C. 1–3. Backed tools AB1, AB6, AC2; 4–5. scrapers CB and CE; 6. awl EB2; 7. "micro-burin" NB; 8–10. backed tools AB1, AB2, AB7; 11–12. retouched flakes/blades HA and HB; 13. "micro-burin" NB; 14–16. backed tools AB1, 2 AB2; 17. scraper CE; 18. awl EB2; 19–20. burins, both BB2; 21–22. backed tools AB11 and AC22a; 23. "micro-burin" NB. All in flint. 1–18 redrawn after McBurney (1959).

Fig. 132 HOYLE'S MOUTH (Pembrokeshire), 1862–72, layer B?. 1–9. Backed tools AA1, AB1, AC2, AC4, AC7, AC17, AC23, AD1, AE1; 10–12. burins BA1, BB2, BD1; 13–14. scrapers CA2, CB; 15. borer EA2; 16–17. awls EB2; 18. saw FB. All in flint except no. 12 in adinole.

Fig. 133 SOUTH-WESTERN WELSH CAVES. 1–3. Backed tools AC7, AE1, AF4; 4. burin BA3; 5. scraper CE; 6–9. backed tools AB1, AC7, 2 AC23; 10. awl EB2; 11–14. backed tools AC2, AC6, 2 AC23; 15. awl EB2. All in flint except no. 9 in Greensand chert.

Fig. 134 KING ARTHUR'S CAVE (Herefordshire), 1871 and 1925–30, c. layers A3c and B. 1–2. Backed tools, both AB2; 3. burin BA1(/NA); 4–8. backed tools AA1, AB2, AC7, AC23, AD1; 9. awl EB2; 10. burin BC1; 11. awl EB2; 12. multiple class GB (BA4/CA2). All in flint.

Fig. 135 HIGH FURLONG (Lancashire), 1970, Zone II layer. 1–2. Barbed points IIFA1 in bone. Estimated radiocarbon age average c. 11,930 B.P.

Fig. 136 KIRKHEAD CAVE (Lancashire), 1968–70, layers B and ?C. 1–3. Backed tools AB1, AB2, AB3; 4–5. scrapers CA2 and CC; 6. backed tool AB2; 7. burin BB4; 8. backed tool AC22a (?Mesolithic layer C). All in flint.

Fig. 137 KINSEY AND VICTORIA CAVES (Yorkshire). 1–2. Backed tools, both AB1; 2. burin BB2; 4. multiple class GE (EB2/FB); 5. bevelled-base point IICBb; 6–7. backed tools AB1 and AB6/12; 8. awl EB2; 9. "harpoon" IIFB2. 1–4 and 6–8 in flint, 5 in *Rangifer* antler and 9 in *Cervus elaphus* antler. 5 redrawn after Garrod (1926).

Fig. 138 BRIGHAM HILL (Yorkshire), 1962–63, surface. 1–12. Backed tools AB1, 3 AB6, AB9, AC22a, 2 AC22b, AC22c, AC23, AD7, AE6; 13–15. awls, all EB2; 16–17. burins BB2 and BC3. All in flint.

Fig. 139 BRIGHAM HILL (Yorkshire), 1962–63, surface. 1. Burin BE5; 2–8. scrapers CA2, CA4, 2CB, 2 CE, CF; 9–10. saws, both FB; 11. "micro-burin" NB; 12. notch FB. All in flint.

Fig. 140 BRIGHAM HILL (Yorkshire), 1962–63, surface. 1. Core-burin IF1; 2. core-burin/scraper IF1/2; 3. axe-sharpening flake NC (with burin); 4. axe-sharpening flake NC. All in flint.

Fig. 141 NORTHERN ENGLISH OPEN-AIR FINDS. 1–2. Backed tools AA1 and AB2; 3. burin BB2; 4. scraper CE; 5–13. backed tools AC22a/AE4, AC23, AE7, AB1, AB2, AB6, AC23, AE1, AE4; 14. "Zinken" DA. All in flint.

Fig. 142 CRESWELL CRAGS CAVES (Derbyshire/Nottinghamshire), 1874–76. 1. Backed tool AB1; 2. burin BA1; 3. rubbed end piece MB; 4–6. backed tools AC6, AC7, AD5; 7. burin BA2; 8. multiple class GB (BA2/CB); 9–11. backed tools AB1, AC2, AC23. All in flint.

Fig. 143 CRESWELL CRAGS CAVES, 1874–76 and 1924–28. 1. Awl IIA; 2. needle IIB; 3. point IICB1 (?/spatula IIGA); 4. ?pendant IIHC1 (?/"flesher"); 5. decorated rib segment IIDB (with human male motif). 1–2 and 4–5 in bone, 3 in antler (*Rangifer*). 1–4 redrawn after Garrod (1926), 5 redrawn after Armstrong (1928).

Fig. 144 MOTHER GRUNDY'S PARLOUR (Derbyshire), 1924, base spit. 1–6. Backed tools 3 AC23, 2 AD1, AD5; 7–13. burins 3 BA1, BA2, BA3, BB2; 14–16. scrapers CA2, CA1, CB. All in flint. All redrawn after Armstrong (1925).

Fig. 145 MOTHER GRUNDY'S PARLOUR (Derbyshire), 1924, base spit. 1–9. Awls, all EB2; 10. notch FB; 11. awl EB2; 12. notch FB. All in flint. All redrawn after Armstrong (1925).

Fig. 146 MOTHER GRUNDY'S PARLOUR (Derbyshire), 1924, lower spit. 1–17. Backed tools 2 AB1, 3 AB4, AB2, AC2, AC1, AC2, 5 AC23, 3 AD1; 18–21. burins 3 BA1, BB2. All in flint. All redrawn after Armstrong (1925).

Fig. 147 MOTHER GRUNDY'S PARLOUR (Derbyshire), 1924, lower spit. 1–9. Scrapers 3 CA1, 4 CA2, CA4, CD; 10–17. awls, all EB2; 18. saw FB; 19. notch FD. All in flint. All redrawn after Armstrong (1925).

Fig. 148 MOTHER GRUNDY'S PARLOUR (Derbyshire), 1924, middle spit. 1–15. Backed tools AB2, AB6, AC1, 2 AC2, AC5, 2 AC7, 5 AC23, AE2, AD1; 16–18. burins 2 BA1, BB2; 19–20. awls, both EB2; 21–23. scrapers CA1, CA4, CE. All in flint. All redrawn after Armstrong (1925).

Fig. 149 MOTHER GRUNDY'S PARLOUR (Derbyshire), 1924, upper spit. 1–13. Backed tools AA1, AB1, 2 AB2, 5 AB6, AB11, AC21, AC22b, AC23; 14–15. burins BA5 and BA2; 16–20. scrapers CA1, 2 CA2, 2 CE; 21–22. awls, both EB2; 23. retouched flake/blade HD2; 24–26. "micro-burins" NB. All in flint. All redrawn after Armstrong (1925).

Fig. 150 MOTHER GRUNDY'S PARLOUR (Derbyshire), 1969, layers SB to D. 1–2. Backed tools, both AA1; 3. burin BA2; 4. retouched blade HD2;

5–6. backed tools AC23 and AE2?; 7. core-scraper IF2; 8–9. backed tools AB2 and AB5; 10. burin BB2; 11. awl EB2; 12. "micro-burin" NB; 13–14. scrapers, both CA2; 15. backed tool AA1; 16–17. scrapers CA1 and CA4; 18. backed tool AB4; 19. scraper CE. All in flint.

Fig. 151 PIN HOLE (Derbyshire), 1924–38, layers 2 and 3. 1–5. Backed tools 2 AC2, AC10, AE4, AE1; 6. "Zinken" DC; 7–8. backed tools AB1 and AC23; 9. awl EB2. All in flint.

Fig. 152 ROBIN HOOD'S CAVE (Derbyshire), 1969, layers A to LSB. 1. Scraper CE (Middle Palaeolithic); 2–4. backed tools AB5, AC1, AC2; 5. awl EB2; 6–9. backed tools AB2, AC2, AC20, AD1; 10. multiple class GA (AC5/BA1). All in flint except no. 1 in quartzite.

Fig. 153 ROBIN HOOD'S CAVE (Derbyshire), 1969, layers LSB to USB. 1–2. Backed tools AC2 and AC10; 2. scraper CE; 4–9. backed tools 2 AA1, AB1, AB2, AD5, AD1; 10. burin BB2; 11. multiple class GA (AC2/BA2); 12. core-scraper IF2?; 13. awl IIA; 14. scraper CD (/NB); 15–16. backed tools AC2 and AC23; 17. burin BA2; 18. scraper CD. All in flint except no. 10 in quartzite and no. 13 in bone. Estimated radiocarbon age c. 10,590 B.P. (layer LSB) and c. 10,390 B.P. (layer OB) (both too young?).

Fig. 154 CRESWELL REGION CAVES. 1–10. Backed tools AC1, 3 AC2, AE1, AB9, AC23, AB5, AC7, AE4; 11. multiple class GB (BA4/CB). All in flint.

Fig. 155 NORTHERN WELSH AND PEAK DISTRICT CAVES. 1–6. Backed tools AC1, AB1, AB2, AD1/AC23, AB6, AA1; 7. saw FA; 8. burin BE5; 9–10. backed tools, both AC1. All in flint.

Fig. 156 EAST ANGLIAN OPEN-AIR FINDS. 1–3. Backed tools AB1, AD1, AF4; 4. barbed point IIFA1; 5–6. backed tools AD1, AB6. All in flint except 4 in bone. 5–6 redrawn after Garrod (1926).

Fig. 157 HENGISTBURY HEAD (Hampshire). 1–6. Backed tools AC2, AD1, 2 AF2, AC2, AC1. All in flint.

Fig. 158 HENGISTBURY HEAD (Hampshire), 1957, layers unknown. 1–11. Backed tools AC1, AD1, AE5, AD1?, AE1, AA1, 2 AB2, AB6, 2 AC23?. All in flint. All redrawn after Mace (1959).

Fig. 159 HENGISTBURY HEAD (Hampshire), 1957, layers unknown. 1–4. Burins BE1, BA2, BA4, BA1; 5. multiple class GB (BA4/CB); 6–8. burins BD1, BC1, BE5. All in flint. All redrawn after Mace (1959).

Fig. 160 HENGISTBURY HEAD (Hampshire), 1957, layers unknown. 1–6. Scrapers CA1, 2 CA2, CA3, 2 CE; 7. awl EB1; 8. saw FA; 9. core-axe IF4?. All in flint. All redrawn after Mace (1959).

Fig. 161 HENGISTBURY HEAD (Hampshire), 1968–69, layers B2/1. 1–9. Backed tools 2 AB1, 2 AB2, 3 AC2, AC4, AF3. All in flint.

Fig. 162 HENGISTBURY HEAD (Hampshire), 1968–69, layers B2/1. 1. Backed tool AE1; 2. multiple class GA (AF2/BA1); 3–4. burins BA1(/NA) and BC3; 5–6. scrapers CA2 and CB; 7–8. awls EB1 and EB2. All in flint.

Fig. 163 HENGISTBURY HEAD (Hampshire), 1968–69, layers B2/1 and A2/A1a. 1. Multiple class GC (CD/FB); 2. awl EB2; 3–4. multiple classes GE (EB2/FB) and GB (BD2/EB1); 5–10. backed tools AB1, 2 AB2, 3 AB6/12. All in flint.

Fig. 164 HENGISTBURY HEAD (Hampshire), 1968–69, layers A2/A1a. 1–5. Backed tools 2 AC22a, AC22b, AC22d, AC23?. 6–10. burins BA1, BA2, BA4, BB4(/NA), BD1; 11. scraper CA1. All in flint.

Fig. 165 HENGISTBURY HEAD (Hampshire), 1968–69, layers A2/A1a. 1–6. Scrapers CA4, 2 CA2, CB, CE, CF; 7–8. saws FA and FB. All in flint.

Fig. 166 SOUTHERN ENGLISH OPEN-AIR FINDS. 1–5. Backed tools AB1, AD2, AE4, AE1, AE4; 6. burin BA3; 7. multiple class GC (CB/FA). All in flint.

Kent's Cavern, Gough's Cave, Hoyle's Mouth, Robin Hood's Cave and Mother Grundy's Parlour are the only five Later Upper Palaeolithic sites to have yielded sufficiently large stone tool samples to permit any sort of simple typological analysis or comparative classification on the basis of percentages of the total stone tools in each of the various samples. The percentages for just the stone tool class headings are plotted for comparison in Figures 167 and 168; the reader is referred to Gazetteer II, sites 1, 15, 25, 64, 65, 66 and 70 for their actual counts and percentage calculations for both these class headings and individual tool-types. It should be noted that Figure 168 presents the tool class frequencies for the entire sequence at Mother Grundy's Parlour only; it is a tripartite figure with the Later Upper Palaeolithic tool classes on the left, Mesolithic on the right for comparison, and what I have termed "Meso./L.U.P." in the middle.

In Figure 167, it will be seen that there is reasonable, broad agreement between the four Later Upper Palaeolithic sites represented in that diagram, although it should be noted that the material from Robin Hood's Cave is only a generalized sample put together from four quite separate assemblages excavated by the present author as well as that found by Dawkins and Mello, this grouping making it possible to include the site in the analysis. The striking feature about all four histograms in Figure 167 is that backed tools always outnumber the other tool classes. A more subtle similarity is that the frequencies of burins and scrapers are always approximately equal, with the burins usually just in the majority. Borers/awls appear to have closely similar frequencies at each site, always less than 10%. Differences include the occurrence of "Zinken" only at Gough's Cave, the apparent importance of saws/notches at Kent's Cavern and Hoyle's Mouth, and the apparent importance of retouched flakes/blades at Kent's Cavern and Gough's Cave. Also the multiple-class tools (i.e. burins/scrapers or other such cross-class combinations) only occur at Gough's Cave and Robin Hood's Cave. However, such differences seem of less significance than the similarities outlined above, except perhaps for the important element of retouched flakes/blades at Gough's Cave: 23.8% or 190 out of 799 stone tools.

The histograms from Mother Grundy's Parlour in Figure 168 present a somewhat different picture, the only strong similarity with the Later Upper Palaeolithic histograms in Figure 167 being an obvious predominance of backed tools right through from the Later Upper Palaeolithic to the Mesolithic. Unlike the assemblages classified in Figure 167, throughout the Mother Grundy's Parlour sequence from Later Upper Palaeolithic to Mesolithic the scrapers always outnumber the burins. Also, the scrapers reach their maximum in the Mesolithic assemblage, and this agrees with my observations of many other British Mesolithic tool samples, including that from my excavation of the A layers at Hengistbury Head Site C2 (see Gazetteer II, site 94). Another feature about the Later Upper Palaeolithic at Mother Grundy's Parlour which helps to set it apart from the sites in Figure 167 is the sharp rise in borers/awls, particularly awls, to 23.9% or 28 out of 117 stone tools. As already noted in section V.I above, these awls are morphologically similar to those at the Earlier Mesolithic site of Star Carr, where they also form an important part of the tool-kit (see Clark, 1954). Later in the Mesolithic, awls drop in importance to about the level they held previously during the height of the Later Upper Palaeolithic, i.e. less than 10%; cf. the Mesolithic sample from Post Glacial Zones VI/VII in Figure 168. The predominance of scrapers over burins, and also of awls over scrapers, in the Later Upper Palaeolithic at Mother Grundy's Parlour strongly suggests that this assemblage is a very late or even final one within the British Later Upper Palaeolithic sequence, perhaps belonging to Late Last Glacial Zone III on the basis of the associated pollen evidence. The middle histogram in Figure 168 is for an assemblage belonging at the interface of layers C and B. Its backed and other tool-forms possess features already discussed above in section V.I where I suggested that it represented an intermediate stage between what we may formally call "Later Upper Palaeolithic" and "Mesolithic"; I have accordingly labelled it "Mesolithic/Later Upper Palaeolithic". Its associated pollen evidence suggests that it dates, as indeed one might expect, from somewhere between Late Last Glacial Zone III and Post Glacial Zone V, which period must indeed include the transition (in whatever terms one views it) between the two formally recognized periods. In future, it is to be hoped that a larger assemblage of this "transitional character" will be found, as this particular one from Mother Grundy's Parlour has only 38 stone tools.

At the other end of the scale, just to complete the picture of the chronological range covered by Figures 167–68, the Kent's Cavern Later Upper Palaeolithic tool sample (Fig. 167) has been shown on the basis of reliable radiocarbon evidence to date from Late Last Glacial Zone I.

Figure 169 provides a comparison of individual backed tool-type frequencies in selected Later Upper Palaeolithic and Mesolithic samples. The backed tools from Gough's Cave, Hoyle's Mouth and Robin Hood's Cave are all Later Upper Palaeolithic and, as may be seen, their type frequencies have a fair measure of broad agreement. Among other things, each of these three has over 15% "Creswell points" (AC1–5), Gough's Cave and Robin Hood's Cave being well over, and each has less than 5% "penknife points" (AC23) and less than 6% obliquely blunted points (AB6). In the Later Upper Palaeolithic backed tool sample of Mother Grundy's Parlour, on the other hand, "Creswell points" (AC1–5) are down to under 10%, while "penknife points" (AC23) exceed 20% and obliquely blunted points (AB6) have also increased; there are also nearly 5% "geometric microliths" (AC22), entirely absent at the other three sites, except for less than 1% at Gough's Cave. As may be seen, the other backed tool-types vary less in these four Later Upper Palaeolithic samples, although it should be noted again that the material from Robin Hood's Cave has been grouped together from a number of definitely separate assemblages. The histogram for the sample of "Meso./L.U.P." backed tool-forms from the C/B interface assemblage shows fairly close agreement with the histogram for the underlying Later Upper Palaeolithic at Mother Grundy's Parlour, and in common with the latter it again shows certain differences from the Later Upper Palaeolithic of Gough's Cave, Hoyle's Mouth and Robin Hood's Cave. The "Meso./L.U.P." assemblage has over 20% "penknife points" (AC23), as did the Mother Grundy's "L.U.P." assemblage, but obliquely blunted points (AB6) show a rise to over 15% and "Creswell points" (AC1–5) show a "return" to just over 20%, though such percentage figures should not be taken too seriously when the sample size is so small. In the fully Mesolithic sample at Mother Grundy's Parlour, obliquely blunted points (AB6) remain at over 15%, while "Creswell points" (AC1–5) drop to about 6%, "penknife points" (AC23) drop to about 5%, "geometric microliths" (AC22) remain close to 5%, and small, straight backed blades (AB2) show a very marked increase to about 40%. "Trapeziform backed blades" or "Cheddar points" (AC6–10) persist in a generally small amount throughout these Later Upper Palaeolithic and Mesolithic histograms, showing their strongest frequency at Hoyle's Mouth, 12%, though the sample is too small for this figure to mean much, the total of backed tools being only 25. Shouldered points of one form or another (AD–E) are present in all of these samples except the Mesolithic one, and they attain their greatest frequency at Robin Hood's Cave, just over 15%, in the generalized sample.

Thus from the above lines of evidence there would appear to be at least two sub-groups within the British Later Upper Palaeolithic: an apparently earlier phase as represented at Kent's Cavern, Gough's Cave, Hoyle's Mouth and Robin Hood's Cave, and an apparently later or final phase as represented at at least Mother Grundy's Parlour. The "earlier phase" has comparatively high frequencies of "Creswell points" (AC1–5) amongst its backed tools, whilst the "later phase" has comparatively high frequencies of "penknife points" (AC23). But both of these "phases" have plenty of common ground in their range of tool-types, even if the frequencies are different: they both have "Creswell points" and "penknife points", as well as some "Cheddar points" (AC6–10), shouldered points (AD–E) and certain other backed tool-forms. The backed tools from Kent's Cavern are not included in Figure 169 as there are only 15 of them; however, I consider it significant that of the 15, 6 are "Creswell points" and none whatsoever are "penknife points". As has already been shown a number of times, the Kent's Cavern Later Upper Palaeolithic assemblage definitely dates from Late Last Glacial Zone I; it may therefore be one of the earliest members of the "earlier" or "Creswell point" phase of the Later Upper Palaeolithic. On the basis of associated pollen evidence, this "Creswell point phase" may survive through Zone II into Zone III at Gough's Cave, where there is an increase in the importance of "penknife points". One might expect the "later", "final" or "penknife point" phase to begin as early as Zone II by comparison with the Tjongerian dates in Holland (see Bohmers, 1956, 1960 and 1963; but also see Noten, 1975 on Belgium) if one assumes a close connection between it and the Tjongerian, but the occurrence at Mother Grundy's Parlour seems to date from Late Last Glacial Zone III. This "penknife point phase" is also characterized by much higher frequencies of awls of EB2 form, than ever exist in the "Creswell point phase". At Mother Grundy's Parlour, this "penknife point phase" of the Later Upper Palaeolithic is then apparently succeeded by a "transitional phase" which I have called "Mesolithic/ Later Upper Palaeolithic" and which is characterized (so far as our present limited evidence goes) by comparatively high frequencies amongst its backed tools of obliquely blunted points (AB6), "Creswell points" (AC1–5) and "penknife points" (AC23). This "transitional phase" dates from some period in the earlier part of the Post Glacial at Mother Grundy's Parlour on the basis of associated pollen evidence, but from the faunal evidence there is perhaps reason to suppose that it may have begun in Late Last Glacial Zone III.

Thus, over the whole of the British Later Upper Palaeolithic there would appear to be at least three phases:

1. a "Creswell point phase" ranging in time from the assumed beginning of the Later Upper Palaeolithic in Zone I to some time in Zone III;
2. a "penknife point phase" ranging in time from an assumed beginning in Zone II to perhaps Zone IV; and
3. a "transitional to Mesolithic phase" ranging in time from an assumed beginning in Zone III to perhaps Zone V (or perhaps beginning in Zone II if one considers the High Furlong barbed points typologically "Mesolithic"; see Clark, 1972; Hallam, Edwards, Barnes and Stuart, 1973).

If we had more and better statistically and stratigraphically viable samples, we would certainly hope to say more than this. Meanwhile, there seems little or no support for Bohmers' (1956) concept of a separate "Cheddarian culture"; rather, the available evidence, scanty though it may be, suggests different overlapping "phases", as outlined immediately above, within the same general British Later Upper Palaeolithic frame of reference, and between that British Later Upper

Palaeolithic and the British Mesolithic. The observed variations may well chiefly represent time trends, but in the present state of the evidence we are simply not in a position yet to test or comment on alternatives such as functional variation or stylistic differences. The industries studied are certainly not all contemporary: of that we can at least be certain.

Various simple metrical analyses also bring out these overlaps and differences between the various assemblages. These analyses, conducted by the present author as simple first tests on selected attributes of certain well-defined tool classes (burins, scrapers and backed tools), are now considered in the following paragraphs. These studies will hopefully be taken further in future work.

Figure 170 presents the results of measurements of burin facet breadths in selected Later Upper Palaeolithic and Mesolithic samples (see. pp. 23–24 and Fig. 3 for technique and location of this measurement; also p. 153 for discussion of significance). The two Gough's Cave burin samples are both Later Upper Palaeolithic; those from spits 11–6 (which might possibly be "Mesolithic/Later Upper Palaeolithic" in part) being from above those from spits 25–12. The Hengistbury Head, Brigham Hill and Flixton Site 1 burin samples are all Mesolithic. Table 50 gives the information on which Figure 170 is based. As may be seen in Figure 170, there is very close agreement between the two Gough's Cave samples, the modal burin facet breadth in both being at 2 mm. The means for these two samples are also in close agreement: 2.2 mm. for the lower burin sample (spits 25–12) and 2.3 mm. for the upper sample (spits 11–6). A mode and mean round about 2 mm. may probably be taken as "normal" for the breadth of burin facets belonging to what I have termed the "earlier" or "Creswell point phase" of the Later Upper Palaeolithic, as I have found similar means in smaller, Later Upper Palaeolithic burin samples which can with some reason be assigned to that particular "phase"; e.g. Flint Jack's Cave: 1.6 ± 0.6 mm. (4 burins), King Arthur's Cave: 2.1 ± 0.7 (4 burins), Robin Hood's Cave: 2.5 ± 1.8 mm. (5 burins), and Hengistbury Head Site C2 (B layers): 2.8 ± 1.0 mm. (5 burins). The maximum breadth of burin facets in what I have termed the "later", "final" or "penknife point phase" of the Later Upper Palaeolithic as represented in the lowest assemblage at Mother Grundy's Parlour is 5.1 ± 2.4 mm. (9 burins), which is certainly of quite a different order. In other words, the burins of the Later Upper Palaeolithic "penknife point phase" would seem to be more robust than those of the earlier "Creswell point phase". As may be seen in Figure 170, this increase in robustness continues into the full Mesolithic as represented at Flixton Site 1, where the mode is at 4 mm. and the mean is close-by at 4.2 mm., and as represented at Hengistbury Head Site C2 (A layers), where the range has two peaks at 4 and 7 mm., with the mean between them at 5.2 mm. The material from Brigham Hill has previously been described as "Creswellian" (Manby, 1966), but as shown above in section V.H. of this chapter, I see no reason to consider it other than Early Mesolithic, and in this I am in complete agreement with Radley (1969).

Its burin facet breadth mode agrees remarkably well with that of Flixton Site 1 (see Fig. 170), and the mean is 5.6 mm., actually rather greater than that at Flixton Site 1 (4.2 mm.). Similarly, the material from Hengistbury Head Site C1 has previously been described as "Late Upper Palaeolithic" (Mace, 1959), but as shown above in section V.L. of this chapter, I see no reason to consider most of it other than Mesolithic, including all of the burins. The mode of the burin facet breadth at Hengistbury Head Site C1 is 5 mm. (see Fig. 170) and the mean is 6.2 mm., even greater than the mean of my own Mesolithic burin sample (5.2 mm.) from Hengistbury Head Site C2 (A layers), and certainly much greater than the already cited mean (2.8 mm.) of my Later Upper Palaeolithic burin sample from the same site (B layers), as well as that of the Gough's Cave samples (2.2 and 2.3 mm.). In summary, even allowing for the smallness of some of the samples, there seems to be a clear trend of increasing robustness in burin manufacture from fairly gracile burins in the Later Upper Palaeolithic to quite robust ones in the Mesolithic. One might be tempted to say this is a swing back to the robust burins of the Earlier Upper Palaeolithic, but there is probably yet no certain reason to suppose the continuity such a statement would imply. The types are generally different, for example the "burin busqué" form certainly does not return in the British Mesolithic. The presence of predominantly gracile burins in the Later Upper Palaeolithic might imply greater concern with delicate bone and antler work in that stage than in either the Earlier Upper Palaeolithic or the Mesolithic, but such would be hard to prove, as all three hunting groups apparently produced some delicate bone work.

Figure 171 presents the results of measurements of the angle of scraper retouch in selected Later Upper Palaeolithic and Mesolithic samples (see pp. 23–24 and Fig. 3 for technique and location of this measurement; also p. 153 for discussion of significance). The Gough's Cave sample is mostly if not entirely Later Upper Palaeolithic and is grouped from all of the relevant spits (26–6) in order to obtain a reasonably sized sample. The Hengistbury Head and Brigham Hill samples are Mesolithic. Table 51 gives the information on which Figure 171 is based. As may be seen in Figure 171, the mode of the angle of scraper retouch in the Later Upper Palaeolithic at Gough's Cave is 75°; its mean is 72.8°. The three selected Mesolithic samples again include Brigham Hill and Hengistbury Head Site C1, following my own interpretation of them. Their modes and means are clearly lower. The relevant material from the A layers in my excavation at Hengistbury Head Site C2 is definitely Mesolithic, and as in previous tests the three assemblages go well enough together and maintain their separation from the Later Upper Palaeolithic at Gough's Cave: the mean of the angle of scraper retouch at Brigham Hill is 62.6°, that at Hengistbury Site C1 is 61.7°, and that at Hengistbury Site C2 (A layers) is 62.7°. This limited experiment thus suggests a shift from higher angled scrapers in the Later Upper Palaeolithic to lower angled scrapers in the Mesolithic. This shift may simply be due to changes in scraper typology, there being more end-of-blade scrapers at Gough's Cave, and more end-of-flake scrapers in the three Mesolithic samples. Such a typological change would probably result in a change in the actual range of angles at which scrapers were held by their makers when in use, but the exact angles of holding might well be hard to determine, although an application of "micro-wear" and related studies might bring such information to light. Of course, the higher angle of scraper retouch apparent in the Later Upper Palaeolithic might also be due to attempts at repeatedly working the same scrapers back in order to re-use them, the Post Glacial Mesolithic hunters perhaps having a greater tendency to toss away unused or only slightly used tools; this subtle possibility might also be checked by "micro-wear" studies. But whatever the exact reason, here we see another criterion of metrical difference between the Later Upper Palaeolithic and the Mesolithic in Britain.

Figure 172 presents the results of measurements of the absolute maximum breadth of backed tools in selected Later Upper Palaeolithic and Mesolithic samples (see pp. 23–24 and Fig. 3 for technique and location of this measurement). Maximum breadth is a good parameter for judging absolute size of backed tools. Length and thickness of backed tools are less satisfactory parameters, length particularly because it is so often affected by a breakage. Breadth, on the other hand, is most frequently a measure of how much the blank was reduced by backing; it is also a measure of when a larger backed tool-form grades into a smaller, "microlithic" form, although, of course, it would be meaningless to propose some precise point of division between the two. But the mean breadth and a frequency graph for a sample of backed tool-forms will often illustrate whether the sample is predominantly "microlithic" or larger. Breadth of backed tool-forms has already been considered in detail further afield by McBurney, for example, in his study of the Upper Palaeolithic and "Epi-palaeolithic" backed tools of the Haua Fteah succession in north-eastern Africa (McBurney, 1967). He also considered their length and thickness to some extent, but it would seem to me that primarily their breadth or "width" was the most informative parameter for demonstrating the widely recognized reduction in size with time.

The information on which Figure 172 is based is given in Table 52. The two samples from Gough's Cave are again Later Upper Palaeolithic, and they both belong mostly to what I have termed the "Creswell point phase", although that from spits 11–6 is younger than that from spits 25–12 and may be in part more "transitional to Mesolithic". The sample from layers B (layers LB and SB, including by correlation, Armstrong's "base and lower middle zones") at Mother Grundy's Parlour is also Later Upper Palaeolithic, but it belongs instead to what I have termed the "penknife point phase", a phase still within the Later Upper Palaeolithic but probably younger than the "Creswell point phase". The sample from layers C–D (including by similar correlation, Armstrong's "middle and upper middle zones") at Mother Grundy's Parlour is a generalized sample representing what has already been shown to be a "transitional to Mesolithic" assemblage followed by a "full Mesolithic" assemblage; I have grouped these two for the purposes of this experiment simply to increase the sample size. The samples from Hengistbury Head Site C1, Brigham Hill and Flixton Site 1 are pure Mesolithic, according to my interpetation of them, except that the sample from Hengistbury Site C1 includes 9 backed tool-forms in the total of 113 which I think might belong to the Later Upper Palaeolithic instead, having been mixed by Mace (1959) with a predominantly Mesolithic assemblage (see Gazetteer II).

As will be seen in Figure 172, there is an apparent trend of reduction in absolute breadth of backed tools from peaks at 11–15 mm. in the Later Upper Palaeolithic of Gough's Cave to peaks at 9–10 mm. in the Mesolithic of Flixton Site 1, Brigham Hill and Hengistbury Head Site C1; the Mother Grundy's Parlour samples have peaks in between and lower at 8–12 mm. This test illustrates, then, as one would expect, that in Britain as elsewhere there is a movement towards increasingly "microlithic" proportions from the Later Upper Palaeolithic through to the Mesolithic. The full Later Upper Palaeolithic (Gough's Cave) and the full Mesolithic (Flixton 1, Brigham, Hengistbury C1) samples have quite different distributions on these graphs, while the scanty evidence from Mother Grundy's Parlour provides some sort of link between them. It will be seen that this visual impression supplied by the modes or peaks in the frequency diagrams finds a proper basis of support in the means and standard deviations for the same samples, which information appears in Table 52.

This apparent trend of reduction in backed tool breadth from Later Upper Palaeolithic to Mesolithic is demonstrated in a different manner in Figure 173. Here the means and standard deviations for backed tool breadths in a number of selected, radiocarbon "dated" assemblages are plotted against their average radiocarbon age in each case. In the sample key, after the names of the relevant sites, the numbers in parentheses indicate first the number of radiocarbon age estimates which have been averaged together, and second, the number of backed tools employed in the calculation of the breadth figures. The actual radiocarbon age estimates may be found in Table 4; their varying reliability is discussed in the section on radiocarbon measurements (pp. 77–84). The actual backed tool breadth means and standard deviations are given in Table 53, together with the associated mean ages and standard deviations of the averaged radiocarbon age estimates. A linear regression has been calculated on the basis of the intersecting points of mean breadth and age for the various samples. The slope of this calculated regression clearly shows the direction and approximate magnitude of the trend of backed tool breadth reduction, directly indicative, as I have said, of reduction in backed tool size, from breadths of over 15 mm. in the earlier part of the Later Upper Palaeolithic to less than 10 mm. in the Mesolithic, the interface or "transition" between the two periods being at about 13 to 9 mm. (see Fig. 173). In addition to this calculated linear regression, I have plotted on Figure

173 an "observed regression" by connecting the age/breadth means in ascending order with a dashed line, plotting that broken line midway between samples 4 and 6 as their apparent ages are parallel. This observed regression is obviously quite zig-zag in shape, but it nevertheless still suggests a trend of reduction with time in absolute breadth of backed tools. With an increase in the number of reliable age estimates and an increase in the quantity of backed tools in each sample employed, this zig-zag shape of the observed regression would probably be reduced. However, in terms of the backed tools themselves, we are dealing with what were, presumably, primarily humanly induced changes in backed tool size; being humanly induced, they would not necessarily be constant changes falling smoothly along a straight trajectory in time. If known environmental changes had any direct or indirect effect on the size of backed tools produced, as has often been suggested, then this too would influence the "shape" of the trajectory. In fact, whether there be any correlation of significance or not between backed tool size and environmental conditions, that is, some sort of ecological interrelationship, it is interesting that the observed regression on Figure 173 does certainly suggest generally higher values for backed tool breadths during the colder climatic phases, and lower values during the warmer phases. This may have something to do with the changes in the preferred and available food animals which existed at different times between about 14,000 and 6,000 years B.P., such changes perhaps demanding alterations in hunting techniques, which would be reflected in the surviving material culture by just such changes as reductions in tool size. However, really thorough exploration of these lines of speculation lies beyond the scope of this present enquiry; many of the hypotheses would be very hard to test and the experimental work would certainly consume more time (and money) than is at present available, although some day such an approach would probably be well worth developing.

Despite the many potential sources of error in the data presented in Figure 173 and Table 53, where for example some radiocarbon "dates" are obviously too young and some backed tool samples much too small, I am still inclined to think that there may be a certain predictive value in the calculated linear regression shown in Figure 173. For example, it is known from the pollen evidence that the Later Upper Palaeolithic assemblage from Hengistbury Head Site C2 (B layers) belongs approximately to the boundary between Late Last Glacial Zones I and II, which for a southern English site would be about 12,000 radiocarbon years ago; now, the mean and standard deviation for backed tool breadth in that assemblage is 16.6 ± 6.3 mm. (15 backed tools) and when that mean is allowed to intersect with the calculated regression line on Figure 173, the "age reading" is about 12,500 years ago, which is not a bad fit at all. Similarly, if one places the mean of the backed tool breadths (13.9 mm., see Table 52) for Gough's Cave spits 25–12 on the regression line, then one obtains an apparent age of about 11,000 years B.P.; the mean of 12.9 mm. for Gough's Cave spits 11–6 yields in the same way an apparent age of about 10,500 years B.P. There is nothing unreasonable about either of these readings if one consults the associated pollen evidence. However, one must not take this line of thought too far as there are still many limitations involved with it. It is certainly very suggestive evidence, but it would require much careful study and refinement, especially by the addition of new material and subtraction of less reliable old material, before it could possibly be employed as any sort of accurate check on age for any given sample from the British Later Upper Palaeolithic and Mesolithic.

Indeed, prolonged testing might eventually reveal a more sensitive measurement or metrical index than backed tool breadth, to plot against time in such a graph. But it may not be too fanciful to envisage the eventual development along these lines of a "calibration curve", in principle not unlike the bristlecone pine calibration curve which is currently so fashionable in the writings of later-period prehistorians, on whose field of study it has undoubtedly had a remarkable effect. What I have in mind would be a less powerful weapon, but it could nevetheless be very useful to the future researcher on British Upper Palaeolithic and Mesolithic material. With some personal regret, however, I must here confine my remarks again to indicating a possibility which the usual exigencies of completing a research project and writing it up within a given period of time have left me unable properly to explore.

Figure 174 presents a comparison of selected Later Upper Palaeolithic and Mesolithic core sample means and standard deviations for maximum length and breadth of flint cores. The numbers in parentheses after the site names in the sample key give the total number of cores measured in each sample. The reader is referred to Figure 3 for the positions of these measurements. It may be seen at a glance in Figure 174 that the Later Upper Palaeolithic cores are generally much larger than those of the Mesolithic, as indeed one might expect. Also, it may be significant that the cores from Gough's Cave spits 11–9, which I have here placed under the label "Meso./L.U.P.", fall in the middle of the Mesolithic cluster, as indeed do those from Mother Grundy's Parlour layers B, which are also lumped under "Meso./L.U.P." for the purposes of this diagram. Gough's Cave spits 23–12, on the other hand, falls more or less midway between the higher Later Upper Palaeolithic cluster and the lower Mesolithic group. Core sizes are discussed above, as well, under section V.L., pages 177 to 178.

I can only conclude this section on the statistical and metrical analysis of my Later Upper Palaeolithic material in the same terms as in the corresponding section of the previous chapter. Essentially, the samples are too few, often too small and of too dubious validity for such treatment, which I find disappointing, though eventually future workers may obtain better material. While I have tried to employ objective methods of study and comparison with a sound statistical basis, I have not felt it worthwhile to proceed very far with such low-quality samples, and I have devoted a higher proportion of my time than originally envisaged to the aspects of my work described in earlier chapters. The use of simple statistics has no magical effect on the tentative conclusions drawn above and summarized below. In fact, these pages are best considered simply as guidelines for future research, research which would hopefully include larger and more reliable samples from Britain as well as continental Europe.

N. CONCLUSIONS

It is, however, always worth making what one legitimately can of any corpus of material, and one does not want to waste too much space lamenting its shortcomings. As regards so-called "type-fossils", it seems reasonable to say that the British Later Upper Palaeolithic material is characterized by numerous generally broad (in British terms) backed tools including notably the "Creswell point" (AC1–5) forms and shouldered point (AD–E) forms, numerous generally gracile burins and pretty numerous steep-angled end-of-blank scrapers, rather scarce multiple-class tools of various kinds and rare antler and bone "harpoons" or barbed points. In the five largest tool samples, Kent's Cavern, Gough's Cave, Hoyle's Mouth, Robin Hood's Cave and Mother Grundy's Parlour (layers B), backed tools are the predominant tool class. In the first four of those assemblages, burins and scrapers occur in about equal frequencies, whilst in the last-named assemblage, scrapers greatly outnumber the burins. The first four assemblages belong to what I have termed the "Creswell point phase" of the Later Upper Palaeolithic and are evidently earlier than that from Mother Grundy's Parlour (layers B), which belongs to what I have termed the "penknife point phase" as it has the greatest proportion of "penknife points" (AC23) and possesses certain other characteristic features such as a high proportion of awls of the EB2 form, similar to those from the Early Mesolithic at Star Carr. At Mother Grundy's Parlour, the "penknife point phase" of the Later Upper Palaeolithic is followed by a closely related "transitional to Mesolithic phase", which I have tentatively labelled "Mesolithic/Later Upper Palaeolithic". It is thought that the "Creswell point phase" is the main and most prolific one of the British Later Upper Palaeolithic, belonging principally to Late Last Glacial Zones I and II, whilst the "penknife point phase" represents the end of the British Later Upper Palaeolithic, lasting into Late Last Glacial Zone III, if indeed not into Early Post Glacial Zone IV.

From the Later Upper Palaeolithic to the Mesolithic in Britain, there is metrically determined evidence for an apparent decrease in the maximum breadth of backed tools (indicating reduction in their absolute size), an apparent increase in the breadth of burin facets at their business end, and an apparent decrease in the angle of scraper retouch. These changes may reflect functional and/or stylistic differences, but this cannot yet be decided. The backed tools are the most important tool class in the Later Upper Palaeolithic from the diagnostic point of view, and they include notably: convex backed blades (AB1), straight backed blades (AB2), a few obliquely blunted points (AB6), the sub-triangular "Creswell points" (AC1–5), a few trapeziform "Cheddar points" (AC6–10), "penknife points" (AC23), shouldered points (mainly AD forms, but some AE forms) and rare tanged points (AF). There is a striking absence of leaf-points of any form whatsoever in the whole of the Later Upper Palaeolithic. Nor are there any other of the characteristic Earlier Upper Palaeolithic tool-forms such as "burins busqués" (BD4), nosed scrapers (CA5) and keeled scrapers (CA6). In fact, there appears to be a complete typological separation, in terms of diagnostic tool-types, between the Earlier and the Later Upper Palaeolithic, the Later Upper Palaeolithic having more in common with the subsequent Mesolithic, particularly through its intervening "Mesolithic/Later Upper Palaeolithic" stage.

If we glance finally at the classic cultural terminology of continental Europe, although the burins, scrapers and "cave harpoons" (i.e. those of Kent's Cavern and Aveline's Hole) of the British Later Upper Palaeolithic are perhaps generally "Magdalenian" in character, the backed tools range from showing certain "Magdalenian" and "Ahrensburgian" characteristics to being generally closer to the "Hamburgian" and "Tjongerian", while preserving also those features of their own which were typically "Creswellian" even in the view of Garrod (1926), though "typical Creswellian" has since rather changed or lost its meaning! In conclusion, I still think it is best for the moment to refer to the relevant British materials as Later Upper Palaeolithic rather than by some value-laden name such as "Magdalenian", "Creswellian", "Cheddarian", "Hamburgian" or "Tjongerian", and in any case, it is hardly for me to pronounce upon more detailed continental connections before I have had a chance to take a longer and more critical look at the relevant continental material than I was able to do whilst digging at La grotte du Coléoptère in Belgium in 1974 with my colleague Monsieur M.C. Dewez, although I did manage to see and discuss quite a lot thanks to much help from him, then and again in 1976.

Within Britain, there seems absolutely no substantial evidence whatsoever to support Bohmers's (1956) recommendation for the recognition of a separate "Cheddarian culture". It would be preferable to say that most or all of the British material (as well as a fair proportion of the Belgian and Dutch material) is "Creswellian" in Garrod's original sense, although her term has since been grossly misused by many other workers, and I have said what I can about divisions within what I regard as essentially a single framework.

According to the available radiocarbon evidence, the time-span of the British Later Upper Palaeolithic certainly gives it contemporaneity with the latter half of the Magdalenian, with the Hamburgian, Ahrensburgian and other "Stielspitzen-Gruppen", and with the Tjongerian and other "Federmesser-Gruppen". Indeed, the Tjongerian might simply be an open-air facies of the Belgian Creswellian (M.C. Dewez, personal communication; cf. Noten, 1975 and Danthine, 1961).

It is now to be hoped that future field-work and laboratory analyses will yield sufficient new information both in Britain and in continental Europe to clarify more firmly the proper relationship of the British Later Upper Palaeolithic to the continental sequence. Work in the critical region of eastern England, northern France, western Germany, Belgium and Holland is happily already well underway (J. Wymer, F. Bordes, J. Hahn, W. Taute, M.C. Dewez, F. van Noten and E. O'Donoghue, personal communications; also see Appendix 4 by Monsieur Dewez in this present book, as well as the following references: Rose, Turner and Wymer, 1973; Wymer, 1971; on East Anglia; Leroi-Gourhan and Brézillon, 1972; Agache, 1971; Schmider, 1971; on northern France; Hahn, Müller-Beck and Taute, 1973; Bosinski and Hahn, 1972; on western Germany; Noten, 1975; Dewez, Brabant *et al.*, 1974; on Belgium; Paddayya, 1973; on Holland), but much further work is still required.

CHAPTER VI

GENERAL CONCLUSIONS AND SUGGESTIONS FOR FUTURE RESEARCH

Based on the foregoing presentation of available evidence obtained by previous workers and by the present author for the nature of the chronology, ecology and cultural material of the British Upper Palaeolithic, the following conclusions are drawn:

1. *Chronology*: The British Upper Palaeolithic is divided between two basic groups, an Earlier and a Later one, which are quite separate when seen in their proper stratigraphic context. The Earlier one, which I simply call Earlier Upper Palaeolithic, has definite dates in the region of 38,000 to 28,000 years B.P., on the basis of available radiocarbon evidence, and may survive until as late as about 18,000 years B.P. Whether the earliest dates fix its first appearance in Britain is not known. There is then an apparent gap in the British Upper Palaeolithic sequence correlated with the maximum ice advances of the Full Last Glacial of 20,000 to 15,000 radiocarbon years B.P. The Later Upper Palaeolithic then follows, with various dates between about 14,500 and 10,000 radiocarbon years B.P. In the typological sense, the Later Upper Palaeolithic or its influence may be said to survive later than this into what may formally be called the Mesolithic period.
2. *Ecology*: The Earlier Upper Palaeolithic is associated with a Middle to Full Last Glacial Sub-Arctic to Arctic environment. The flora is that of a treeless steppe-tundra to tundra, with scattered shrubby forms of willow and birch. The fauna suggests similar ecological conditions and its land mammals include notably hyena, mammoth, woolly rhinoceros, wild horse, giant deer and reindeer, the main food animals being wild horse (*Equus przewalskii*) and reindeer (*Rangifer tarandus*). The Later Upper Palaeolithic is associated with a Late Last Glacial environment varying between Boreal and Sub-Arctic. The flora is that of a nearly treeless steppe-tundra, with shrubby forms of juniper, willow and birch, and, particularly during Zone II, occasional coppices of tree birches and just possibly some pine. The fauna suggests similar ecological conditions and its land mammals include notably brown bear, woolly rhinoceros, wild horse, red deer, giant deer and reindeer, the main food animals again being wild horse (*Equus przewalskii*) and reindeer (*Rangifer tarandus*), with perhaps some sort of "special relationship" evolving between man and his "prey", horse and reindeer. The maximum cold of the British Last Glacial definitely occurs during the

latter part in the Full Last Glacial, and this approximately coincides locally with maximum ice advances and generally with a maximum fall in world-wide sea level to below −100 metres. The Full Last Glacial has an Arctic to High Arctic environment with little flora and fauna, apparently not sufficient to support a human population, at least throughout the year. The British shoreline for both the Earlier and the Later Upper Palaeolithic hunters would be generally at about −50 metres, providing a land-bridge with continental Europe, but not with Ireland. As for human population sizes, one might suggest that the Earlier Upper Palaeolithic level was of the order of 100–500 people, whilst that of the Later Upper Palaeolithic saw an increase from 500 to possibly 5,000.

3. *Cultural Material*: The artifacts of the British Upper Palaeolithic are divided typologically between the two basic, stratigraphic groups, the Earlier Upper Palaeolithic and the Later Upper Palaeolithic. The finds of both these groups in Britain are entirely confined so far to England and Wales, but those of the Later Upper Palaeolithic extend slightly further north. These finds are in both cases primarily concentrated at cave-sites along the contact zone between highlands and lowlands, but there are more open-air sites now known for the Later Upper Palaeolithic. This distribution pattern may partly reflect modern collecting bias, but it seems likely also to reflect ancient hunting practice: the exploitation of two or more viable ecological zones at the same time for greater economic yield and balance. Also, in the Later Upper Palaeolithic there may have evolved a sort of socio-economic/ecological "buffer zone" between the northern and southern sites, acting presumably as a food (i.e. game) reservoir. The surviving artifacts of these hunters, both Earlier and Later, are predominantly in flint, but there are also some in various cherts, adinole, quartzite, etc. as other inorganic raw materials, and some in antler, ivory, teeth, bone and shell as organic raw materials. The use of cherts and ivory is more common in the Earlier Upper Palaeolithic, whilst the use of antler is more common in the Later Upper Palaeolithic. By ethnographic analogy, many other organic materials may have originally been employed, particularly fur, skin, gut and various plant materials, though no traces survive. The extant tool-forms of the Earlier Upper Palaeolithic are characterized by generally robust burins including the "burin busqué" form; by various stout scrapers including the nosed form; by various unifacial leaf-points; and by rare bifacial leaf-points. In the two largest assemblages, scrapers are the predominant tool class. The "cultural affinity" of the Earlier Upper Palaeolithic is probably mainly with the continental Aurignacian, but as the leaf-points and various other features distinguish it from the latter, it is thought best to use for the present the suggested name of the former for the relevant British material. Some connection with such continental "leaf-point cultures" as the "Altmühlgruppe", the Jerzmanovician and the Solutrean may well be possible; chronology and access to Britain would allow it. Useful sub-division of the Earlier Upper Palaeolithic in Britain is not really possible in the present state of the evidence, although it

might be added that Kent's Cavern and Paviland Cave could just possibly represent different sub-groups. The extant tool-forms of the Later Upper Palaeolithic are characterized by relatively broad (in local terms) backed tools including abundant sub-triangular backed blades or "Creswell points", rare trapeziform backed blades or "Cheddar points", fairly common "penknife points" and shouldered points; by various generally gracile burins; by various comparatively steep-angled end-of-blank scrapers; by various multiple-class tools; and by rare antler and bone "harpoons" or barbed points and other occasional bone work. In the five largest assemblages, studied in some detail, backed tools are the predominant tool class. Four of these assemblages belong to what I have termed the "Creswell point phase" of the Later Upper Palaeolithic and are generally earlier than the fifth assemblage, which belongs to what I have termed the "penknife point phase". The "Creswell point phase" has burins and scrapers occurring in about equal frequencies and is particularly characterized by high frequencies of "Creswell points". The younger "penknife point phase" has more scrapers than burins, possesses particularly a high frequency of awls somewhat similar to those in the British Early Mesolithic, and is particularly characterized by a high frequency of "penknife points". This "penknife point phase" gives way to an as yet poorly defined "transitional to Mesolithic phase", which I have labelled "Mesolithic/Later Upper Palaeolithic". It is thought that the "Creswell point phase" represents the major part of the British Later Upper Palaeolithic. The "cultural affinities" of the Later Upper Palaeolithic very likely lie mainly with the continental "Federmesser-Gruppen" and partly with the "Stielspitzen-Gruppen" and the Magdalenian of the European Late Last Glacial, but this question calls for further research and is here left open. As the British material certainly has its own distinctive features, it is thought best for the present simply to use for it the term Later Upper Palaeolithic. The term "Creswellian" is no longer satisfactory, due to its frequent misuse, and the term "Cheddarian" is considered completely meaningless. Finally, the available British Upper Palaeolithic cultural material, both Earlier and Later, appears generally unsuitable for elaborate statistical treatment. It is to be hoped that future work will yield sufficient new information to improve this situation and eventually to clarify the relationship of both main groups in the British Upper Palaeolithic to the continental European sequence.

As regards suggestions for future research on the British Upper Palaeolithic, the following are but a few of the more prominent needs. Many more radiocarbon age estimates are required, particularly for the Earlier Upper Palaeolithic. New and more detailed ecological information would certainly be useful, as would economic information on hunting and gathering practices; more detailed analyses of available faunal collections than I have attempted would certainly be worthwhile and are indeed urgently required. Simulation of British Upper Palaeolithic ecosystems would certainly seem worth attempting. More detailed artifact attribute analyses than those carried out so far by the present author might be conducted on some at least of the artifact samples, such as those from Kent's Cavern, Gough's Cave, Paviland Cave, Hoyle's Mouth, Mother Grundy's Parlour, Robin Hood's Cave and Hengistbury Head. Intensive field surveys for new open-air sites should be begun as soon as posible. For chronological and ecological control, renewed excavations would be quite worthwhile at Kent's Cavern (Great Chamber), Badger Hole (blocked east entrance), Cae Gwyn Cave (west entrance), Mother Grundy's Parlour (south-western part of entrance platform), Robin Hood's Cave (between entrances and road) and Hengistbury Head (peat-bog in depression north of Site C2). New and undisturbed cave-sites should be urgently sought, as an encouraging number of presumably virgin and worthwhile ones have recently come to light, e.g. Rectory Cave (Torbryan, Devonshire) and Ogof-y-Pebyll (Bridgend, Glamorganshire), both of which possess very promising entrance platforms. In short, the British Upper Palaeolithic is ripe for further research, and, indeed, I certainly hope to take part in some of that work myself. Finally, future research projects should be extended to the once linked parts of continental Europe, particularly northern France, Belgium and Holland. "Proto-Solutrean/Aurignacian" sites quite similar to the British Earlier Upper Palaeolithic are already known in Belgium (Eloy, 1956; Smith, 1966; Otte, 1974b and personal communications), and "Creswellian" sites quite similar to the British Later Upper Palaeolithic are now known in northern France, Belgium and Holland (Schwabedissen, 1954; Bohmers, 1956, 1960 and 1963; Danthine, 1961; Noten, 1975; Prof. F. Bordes, M.C. Dewez and F. van Noten, personal communications). It is not even impossible that underwater archaeology might be employed for the intervening, now flooded, regions (e.g. see Bonifay, 1970 on underwater caves off southern France). Meanwhile, I hope that this book will at least have succeeded in bringing a degree of order out of chaos, establishing a sound basis for future work, and suggesting a number of fresh starting points which will prove profitable.

APPENDIX 1

Charred Wood Fragments from Kent's Cavern, 1866: Later Upper Palaeolithic Ecology and Ethnobotany

by

JOHN B. CAMPBELL

(Department of Behavioural Sciences, James Cook University, Queensland, Australia)

with identifications by

J. D. BRAZIER

(Department of the Environment, Building Research Establishment, Princes Risborough Laboratory, Princes Risborough, Aylesbury)

In the winter of 1866–67 during the course of his long-term systematic excavations at Kent's Cavern (Torquay, Devonshire), William Pengelly discovered and recorded a number of concentrations of charred and/or burnt materials in the Vestibule. These dark horizons occurred within the first foot spit (first 30 cm.) of the stony "cave earth" (i.e. thermoclastic scree to judge from the surviving sections) *beneath* an undisturbed one foot (30 cm.) layer of "granular stalagmite". He labelled these dark horizons the "Black Band", presumably for convenience, as he also quite clearly realized and so recorded his view that these were in fact a series of superimposed hearths which had accumulated over a number of centuries and which were due entirely to the activities of prehistoric man (Pengelly, 1865–80, Vol. I, pp. 287–359; Pengelly, 1884). He described the "Black Band" as a variable three to nine inches (c. 8–23 cm.) in thickness, sometimes separating into at least two distinguishable bands, and he recorded in great detail its horizontal distribution in his grid system.

New sections and plans based on a careful analysis of Pengelly's Diary (Pengelly, 1865–80) and of the numbers written by him on his still extant finds are presented elsewhere in this book (see Figs. 4 to 9, especially 6 and 9). The artifacts and faunal remains found by him in association with the "Black Band", or more specifically within, just above and just below it, demonstrate that it clearly belongs to the Later Upper Palaeolithic and the Late Last Glacial. Radiocarbon age estimates on bones from within and near the "Black Band" suggest time ranges of 14,275 ± 120 years B.P. (GrN–6203) and 12,180 ± 100 years B.P. (GrN–6204), respectively (also see pp. 41–42). In other words, the "Black Band" is not only Late Last Glacial in age, but it probably belongs entirely to Zone I.

The charred wood fragments with which we are concerned are from Pengelly's find-number 1847. They were found 4th December, 1866. Their position is given as the second series datum, 59th parallel, first yard left, first foot level of "cave earth/Black Band" (Pengelly, 1865–80, Vol. I, p. 294; also see pp. 37–42 in this present book on his excavation method and the site's stratigraphy). Further, they were found in the same prism (i.e. same parallel, yard and foot) as the famous eyed bone needle, at the westernmost edge of the "Black Band" (i.e. at a point furthest from any known disturbance by either badgers or earlier excavators).

According to Dr. J.D. Brazier, who has kindly identified these specimens, there are three timbers represented in this series from find-number 1847, namely:

'*Quercus* sp. OAK. This is somewhat unusual; it is of very vigorous growth with a structure resembling an evergreen (currently southern European) oak. However, the fragment is very small and there must also be the possibility of its being cut from a very young growth of a northern European "white" (more temperate) oak, which can sometimes have a structure somewhat like the sample.

Ulmus sp. ELM. There are two fragments. One is from a very young twig or stem; the other is apparently of fairly mature and comparatively slow growth.

?*Rhamnus catharticus*. Possibly COMMON BUCKTHORN. This is a very young, seven-year old stem. The structure is very distinctive and matches that of common buckthorn, but we cannot rule out entirely the possibility of its being another woody plant which reaches a small size only and has a similar structure. *Hippophaë* and *Empetrum* can be eliminated, but *Helianthemum* is not represented in our collection; however, from the little about it in Metcalfe and Chalk (1950), this too seems unlikely' (J.D. Brazier, personal communications).

The problem posed by the fact that these Temperate type wood fragments were found in direct association with other items of a more Arctic to Boreal character is indeed quite intriguing, both from palaeoecological and palaeoethnobotanical points of view. It is, perhaps, pertinent that today each of the three timbers identified occurs to quite northerly latitudes, *Rhamnus catharticus* to 62° North, *Quercus* to 63° North and *Ulmus* to 67° North (i.e. over 66°32′N., or north of the Arctic Circle), admittedly all on the western coast of Norway. But this may provide some analogy with ecological conditions in or near south-western Britain during the Later Upper Palaeolithic and the Late Last Glacial, with the Gulf Stream presumably at least having an influence on the coastal vegetation (Brazier, suggested in personal communication).

Unfortunately, most traces of that coastal vegetation would be submerged now under about 50 to 100 metres of seawater. This might, of course, help to explain in part why definite occurrences of *Rhamnus, Quercus* and *Ulmus* have not previously been recorded for the British Last Glacial, or for that matter, for any of the British glacials, the times of generally lowered sea levels.

It is thus apparently thanks to human collecting behaviour during the Later Upper Palaeolithic that we have managed to find any trace at all of these three thermophilous genera. Whether these plants in fact grew in some sort of more congenial, periglacial maritime refugia somewhere to the south and west of the present British Isles, one simply cannot as yet say. None of the Kent's Cavern timbers appears to be the result of gathering driftwood (Brazier, personal communication), but though fresh in condition, it remains uncertain just how far Later Upper Palaeolithic people really transported them. The nearest known potential sources during the Late Last Glacial are probably those which existed in Spain and southernmost France (e.g. see Butzer, 1971). But it seems just as conceivable that they came from sources nearer at hand but still unknown to us today.

Remains of *Quercus* and *Ulmus* are well known from the Post Glacial, and they are also known from earlier interglacials in Britain (e.g. see Godwin, 1956; Pennington, 1969; Walker and West, 1970; Sparks and West, 1972). However, even though the known floras of the glacials and their interstadials are predominantly Boreal to Arctic/Alpine in character, as suggested above, there is as yet far too little known to be absolutely certain what the now submerged coastal floras may have really included amongst their presumably broader vegetational spectra. From off the present coast of southern Ireland down to the Iberian Peninsula there may at times have been reasonably well-established plant communities, which if found today, one might interpret as Temperate to Boreal.

When found during the glacials, the presence of pollen of *Quercus* and *Ulmus*, or for that matter the pollen of most any thermophilous tree or shrub not represented by macrofossils, is normally interpreted as due to derivation on the one hand or long-distance wind-transport on the other. Nevertheless, such pollen appears in many British and continental European pollen diagrams (see examples in this present book, as well as in Godwin, 1956; Leroi-Gourhan, 1966 and 1968; West, 1968; Paquereau, 1969; Bastin, 1970; Walker and West, 1970; Campbell and Pohl, 1971; Sparks and West, 1972; Damblon, 1974). In France and Belgium this is sometimes interpreted as indicative of a real, albeit local occurrence of thermophilous tree and shrub genera during interstadials (e.g. see Leroi-Gourhan, 1964; Campo, 1969; Renault-Miskovsky, 1972; all on France; and Bastin, 1970; Damblon, 1974; both on Belgium). But the frequency of such pollen types both in Britain and the rest of north-western Europe seems always to be comparatively low, suggesting presumably that when present at all in the actual vegetation, these plants would comprise only a small component in what would otherwise have been a fairly open landscape.

During the Last Glacial variegated Temperate, Boreal and Arctic natural vegetation zones along the coast of western Europe may in fact have been much more altitudinally, latitudinally and longitudinally compressed than they generally are today, perhaps more or less as they still are within northern coastal regions such as southern Alaska, western British Columbia and western Norway, all examples with an interface between warm oceanic and cool continental climatic régimes. If this sort of compression of vegetation, or rather ecological, zones had in fact occurred in Last Glacial western Europe, the zone we would be least likely to find evidence for would be the Temperate zone, which particularly to the north-west and west of France would have been entirely below present sea level. The northern limit of this Temperate zone would have been to the west and south of the British Isles, an assumption which might also help to explain its absence in those regions on Frenzel's otherwise highly instructive reconstruction maps of Eurasian vegetation zones during the Last Glacial (Frenzel, 1968a and b). He does, however, show that maritime thermophilous forests did exist in certain parts of south-western Europe.

But really unless an intensive search were undertaken for palaeobotanical sites on the continental shelf of the "Franco-Hibernian" (French-Irish) region, this problem would most likely remain unresolved. Amongst other possibilities, a series of borings would certainly be worth doing, perhaps particularly along lines drawn from Cherbourg to Plymouth and from Brest to Cork. The middle of this region is in fact shown on Map 21 in this present book with the tentative suggestion that a "maritime shrub tundra with some trees" existed just above the 100 metre submarine contour during Late Last Glacial Zone I. As shown on Maps 19 to 23 as a whole such a maritime zone would probably have had a constantly varying and somewhat precarious distribution, perhaps even retreating completely to off the south-west of France or at least well off the south-west of Britain during the height of the Full Last Glacial (i.e. off the area shown on Map 20), though its distribution may often have centred between about the present 100 metre and 25 metre submarine contours. It could have included refugia for the more hardy thermophilous species, refugia which one would hardly term "tundra", rather a "maritime steppe/forest mosaic".

In any case, some Temperate elements do in fact occur in British Last Glacial deposits found above present sea level, but these are mostly "close-to-the-ground" life forms (e.g. see Kerney, 1963; Evans, 1972; on snails; Bell, 1969 on herbs; Coope, Morgan and Osborne, 1971; Coope, 1975; on beetles). Such "lowly" life forms are presumably indicative of microenvironmental improvements and/or they imply they are better able to migrate, colonize and propagate during slight climatic ameliorations than are Temperate trees, or even Boreal ones. In fact, the only fully Boreal interstadial known thus far for the "land-locked" Last Glacial is the so-called "Chelford" Interstadial (?="Brørup") of about 60,000 radiocarbon years B.P., which had extensive forests of *Betula*, *Pinus* and *Picea* (Simpson and West, 1958; also see West, 1968 and 1970; Frenzel, 1968a and b; Sparks and West, 1972). The only other interstadial which approached but did not achieve botanically Boreal conditions during the Last Glacial of the British Isles is the so-called "Allerød" Interstadial (= Late Last Glacial Pollen Zone II) of about 11,000 radiocarbon years B.P., which had woodlands of *Betula* in the more congenial parts of England and Wales, and possibly to some extent in Scotland and Ireland as well, but which still had much open steppe "parkland" and "Alpine" meadows (West, 1968; Pennington, 1969; Sparks and West, 1972).

Obviously, it is the "Allerød" Interstadial which is of more relevance to our immediate problem, or rather it is the whole ecosystemic complex of the Late Last Glacial in general and Zone I in particular. Detailed analyses of fossil assemblages of beetles suggest that the greatest summer warmth was attained not during "Allerød" but during the immediately preceding Zones Ib–c (Coope, Morgan and Osborne, 1971; Coope, 1975). If this be so, then presumably the reaction of *Betula* to this increased warmth was slower than that of the beetles, at least as far as the evidence goes for the areas of Britain above present sea level. Below present sea level the response of *Betula*, and perhaps *Pinus, Corylus, Quercus, Ulmus*, etc., in what might also be termed "maritime arboreal refugia", to this Late Last Glacial thermal maximum would have been, one might assume, more dramatic. Thus, it may not simply be a coincidence that the apparent age of the Kent's Cavern charred wood fragments (between c. 14,275 and c. 12,180 radiocarbon years B.P.) agrees well with the many age estimates now available for that thermal maximum, estimates which centre round 13,000 to 12,500 radiocarbon years B.P. (Coope, 1975; also see Fig. 83 in this present book).

The "maritime arboreal refugia" hypothesis presented here might help partly to explain, at least indirectly, the enigmatic so-called "Lusitanian-Mediterranean" elements of the modern natural Irish flora. Different types of refugia would have been required for these elements to survive, but they would nonetheless also have needed to have been warm. Mitchell and Watts (1970) postulate the possibility of now submerged "maritime steep slope refugia", embanked heathlands of a sort, off the south-western coast of Ireland for the Ericaceae. In the more sheltered valley bottoms of such refugia Boreal/Temperate species of trees may also have survived. The annual, monthly and even daily ranges of actual air and ground temperatures in these coastal valleys may have been much more limited and much less severe than what was being experienced elsewhere to the north and east in Ireland and Britain. In fact, in the more favoured of these valleys during most if not all of the Last Glacial, there may have been no permafrost or related periglacial phenomena (also see Lindroth, 1970 on the general nature of ice-free refugia). Such conjectures are, of course, still very difficult to test. West (1968 and 1970) correctly cautions that the individual species concerned in the Irish flora probably had separate histories, that some might have been better able to withstand actual periglacial conditions than we might normally assume, and that they should not be considered in groups (e.g. such as "Lusitanian").

Although the species of *Quercus* identified from Kent's Cavern remains uncertain, it is of considerable interest that it might possibly be *Q. ilex*. *Q. ilex* is a native of the Mediterranean region today, but it also extends north to Brittany and it is sometimes naturalized in southern England (Clapham, Tutin and Warburg, 1962). The two fragments of *Ulmus* identified by Dr. Brazier belong presumably to *U. glabra*, which is native today throughout the British Isles, being commoner in the west and north (Clapham, Tutin and Warburg, 1962).

If the third timber represented in the Kent's Cavern series did in fact belong to *Rhamnus catharticus*, then it is possibly the earliest record so far for this species in the British Isles (for younger records see Godwin, 1956; Pennington, 1969; Walker and West, 1970). Today it occurs as a native in England, Wales and Ireland, being found on fens and in ash and oak woods. It prefers calcareous soils and thus is not generally found in Cornwall, Devon, central Wales and Scotland (Clapham, Tutin and Warburg, 1962). Together with hawthorn (*Crataegus*), juniper (*Juniperus*), dogwood (*Cornus*) and privet (*Ligustrum*) it sometimes colonizes chalk grasslands today (Pennington, 1969). If it survived the Last Glacial in as yet unproved "maritime refugia", then it may have found the appropriate calcareous soils in certain parts of the south-western British and southern Irish continental shelf, at least according to the geological data in Donovan (1968).

Of course, in the case of Kent's Cavern one is not dealing simply with a palaeobotanical site, rather it is a palaeoethnobotanical site, the nature of which is due almost entirely to prehistoric human behaviour. But having said that, the cultural-ecological patterns involved, unfortunately, remain obscure. For example, whether any elaborate exchange or trade networks were involved in transferring the materials which ultimately became these charred wood fragments, we simply cannot as yet tell. Similarly, it is uncertain whether these fragments once belonged to intentionally shaped artifacts, such as baskets or other food containers (there is no evidence whatsoever on these tiny burnt bits for any sort of workmanship).

Returning to the question of trade or exchange, it is worth looking at some ethnographic examples of certain aspects of trading amongst hunter/gatherers. First let us look at the nomadic Walmadjeri and Gugadja Aboriginal groups of the Balgo region in the Western Desert of Australia. According to Berndt:

'Trade goods are passed, so to speak, from one interactory zone to the next. When large ceremonies and rituals are held, some of the participants come from places a great distance apart; they provide, therefore, an ideal opportunity for bartering. Trade takes place within the context of ritual and often is not seen as being something separate. Items that change hands at such times include red ocher, spears, and native tobacco . . . But the most important "goods" are pearl shells and sacred boards.' (Berndt, 1972, p. 188)

This suggests, amongst other possibilities, that substantial distances and close association with rituals can be closely interwoven in hunter/gatherer trade systems. It is certainly distance which would be required to get the Kent's Cavern timbers to Kent's Cavern, and once there it may have been ritual which eventually saw them burnt.

Now let us look at the North American Indians of the North-west Coast, certainly amongst the most famous of hunter/gatherers owing to their comparative wealth, densely populated permanent villages, elaborate social stratification, and extensive trade and display systems, including the so-called "potlatch" which is a way of converting surplus wealth into glory and prestige. According to Suttles:

'The areas where commercial transactions were most open and honorable seem to have been the north and, most especially, the lower Columbia. But even here commerce was not wholly free for it is reported that chiefs . . . held monopolies over trade in their territories. It seems possible that such monopolies may have had their origins in exchanges between affines in areas of different resources. In both cases the most important trade was between the coast and interior – the Tlingit with the Athapaskan hinterland and the Chinookans with the Plateau Sahaptins.' (Suttles, 1968, p. 67)

This brings to mind the possibility that different groups may have lived in Later Upper Palaeolithic Britain, perhaps "wealthier" groups on the now submerged coast and more "nomadic" groups in the interior. Spots such as Kent's Cavern could have been at the crossroads between such groups, sharing in the resources from both areas. But to suggest that the charred wood fragments from Kent's Cavern and the "Black Band" as a whole might be the result of a series of "potlatches" would probably be too much for the moment. However, in this vein it is well worth quoting Suttles' final remarks:

'Finally, I would ask, when we find archeological evidence of unexpected cultural complexity and population density, is it altogether hopeless to seek ethnographic parallels from the Northwest coast? This is not to say that I believe the Upper Paleolithic Europeans held slaves and gave potlatches. But I expect that some day we will be in a position to say whether they possessed the functional equivalents.' (Suttles, 1968, p. 68)

Of course, in terms of the Upper Palaeolithic Suttles is referring mainly to the remarkable character of south-western France in those times. But if I may stretch the point further, if such "functional equivalents" did ever evolve in France, they may also have been strung out in a long thin line along the Last Glacial coast of western Europe, extending right up to near south-western Britain in a manner not terribly unlike their much more recent distribution in British Columbia.

There is now obviously the need to search for additional samples of charred wood from the Kent's Cavern "Black Band", specimens of which might still be extant in one or another of the many museums to which Pengelly's finds were distributed. There is a list of these museums in his Diary (Pengelly, 1865–80, Vol. V). From his list and from what is known of the various collections, it would seem that the most promising places to look first would include the British Museum, the British Museum (Natural History), the Oxford University Museum and the Torquay Natural History Society Museum. It is at the latter Museum that a copy of Pengelly's Diary is kept, and it is from that Museum that the charred wood fragments reported upon here came.

APPENDIX 2

Land Snail Faunas from Cathole, 1968 and Long Hole, 1969

by

JOHN G. EVANS

(Department of Archaeology, University College, Cardiff)

There have been few studies of subfossil land snails from cave deposits in the British Isles, and none of a detailed nature. The samples from Cathole and Long Hole were looked at specifically to see if a Late Pleistocene cold-climate fauna like that from open-sites in southern England could be extracted. In this, the analyses were unsuccessful. The faunas from both sites, where present, were Post Glacial and considered to be largely intrusive, the snails exploiting the cavities of the loose rock rubble which comprised the bulk of the deposits.

The following samples were analysed:

Cathole, 1968

Layer	Age	Weight (kg.)
F	Modern soil	0.20
E	Medieval to 19th century	0.55
D	Mesolithic to Bronze Age	0.25
C	Early Post Glacial (Mesolithic)	0.25
LSB–USB	Late Last Glacial (L.U.P.)	1.00
A1–A3	Middle to Full Last Glacial	1.20

The A layers were devoid of shells. Samples from the B layers were combined.

Long Hole, 1969

A combined sample from layers C/B to F. The A layers and B layers were devoid of shells.

The weights of the samples are after the removal of stones greater than 1.5 cm. A number of larger snails extracted by sieving during the excavations were also provided. The earth from these was removed and analysed for smaller shells, and both groups were combined with the totals from the analyses of the bulk samples. The final lists (Table 27) must be regarded therefore only as approximations to the actual faunas once living on the site, although the low abundance of the larger species, particularly *Cepaea*, suggests that little distortion has resulted from this procedure.

The analyses were done as described by Evans (1972), all shells greater than 0.5 mm. being extracted and counted. Apart from the usual difficulties of separating the juveniles of the two species of *Cochlicopa* and those of *Cepaea*, and the virtual impossibility of identifying the various species of limacid slug from their internal shells, the Zonitidae (*Vitrea* and *Oxychilus*) presented certain problems. The majority of *Vitrea* shells were *V. contracta*, the form with the open, parallel-sided umbilicus. A number, however, in having an umbilicus somewhat intermediate between that of *V. contracta* and that of *V. crystallina*, were confusing, particularly as there were some undoubted specimens of *V. crystallina* present. It was eventually decided, however, that the majority were *V. contracta*, and there was no question of there being a graded series between the two species. No examples of the newly recognised British species, *Vitrea diaphana*, were present. With *Oxychilus*, four possible species were involved. *O. cellarius*, the commonest molluscan denizen of underground places such as cellars, coffins and caves, was the most abundant. Some of the specimens were very large, being up to 15 mm. in diameter, far beyond the usual 10 mm., and approaching *O. draparnaldi* in size. However, there was only one possible example of this species, the others all being confidently identified as the large form of *O. cellarius* recorded from other sites in the west of Britain. Two other species, *O. alliarius* and *O. helveticus*, were probably present, but it was impossible to be certain of the identifications.

The results have been presented in tabular form (Table 27); the order of species is a taxonomic one, this and the nomenclature being after Ellis (1951). The results from Cathole have also been presented in the form of a histogram (Fig. 80) showing the relative abundance of species or groups of species as a percentage of the total fauna.

Four broad ecological groups may be recognised – troglophile, "woodland", catholic and open country. The troglophile species are those which are frequent in caves and other underground places by virtue of their ability to eat animal food; but they are not confined to such places and can live just as readily on vegetable matter. The "woodland" species include those which require a certain degree of moisture and shade, and a lack of disturbance in the habitat. They are generally, although not always, found in woodland. *Retinella nitidula* and *R. pura* probably feed exclusively on vegetable matter. The catholic group occur in a much wider variety of places and can tolerate a degree of impoverishment and dryness in their environment which the "woodland" species cannot. The open-country species are those which are virtually restricted to dry and unshaded habitats.

The faunas from both sites are dominated by the troglophile group. At Cathole the "woodland" species become increasingly important in layers E and F, suggesting an increase in the vegetable mould content of the habitat. Catholic species are sparsely represented; open-country species hardly at all.

In the past one has tended to consider subfossil shell assemblages as being broadly contemporary with the deposits in which they occur. Derivation of shells from older deposits and their incorporation into those of more recent origin undoubtably occurs in some cases; so too does intrusion of shells into older deposits via root channels, etc. But on the whole the evidence of former environments as inferred from molluscan assemblages and that of other data such as the lithology of the deposits and their small mammal faunas has been remarkably consistent. But in the case of Cathole and Long Hole, and of other sites in which rock rubble constitutes a major component of the deposits, this is not so (Evans and Jones, 1973).

Instead, what appears to have happened is that the snail species which are able to colonise underground cavities – the troglophile group – have crawled in amongst the rock rubble after its deposition. This is shown particularly by the fauna from the B layers at Cathole. The lithology, pollen, small mammal fauna and archaeology of this layer complex indicate a cold climate, an open-country landscape and a Late Last Glacial date, but the molluscan fauna is unequivocably of warm temperate character. Not only does it lack the open-country species – there being only one *Pupilla*, two *Vallonia* and no *Helicella* (see Fig. 80 and Table 27) – so characteristic of the Late Last Glacial period, but many of its predominant elements – *Discus rotundatus, Oxychilus cellarius* and *Pomatias elegans* – apparently did not enter Britain prior to late Boreal times (Kerney, 1966; Evans, 1972).

The complete absence of shells from the A layers at Cathole and the A and B layer complexes at Long Hole is odd. It is possible that there was no molluscan fauna present in the Gower Peninsula at this time, but this is unlikely in view of the existence of cold climate faunas of Full or Late Last Glacial date at Brean Down in Somerset (ApSimon, *et al.*, 1961) and Cainscross near Stroud in Gloucestershire (Large and Sparks, 1961). No cave faunas from sites in the British Isles which are convincingly pre-Flandrian are known, all those which have been claimed as Pleistocene being characterised by *Oxychilus, Discus* and other fully temperate species (Evans, 1972; Evans and Jones, 1973). Lozek (1971) describes a profile from Moravia very similar to that from Cathole in deposits at the foot of a limestone cliff. Rock rubble predominates and the fauna in the bulk of the deposits is Post Glacial. But in the lowest layer a Late Last Glacial fauna is clearly present with species of *Pupilla, Vallonia* and *Helicella* common. The absence of such a fauna from Cathole and Long Hole is more likely to have a local ecological rather than general origin, but just what this may be is difficult to say.

The study of subfossil land mollusc faunas from caves has been much neglected and there is a considerable need for some detailed work.

APPENDIX 3

Deux coquilles, probablement d'origine anglaise, découvertes à Spy, Belgique

par

MARCEL OTTE

(Service d'Archéologie Préhistorique de l'Université de Liège)

Deux coquilles fossiles ont été trouvées en 1958 par M. Dewez, lors de fouilles de remblais de la grotte de Spy. Elles sont actuellement conservées au Service d'Archéologie Préhistorique de l'Université de Liège.

Elles présentent toutes deux un émoussé prononcé sur toute leur surface et contiennent des traces de sédiment rouge. Elles ont chacune une perforation provoquée par une cassure et donc due, vraisemblablement, à une action naturelle.

Les fouilles les plus importantes de la grotte de Spy sont dues à M. de Puydt et M. Lohest, en 1885–1886. Ils reconnurent trois niveaux paléolithiques. Le second les avait frappé par sa forte coloration en rouge, due à un apport humain d'oxyde de fer (Puydt et Lohest, 1886, pp. 209 et 213). Cette observation a été refaite lors des fouilles des Musées Royaux du Cinquantenaire, en 1906 et 1909 (Loë et Rahir, 1911, p. 46). On a, par la suite, interprété les industries découvertes à Spy: premier niveau: Périgordien supérieur à Font-Robert; second niveau: Aurignacien, vraisemblablement mêlé à une industrie moustérienne (c'est le "niveau rouge"); niveau intermédiaire contenant les inhumations des hommes de néanderthal, accompagnés semble-t-il par une industrie moustérienne de type Quina; troisième niveau: Moustérien de tradition acheuléene (Breuil, 1922b, p. 129; Angelroth, 1953, pp. 177–179; Bordes, 1959, p. 157; Sonneville-Bordes, 1961, p. 427; Otte, 1969–1970, pp. 86–97). Les niveaux tout à fait supérieurs ont livré du matériel en faible quantité, appartenant à des périodes extrêmement diverses à partir de la fin du Paléolithique.

Etant donné l'extrême abondant du matériel paléolithique, il est très probable que ces coquilles appartiennent à ces périodes. Plus particulièrement au Paléolithique supérieur, comme on en a plusieurs fois noté la présence (Loë et Rahir, 1911, p. 46). Le sédiment rouge, à l'intérieur de ces coquilles, tend à confirmer cette attribution. On a également remarqué l'utilisation d'oxyde de fer dans le niveau supérieur (Loë et Rahir, 1911, p. 46). Il peut donc s'agir, soit du premier niveau (Périgordien), soit du second niveau (Aurignacien). Les coquilles semblent tout au moins appartenir à une industrie du début du Paléolithique supérieur.

Déterminations

La première coquille a été determinée par M.P. Norton (Département de Zoologie de l'Université de Glasgow) comme *Trivia monacha* ou *T. europaea* et par M. Glibert (Institut Royal des Sciences Naturelles de Belgique) comme *T. coccinelloides* (Sowerby), ce qui, selon M. Glibert, est équivalent à *T. europaea* (Harmer, 1914–19, p. 49). Elle peut appartenir à des dépôts soit miocènes soit pliocènes soit pléistocènes (idem).

La seconde coquille a été déterminée par M.P. Norton comme *Nassarius reticosus*. M. Glibert, qui l'avait attribuée à *N. reticulatus* (Linné), a bien voulu reprendre la question de cette attribution et a confirmé sa détermination, sur bases de comparaisons avec des pièces de son laboratoire.

Origine géographique (Map 47)

Selon M. Glibert, ces fossiles proviennent des affleurements de Grande-Bretagne, dans l'East Anglia (Harmer, 1914–19, p. 49, *Trivia coccinelloides* et p. 319, *Nassarius reticulatus*; "British Caenozoic Fossils", 1963). M. Norton a bien voulu nous préciser que ces espèces sont connues dans les séries de Red Crag et Norwich Crag.

Comme l'ont fait remarquer Messieurs Norton et Glibert, certaines de ces formes peuvent apparaître dans les dépôts du Sud de la Hollande, mais elles ne sont connues que par des sondages et étaient, selon M. Glibert, inaccessibles à l'époque würmienne.

Les limites des zones de contact entre l'Angleterre et le Continent sont actuellement bien établies pour la période würmienne (Veenstra, 1970). Une hypothèse de passage humain est donc tout à fait possible. On a déjà noté des formes communes entre l'outillage du début du Paléolithique supérieur anglais et belge (Smith, 1966, p. 286). Un récent travail de synthèse sur les cultures anglaise du Paléolithique supérieur a soulevé à nouveau la question (Campbell, 1971, p. 300). Les observations que nous avons pu faire, à la fois sur le matériel belge et sur le matériel anglais, semblent confirmer cette idée et elles seront développées dans notres travail de doctorat (Otte, 1974b).

On a déjà noté la fréquence des mollusques dans les niveaux d'occupation du Paléolithique supérieur et l'intérêt qu'offre leur étude a été souligné (Breuil, 1950, pp. 191–240). Les hypothèses de leur utilisations ou de leur signification pour l'Homme préhistorique sont diverses (rites, commerce, jeux, etc. . .). Mais l'étude de leur lieu d'origine semble fractueuse pour la détermination des déplacements et la reconnaissance des contacts éventuels (Hahn, 1972, pp. 260 et 261). Généralement, les coquilles découvertes en Belgique ont été reconnues comme d'origine française (Champagne, Reims) mais elles appartenaient le plus souvent à des cultures plus tardives dans le Paléolithique supérieur (Dupont, 1865, pp. 19 et 20). Les exemplaires présentés ici en sont d'autant plus intéressants et peut-être ces coquillages ont-ils une signification culturelle.

Résumé

Deux coquilles fossiles, *Nassarius reticulatus* (Linné) et *Trivia coccinelloides* (Sowerby), proviennent des découvertes, sans stratigraphie, des dépôts de la grotte de Spy. Elles appartiennent très vraisemblablement à une phase du Paléolithique supérieur initial (Périgordien ou Aurignacien). Leur affleurement le plus proche se trouve en Grande-Bretagne, dans le sud-est de l'East Anglia.

Remerciements

Nous tenons à remercier Monsieur Glibert de l'Institut Royal des Sciences Naturelles à Bruxelles ainsi que Monsieur P. Norton du Département de Zoologie de l'Université de Glasgow, pour leur déterminations et les commentaires qu'ils ont bien voulu nous fournir. Nous remercions Monsieur le Professeur J. de Heinzelin, de l'I.R.Sc.N., pour les conseils qu'il nous a fournis et Monsieur J. Campbell pour l'invitation qu'il nous a faite de publier cette note.

Datation d'os par carbone–14

22,105 ± 500 B.P. (IRPA–132) par Messieurs M. Dauchot-Dehon et J. Heylen (1975). Les niveaux du Paléolithique supérieur initial (Périgordien ou Aurignacien) à la grotte de Spy.

APPENDIX 4

Les groupes du Tardiglaciaire et le problème du Creswellien en Belgique

par

MICHEL C. DEWEZ

(Centre Interdisciplinaire de Recherches Archéologiques, Université de Liège)

Introduction historique

La recherche en archéologie préhistorique débuta très tôt en Belgique et, par hasard, les sites de grottes occupés par les peuples du Tardiglaciaire furent les premiers à être fouillés avec méthode.

Vers 1833, Schmerling (1833–34) avait déjà publié un fragment d'outil en bois de renne, à base en double biseau, provenant du site malheureusement perturbé des grottes des Fonds-de-Forêt. C'est le premier document archéologique osseux du Tardiglaciaire publié avec une interprétation correcte.

En 1864, le géologue E. Dupont (1867 et 1873) entreprit une campagne de fouilles subsidiée par le gouvernement. Les premiers travaux concernèrent les grottes magdaléniennes de la vallée de la Lesse (Trou des Nutons, Trou du Frontal, Trou de Chaleux). Il s'agissait de grottes n'ayant connu qu'une seule occupation, au Paléolithique; d'autre part, la stratigraphie avait été très bien relevée, comme c'était généralement le cas dans les fouilles de Dupont, aussi, le matériel récolté reste-t-il toujours scientifiquement exploitable.

D'autres grottes, comme le Trou Magrite, les Grottes de Goyet ou même la célèbre Grotte de Spy, ont été fouillées avec moins de méthode, et les recherches ne peuvent souvent y aborder les problèmes d'occupation que par le biais de la typologie.

Au cours de ce siècle, les recherches se poursuivirent, souvent d'une manière sporadique, mais les travaux exécutés avant la seconde guerre mondiale sont d'une qualité inférieure à la bonne période des fouilles de Dupont.

Après la guerre, Mlle H. Danthine (1961) reprit des fouilles dans la Grotte de Presle, tandis que notre collègue F. van Noten (1967 et 1975) entreprit l'étude des sites de plein air du Tjongérien.

Pour notre part, nous avons repris des fouilles à la Grotte de Remouchamps d'abord, et ensuite à la Grotte du Coléoptère à Bomal-sur-Ourthe, ces dernières entreprises avec la collaboration de notre collègue et ami J. Campbell (Dewez, 1974 et 1975).

Groupes culturels

(1) *Magdaléniens et Ahrensbourgiens*

Nous comprenons sous le terme "Paléolithique supérieur final" ("Later Upper Palaeolithic") les groupes de chasseurs nomades qui parcoururent notre pays pendant le Tardiglaciaire (Map 48).

Les Magdaléniens paraissent n'avoir fréquenté que la zone de grottes des collines calcaires situées en Wallonie (sud-est de la Belgique), avec une certaine prédilection pour les petites vallées, comme celle de la Lesse ou de l'Ourthe.

Jusqu'à présent nous n'avons pas pu identifier du Magdalénien ancien, mais il est possible qu'il en existe dans les Grottes de Goyet, pour lesquelles nous ne possédons que des indications d'ordre typologique. Les datations des occupations magdaléniennes vont de la fin du *Dryas* I ("Late Last Glacial Zone Ia", Grotte de Verlaine, vallée de l'Ourthe) jusqu'au *Dryas* II ("Late Last Glacial Zone Ic", Grotte du Coléoptère, Bomal-sur-Ourthe).

Plusieurs grottes ont livré des oevres d'art mobilier, parmi lesquelles on peut citer: figure humaine à Verlaine, bison gravé au Trou des Nutons, bovidé, cheval, bouquetin et cervidé à Chaleux, bovidé au Trou du Frontal, bouquetin à Goyet.

L'outillage osseux est très riche dans les sites magdaléniens, et parmi la documentation non utilitaire, il faut remarquer la présence de coquilles fossiles provenant du Tertiaire du Bassin de Paris.

Les Ahrensbourgiens paraissent s'être répandus entre les vallées de la Vesdre et de l'Ourthe, dans la partie est de la Wallonie, où ils occupèrent des grottes comme Remouchamps, Fonds-de-Forêt, la Préalle, et aussi la Grotte du Coléoptère à Bomal (couche 6). Plus au nord, ils remontèrent la Meuse pour installer des camps de plein air, notamment en Campine dans la zone de partage entre les bassins de la Meuse et de l'Escaut.

Leur répartition dans le temps nous semble bien limitée à la dernière séquence froide du *Dryas* III ("Late Last Glacial Zone III"). Leur documentation non utilitaire comprend des coquilles fossiles provenant du Bassin de Paris.

(2) *Creswelliens*

La première identification de ce groupe est due à Mlle H. Danthine (1961), à propos du site de Presle (Map 48). Ce site, dont la fouille n'est pas achevée, a cependant déjà fourni un matériel archéologique assez riche.

A côté des pointes qui associent une troncature à un bord abattu, comme les pointes de Creswell, on y rencontre des pointes à cran, des lames à bord abattu courbe et des lamelles à dos rectiligne. Les burins sont très nombreux, et jusqu'à présent les grattoirs paraissent extrêmement rares.

L'outillage osseux est très pauvre. On doit cependant signaler la présence d'un métapode de renne, gravé de signes plutôt énigmatiques, et d'une figuration féminine schématisée du type de Gönnersdorf.

Il est curieux de constater que dans ce site, les coquilles fossiles recueillies ne proviennent plus du Bassin de Paris, mais des côtes de la Mer du Nord et probablement d'Angleterre, ce sont en particulier des *Nucella lupillus* (Map 48).

Le matériel comprend aussi un certain nombre de bois de chute de renne femelle. Ce phénomène de stockage de bois de chute de renne femelle, dont on ignore l'utilisation puisque les outils sont toujours en bois de mâle, a déjà été constaté dans plusieurs sites magdaléniens (Trou des Nutons à Furfooz, par exemple).

Nous pensons qu'un certain nombre d'autres sites que l'on peut dater du Tardiglaciaire doivent être associés au groupe Creswellien. Malheureusement, peu d'entre eux ont été complètement fouillés, et nous ne possédons souvent qu'une documentation quantativement faible.

Le site de la Grotte de Martinrive, que nous avons récemment contrôlé, montre une industrie plus riche. Le matériel lithique, avec ses lames à dos courbe, ses troncatures, ses pointes à cran, ses nombreux burins et une quasi absence de grattoirs, ne peut être magdalénien. D'autre part, comme à Presle, le matériel osseux est très pauvre; il se réduit à un seul poinçon. A notre avis, le matériel lithique ressemble à celui de la Burckardtshöhle dans les Alpes souabes (Riek, 1959). H. Schwabedissen (1954) avait déjà isolé ce faciès sous le nom de faciès de Probstfels, mais ce site présente malheureusement un mélange de couches.

S'il y a une similitude entre l'industrie de Martinrive et le faciès creswellien bien clair de Presle, il y a également des différences. Aussi, nous proposons de distinguer un second faciès où les pointes de Creswell sont plus rares, et que l'on pourrait dénommer "Creswellien continental".

Une série de grottes, comme l'Abri de la Poterie à Furfooz, les Grottes des Fonds-de-Forêt, l'Abri de Mégarnie à Engihoul, le Trou Dubois à Moha et peut-être aussi le Trou des Blaireaux à Vaucelles, pourraient appartenir à l'un ou à l'autre de nos faciès creswelliens. Les recherches futures permettront seules d'éclaircir ce problème.

Par ailleurs, des sites de plein air appartiennent aussi aux faciès creswelliens. L'un des seuls sites dont la documentation n'est pas mélangée est un bel atelier de débitage découvert au Bois-St-Macaire près de Mons, par L. Letocart (1970). Dans le nord du pays, par exemple à Lommel en Campine, on peut remarquer des traces de nombreuses présences creswelliennes. La fréquence des "penknife points" augmente alors que dans les sites de grottes, elles sont très rares. On y trouve également beaucoup plus de grattoirs.

Malheureusement, beaucoup de ces sites ont été fouillés anciennement, et il est devenu impossible de dissocier dans les séries actuellement mélangées les ensembles vraiment creswelliens et ceux de leur faciès épigone "tjongérien". La parenté du faciès tjongérien avec le Creswellien au sein du techno-complexe des pointes à bord abattu courbe a bien été établie dès 1954 par H. Schwabedissen, et nous ne pouvons que confirmer ce point de vue.

APPENDIX 5

Bird Faunas from Cathole, 1968 and Robin Hood's Cave, 1969

by
DONALD BRAMWELL
(Bakewell, Derbyshire)

The bird remains from Cathole and Robin Hood's Cave are listed in Tables 29 and 30, respectively. The system of classification is that of James Fisher (1966). The remains are rather fragmentary and so my identifications entail a lot of queries, but the apparent groups matter more than individual species, whether of open ground, heath, marsh, water, woodland etc., as birds are generally more adaptable than mammals. The column on "possible seasonal behaviour" in Tables 29 and 30 is offered simply as a suggestion. It is based on my own first-hand knowledge of avian behaviour, as well as the information contained in Voous (1960), in conjunction with a careful reading of the pollen and mammalian evidence for the changing ecological settings at Cathole and Robin Hood's Cave (see Chapter III), and many hours of discussion with Dr. Campbell.

Comments on Cathole (Table 29)

The bird remains from Post Glacial layers C to F are dark in colour and often eroded due to the bad preservative qualities of soil rich in humus, whereas those bones from the Last Glacial levels (layers A2 and LSB to USB) are lighter and well preserved.

The species from layers D, E and F are mainly woodland, cliff-dwelling ones and include one notable bird of open ground, skylark, in layer E.

The aspect of the species from layer C is very close to that of the Last Glacial layers, a feature which agrees with certain Peak District caves, suggesting that the climate was still fairly cold and that there was still a lot of heathland about. The species of birds at Cathole for this epoch (Mesolithic, ?Pre-boreal) would all be desirable food for a hunting people: goose (at the interface of layers C and USB), grouse, stint, curlew and little auk. The last named species normally nests on Arctic cliffs and seldom reaches the Bristol Channel water today. It has also been found at several levels in Chelm's Combe Shelter, Somerset among Late Last Glacial mammals (Bramwell, 1960; also see Gazetteer II, site 12), and in disturbed but assumed Late Last Glacial contexts at Merlin's Cave, Herefordshire (Newton in Hewer, 1926) and at Chudleigh Fissure, Devonshire (Bramwell, 1960).

The environment suggested by the avifauna of layers A2 and LSB to USB (as well as C) is typically Last Glacial, requiring heath, marshy ground and water. The woodcock (layer USB) is a little doubtful among these assemblages, but the determination is based on a very imperfect femur shaft, and so it may belong to some other large wader. The woodcock does, however, frequent pine as well as deciduous forest glades. Most warblers (layer LSB) are woodland species, but the willow and Arctic warblers reach the far north of Scandinavia where virtually only tundra exists. Some of the species of birds present during the Later Upper Palaeolithic occupations (layers LOB and MSB) would certainly be desirable food for a hunting people:

goose and grouse.

The bird remains from Cathole were very fragmentary, but even so they have provided quite interesting additional information on the conditions existing in South Wales at different stages of the Last Glacial and Post Glacial.

Comments on Robin Hood's Cave (Table 30)

The general aspect presented by this collection of bird species is one of open heath with crowberry, other dwarf shrubs and lichens. Black grouse (layer OB) suggests some pine, willow and birch, and the goshawk (layers OB and USB) usually nests in pine in the northern limits of its range. Some of the birds demand marshy ground and open water, e.g. duck (layer OB) and plover (layer B/A) species. Short-eared owl (layers LSB and USB) feeds largely on voles in Britain today but on lemmings in Scandinavia; both voles and lemmings were available at Robin Hood's Cave (see Chapter III and Fig. 82). Jackdaw (layers LSB and OB) and kestrel (layers A, B/A and OB) probably nested in the cliffs of the gorge at Creswell, and kestrel was no doubt responsible for the ring ousel, fieldfare, bunting and finch type bones at Robin Hood's Cave.

Limitations of my reference material have probably led to the non-recognition of other interesting species, but none of those recognised would be out of place in the Boreal forest and heathlands of north-western Europe, and some such as ptarmigan would extend to the Arctic tundra. The species represented in layer OB in particular agree quite well with the age and ecological setting suggested by the associated pollen and mammals, namely the so-called "Allerød Interstadial" or Zone II (see Figs. 77 and 82 and compare with Table 30).

None of the bones show signs of charring, but a few bear sharp fractures (e.g. mallard, goldeneye, grouse, ptarmigan and plover) similar to those of the varying hare, *Lepus timidus*, at this cave, and so these may be the result of human activity rather than that of predatorial birds.

APPENDIX 6

Human Remains from Robin Hood's Cave, 1969: Later Upper Palaeolithic Osteology

by

ROSEMARY POWERS

(Sub-Department of Anthropology, Department of Palaeontology, British Museum Natural History, London)

and

JOHN B. CAMPBELL

(Department of Behavioural Sciences, James Cook University, Queensland, Australia)

The human material described here is from the July 1969 excavations at Robin Hood's Cave (Creswell Crags, Derbyshire). Even though only one bone, a frontal fragment, is from an undisturbed context (layer OB, Later Upper Palaeolithic horizon 3) and the rest of the human remains are from the nineteenth-century tip (layer E), the human material as a whole nonetheless appears to represent a single individual. This assumed individual belongs at least definitely to *Homo sapiens sapiens*. It seems to represent an adult male skull, lacking the vault, but originally having the upper cervicals attached. This male was fully adult but comparatively young, probably having been between 23 and 30 years old at death. No osteometric (craniometric) measurements are possible, but the parts preserved are shown in black on the diagram of a skull in normae in Figure 175.

The dental formula may be reconstructed as follows:

– 7 6 5 4̸ 3̸ – –	1̸ 2̸ 3̸ 4 5̸ 6 7̸ 8̶
8̸ 7 6 5̸ – – – –	– – – – – – – –

8̶ = bone missing
8̸ = tooth missing
– = all missing

Attrition has reached the dentine on the upper first molars and the lower first and second molars.

By layer and bone the human remains are as follows:

From Layer OB (undisturbed Late Last Glacial thermoclastic scree), definitely Later Upper Palaeolithic:

Frontal (find 465, square D3, 238 cm. below 0 m. datum). This is the lower part of a male frontal bone. The upper part is detached by an old, stained break, but the left orbital corner is detached by a fresh, clean break, an unfortunate "mark of discovery" made during the 1969 excavation. There is no trace of a metopic suture.

From layer E (nineteenth-century tip), possibly Later Upper Palaeolithic:

Temporal (finds 130 and 153, squares D3 and C1, 148 cm. and 151 cm. below 0 m. datum, respectively). The two fragments comprise the anterior and posterior halves, respectively, of an adult male right temporal bone, with part of the sphenoid wing attached. The tympanic plate is perforated, probably post-mortem, but the bone is otherwise in a remarkably good state of preservation.

Occipital (find 131, square D3, 134 cm. below 0 m. datum). This is the basioccipital, belonging to the same skull as finds 130 and 153 and joining onto them.

Sphenoid (find 128, square D3, 138 cm. below 0 m. datum). Another piece of the skull, being part of the right wing of the sphenoid adjoining the orbit. It does not, however, contact any other fragments.

Maxillae (finds 133 and 466, squares B2/C2 and E2, c. 150 cm. and 170 cm. below 0 m. datum, respectively). These fragments include the rear part of the right maxilla and the main parts of the left and right maxillae. The left and right maxillae do not join, as a portion of the right just below the nasal aperture is missing. But the two surviving fragments of the right maxilla do fit together and are now reconstructed.

Molar (find 230, square C1, 192 cm. below 0 m. datum). This is the upper left third molar. It is from the left maxilla described above.

Mandible (find 121, square E3, 135 cm. below 0 m. datum). This is the right angle of a mandible, with two molars in place. It is possibly from the same skull as the finds described above, but unlike those pieces this bone shows surface pitting from soil action (though it should be noted also that this specimen was found in the most humic part of the tip amongst many active plant roots). Soil action on the dental enamel has also altered the apparent wear pattern.

Vertebra (find 132, square D3, 134 cm. below 0 m. datum). This is a complete third cervical vertebra.

These finds are housed in the Sub-Department of Anthropology at the British Museum (Natural History), where they may be examined. Further analytical tests might be worthwhile, but the material is fragmentary and is mostly from a derived context, excepting the frontal.

By association the age of the frontal and perhaps that of the whole "skull" would be Late Last Glacial. Faunal (mammals and birds) and floral (pollen) evidence from layer OB suggests that this layer belongs mostly, if not entirely, to Zone II, or the so-called "Allerød" Interstadial. A radiocarbon age estimate obtained on a metacarpal of *Equus przewalskii* from layer OB has come out, however, at only 10,390 ± 90 years B.P. (BM–603), though that sort of age, if true, would fall at least in Zone III. For further details on the associated palaeoecological and archaeological evidence, the reader is referred to the main text of this book (see especially pp. 64–69, 92, 105, 125–31, 174–75 and Figs. 27–36; also see Appendix 5). The stratigraphic position of the frontal is shown on Figures 30 and 35.

There is no surviving evidence for an intentional burial in the conventional sense, but after death the "skull" of the "victim" represented by the frontal may just conceivably have served for a time as some sort of trophy. If this were so and if all of the fragments found (both layers E and OB) were to represent the single individual depicted above, then it may have been that the "head" was intentionally severed from the missing "body" just below the third cervical vertebra somewhere beyond the 1969 excavations, or possibly even well away from Robin Hood's Cave, the "head" then having been carried "home". Eventually, this "skull/trophy" may have been ceremonially shattered, the various bits being scattered over the site in the process, perhaps after it had been suspended or mounted for a time as a display to ward off the "enemy", be he feuding kin or another band of hunter/gatherers.

The absence of the vault to the skull might also suggest the possibility of cannibalism, but again there is no clear evidence for this either. However, it is worth noting once more that the surviving lower part of the frontal is detached by an old break, and further that the tympanic plate of the surviving temporal bone is perforated, the perforation probably being post-mortem. The ultimate destruction of the skull may also have been meant to protect the destroyers from the "spirit", or whatever, of its original owner, as well as possibly to consume that "spirit" to share its strength.

On the basis of what little evidence there is, it is almost impossible to make a more complete ethnographic reconstruction at this stage. However, to follow the vein of capture and cannibalism further, it may be worth quoting Forde on the Boro of the Japura region of the western Amazon forest in South America.

'Neither the Boro nor any other group have any effective tribal organization, and conflicts are between small bands from one or a few temporarily combined communities. . . Prisoners are taken and are carried off by the victors in these encounters; the adults are killed and eaten at the dance feast which celebrates the event. Captured children too young to escape and so betray the settlement to the enemy are, however, handed over to the chief, who brings them up as members of his household in which they serve.' (Forde, 1934, p. 146)

Turning to the Jivaro, who also live in the western Amazon, we may quote Meggers' more recent work:

'The Jivaro have elaborated blood revenge and warfare to a point where these activites set the tone for the whole society. . . . . . . The consequence is a high male death rate and an adult sex ratio of one male to two females. . . . . . . All Jivaro men are under heavy social pressures to execute blood revenge or to take a head, and thereby to risk their own lives.' (Meggers, 1971, pp. 108–10)

Although the Amazon, from socio-economic and ecological points of view, may not offer the best ethnographic analogies for Later Upper Palaeolithic Britain, the above quotes may at least help to underscore the fact that it may be significant that a young adult male seems to be represented by the assumed "skull-bashing" activity at Robin Hood's Cave, a male who may also have once been a "warrior", before his head was "taken". Of course, on general cultural-ecological grounds, the Eskimos of the Arctic and the Athapaskans of the Sub-Arctic and Boreal regions of North America would normally offer better ethnographic analogues. But then we cannot for a moment even assume that the Late Last Glacial/Later Upper Palaeolithic people of Britain behaved precisely like their more recent counterparts in North America. In any case, "head-hunting" was not normally practiced by the Eskimos and Athapaskans, and cannibalism was comparatively rare amongst both groups, although it did occur under times of stress or famine, though its occurrence depended very much on the attitude of the individuals or band concerned at the time (e.g. see Balikci, 1970, p. xxi on the Ukjulingmiut Eskimo; and VanStone, 1974, p. 27 on the Kutchin and other Athapaskans). To judge by the food remains (i.e. animal bones) found in layer OB, it does not seem likely that famine was a serious problem at Robin Hood's Cave (there is at least 3,276 kg. of meat represented in layer OB, see Table 43). Of course, from the evidence available one cannot tell what sort of seasonal shortages there may at times have been, but presumably meat was normally frozen for the winter and dried for the summer. Anyway, the individual human represented by the Robin Hood's Cave remains certainly seems to have been quite healthy at the time of his death.

In concluding it should be noted that no human remains are recorded from the 1874–76 excavations by Dawkins and Mello, but they or a contemporary (perhaps the infamous Dr. Laing) removed part of layer OB and then redeposited some of its contents in their tip or our layer E (see Figs. 30 and 35).

BIBLIOGRAPHY

N.B. This is a comprehensive bibliography for all references in the book, be they in tables, text, figures, maps or the gazetteers.

Adovasio, J.M. and Lynch, T.F. (1973). Preceramic Textiles and Cordage from Guitarrero Cave, Peru, *Amer. Antiquity* **38**, 84–90.

Agache, R. (1971). Informations archéologiques: circonscription de Nord et Picardie, *Gallia Préhist.* **14**, 271–310.

Alexander, E.M.M. (1964). Father John MacEnery: Scientist or Charlatan?, *Trans. Devons. Assoc.* **96**, 113–46.

Angelroth, H. (1953). Le Périgordien et l'Aurignacien, *Bull. Soc. Anth. Bruxelles* **64**, 163–83.

ApSimon, A.M. (1955). King Arthur's Cave Plan and Sections, unpubl. Univ. Bristol Spel. Soc. Museum.

— — (1957). A Creswellian Implement from the Chew Valley, North Somerset, *Proc. Univ. Bristol Spel. Soc.* **8**, 46–48.

ApSimon, A.M., Donovan, D.T. and Taylor, H. (1961). The Stratigraphy and Archaeology of the Late-glacial and Post-glacial Deposits at Brean Down, Somerset, *Proc. Univ. Bristol Spel. Soc.* **9**, 67–136.

Armstrong, A.L. (1923). Exploration of Harborough Cave, Brassington, *J. Roy. Anth. Inst.* **53**, 402–16.

— — (1925). Excavations at Mother Grundy's Parlour, Creswell Crags, Derbyshire, 1924, *J. Roy. Anth. Inst.* **55**, 146–78.

— — (1928). Pin Hole Cave Excavations, Creswell Crags, Derbyshire; Discovery of an Engraved Drawing of a Masked Human Figure, *Proc. Prehist. Soc. E. Ang.* **6**, 27–29.

— — (1931a). Excavations in the Pin Hole Cave, Creswell Crags, Derbyshire, *Proc. Prehist. Soc. E. Ang.* **6**, 330–34.

— — (1931b). A Late Upper Aurignacian Station in North Lincolnshire, *Proc. Prehist. Soc. E. Ang.* **6**, 335–39.

— — (1932). Upper Palaeolithic and Mesolithic Stations in North Lincolnshire, *Proc. Prehist. Soc. E. Ang.* **7**, 130–31.

— — (1939). Palaeolithic Man in the North Midlands, *Mem. & Proc. Manchester Lit. & Phil. Soc.* **8**, 87–116.

— — (1956). Prehistory: Palaeolithic, Neolithic and Bronze Ages, *Sheffield and its Region (Brit. Assoc.)*, pp. 90 ff.

— — (1957). Report on the Excavation of Ash Tree Cave, near Whitwell, Derbyshire, 1949 to 1957, *J. Derbys. Arch. & Nat. Hist. Soc.* **76**, 57–64.

Arnold, J.R. and Libby, W.F. (1951). Radiocarbon Dates, *Science* **113**, 111–20.

Baden-Powell, D.F.W. (1949). Experimental Clactonian Technique. *Proc. Prehist. Soc.* **15**, 38–41.

Baden-Powell, D.F.W. and Moir, J.R. (1944). On the Occurrence of Hessle Boulder Clay at Happisburgh, Norfolk, containing a Flint Core, *Geol. Mag.* **81**, 207–15.

Balch, H.E. (1914). *Wookey Hole: Its Caves and Cave-Dwellers*, Oxford.

— — (1928). Excavations at Wookey Hole and Other Mendip Caves, 1926–7, *Antiquaries J.* **8**, 193–210.

— — (1935). *Mendip:- Cheddar: Its Gorge and Caves*, Wells.

— — (1938–53). Exploration of the Badger Hole at Wookey Hole, 2 vol. unpubl. Diary of excavations, Wells Museum.

Balikci, A. (1968). The Netsilik Eskimos: Adaptive Processes, *in* R.B. Lee and I. Devore (eds.), *Man the Hunter*, Chicago, pp. 78–82.

— — (1970). *The Netsilik Eskimo*, New York.

Barendsen, G.W., Deevey, E.S. and Gralenski, L.J. (1957). Yale Natural Radiocarbon Measurements III, *Science* **126**, 908–19.

Barker, H., Burleigh, R. and Meeks, N. (1969). British Museum Natural Radiocarbon Measurements VI, *Radiocarbon* **11**, 278–94.

— — (1971). British Museum Natural Radiocarbon Measurements VII, *Radiocarbon* **13**, 157–88.

Barker, H. and Mackey, C.J. (1960). British Museum Natural Radiocarbon Measurements II, *Radiocarbon* 2, 26–30.

— — (1961). British Museum Natural Radiocarbon Measurements III, *Radiocarbon* **3**, 39–45.

Barnes, B., Edwards, B.J.N., Hallam, J.S. and Stuart, A.J. (1971). Skeleton of a Late Glacial Elk Associated with Barbed Points from Poulton-le-Fylde, Lancashire, *Nature* **232**, 488–89.

Bartley, D.D. (1962). The Stratigraphy and Pollen Analysis of Lake Deposits near Tadcaster, Yorkshire, *New Phytol.* **61**, 277–87.

Bastin, B. (1970). La chronostratigraphie du Würm en Belgique, à la lumière de la palynologie des loess et limons, *Ann. Soc. géol. Belg.* **93**(3), 545–80.

Baxter, M.S., Ergin, M. and Walton, A. (1969). Glasgow University Radiocarbon Measurements I, *Radiocarbon* **11**, 43–52.

Beaulieu, J.L. de (1969). Analyse pollinique des sédiments du sol de la cabane acheuléenne du Lazaret, *in* H. de Lumley *et al.*, Une Cabane Acheuléene dans la Grotte du Lazaret (Nice), *Mém. Soc. Préhist. Franç.* 7, 125–26.

Beck, C.W. (1965). The Origin of the Amber found at Gough's Cave, Cheddar, Somerset, *Proc. Univ. Bristol Spel. Soc.* **10**, 272–76.

Becker, C.J. (1971). Late Palaeolithic Finds from Denmark, *Proc. Prehist. Soc.* 37, 131–39.

Bell, F.G. (1969). The Occurrence of Southern Steppe and Halophyte Elements in Weichselian (Last-glacial) Floras of Southern Britain, *New Phytol.* **68**, 913–22.

Benirschke, K. (1969). Cytogenetics in the Zoo, *New Scientist*, 16 January, 132–33.

Berger, R. and Libby, W.F. (1969). UCLA Radiocarbon Dates IX, *Radiocarbon* **11**, 194–209.

Berndt, R.M. (1972). The Walmadjeri and Gugadja, *in* M.G. Bicchieri (ed.), *Hunters and Gatherers Today*, New York, pp. 177–216.

Beug, H.–J. (1961). *Leitfaden der Pollenbestimmung für Mitteleuropa und angrenzende Gebiete*, Lief. 1, Stuttgart.

Beynon, F. (1934). The Cow Cave, Chudleigh, *Trans. & Proc. Torquay Nat. Hist. Soc.* **6**, 127–32.

Beynon, F., Dowie, H.G. and Ogilvie, A.H. (1929). Report on the Excavations in Kent's Cavern, 1926–9, *Trans. & Proc. Torquay Nat. Hist. Soc.* 5, 237–42.

Binford, L.R. (1962). Archaeology as Anthropology, *Amer. Antiquity* **28**, 217–25.

— — (1964). A Consideration of Archaeological Research Design, *Amer. Antiquity* **29**, 425–41.

— — (1965). Archaeological Systematics and the Study of Cultural Process, *Amer. Antiquity* **31**, 203–10.

— — (1972). *An Archaeological Perspective*, New York.

Binford, L.R. and S.R. (1966a). The Predatory Revolution: A Consideration of the Evidence for a New Subsistence Level, *Amer. Anth.* **68**(2), pt.1, 508–12.

— — (1966b). A Preliminary Analysis of Functional Variability in the Mousterian of Levallois Facies, *in* J.D. Clark and F.C. Howell (eds.), Recent Studies in Palaeoanthropology, *Amer. Anth.* **68**, 238-95.

— — (eds.) (1968). *New Perspectives in Archeology*, Chicago.

Bishop, W.W. and Dickson, J.H. (1970). Radiocarbon Dating of Scottish Late-glacial Sea,

Nature **227**, 481–82.

Blackburn, K.B. (1952). The Dating of a Deposit Containing an Elk Skeleton Found at Neasham, near Darlington, Co. Durham, *New Phytol.* **51**, 364–77.

Boas, F. (1901). The Eskimo of Baffin Land and Hudson Bay, *Bull. Amer. Mus. Nat. Hist.* **15**, 1–370.

Bohlken, H. (1961). Haustiere und zoologische Systematik, *Zeitschr. Tierzücht. Züchtungsbiol.* **76**, 107–13.

Bohmers, A. (1951). Die Höhlen von Mauern, *Palaeohistoria* **1**, 1–107.

– – (1956). Statistics and Graphs in the Study of Flint Assemblages, *Palaeohistoria* **5**, 1–25.

– – (1960). Statistiques et graphiques dans l'étude des industries lithiques préhistoriques, *Palaeohistoria* **8**, 15–37.

– – (1963). A Statistical Analysis of Flint Artifacts, *in* D. Brothwell and E. Higgs (eds.), *Science in Archaeology, a Comprehensive Survey of Progress and Research*, London.

Bökönyi, S. (1974). *The Przewalski Horse*, London.

Bonifay, M. (1970). Antiquités préhistoriques sous-marines: Côtes françaises de la Méditerranée, *Gallia Préhist.* **13**, 585–92.

Bonnichsen, R. (1973). Millie's Camp: an Experiment in Archaeology, *World Arch.* **4**, 277–91.

Bordes, F. (1959). Le contexte archéologique des Hommes du Moustier et de Spy, *L'Anthropologie* **63**, 154–57.

– – (1961). *Typologie du Paléolithique Ancien et Moyen*, vols. 1 and 2, Bordeaux.

– – (1968). *The Old Stone Age*, London.

– – (1969). Reflections on typology and technology in the Palaeolithic, *Arctic Anth.* **6**, 1–29.

– – (1973). On the Chronology and Contemporaneity of Different Palaeolithic Cultures in France, *in* C. Renfrew (ed.), *The Explanation of Culture Change: Models in Prehistory*, London, pp. 217–26.

Bosinski, G. (1967). *Die mittelpaläolithischen Funde im westlichen Mitteleuropa*, Köln (Cologne).

– – (1972). Late Middle Palaeolithic Groups in North-western Germany and their Relations to Early Upper Palaeolithic Industries, *in* F. Bordes (ed.), *The Origin of* Homo sapiens, *Ecology and Conservation* 3, Paris: Unesco, pp. 153–60.

Bosinski, G. and Hahn, J. (1972). Der Magdalénien-Fundplatz Andernach (Martinsberg), *Rheinische Ausgrabungen* **11**, 81–257.

Boughey, A.S. (1971). *Fundamental Ecology*, London.

Bowen, D.Q. (1966). Dating Pleistocene Events in South-west Wales, *Nature* **211**, 475–76.

– – (1970). The Palaeoenvironment of the 'Red Lady' of Paviland, *Antiquity* **44**, 134–36.

Bramwell, D. (1955). Third Report on Excavations at Ossum's Cave, *Peakland Arch. Soc. Newsletter* **13**, 7–8.

– – (1959a). The Excavation of Dowel Cave, *Derbys. Arch. J.* **79**, 97–109.

– – (1959b). Report on a Collection of Bird Bones from the 1929 Excavations at Soldier's Hole, Cheddar, *Proc. Soms. Nat. Hist. & Arch. Soc.* **104**, 87–90.

– – (1960). Some Research into Bird Distribution in Britain during the Late Glacial and Post Glacial Periods, *Bird Report 1959–60 (Merseyside Nat. Assoc.)*, pp. 51–58.

– – (1962). The Nature of the Upper Palaeolithic Cultures in the Western Areas of the Peak District, *Peakland Arch. Soc. Newsletter* **18**, 17.

– – (1964). The Excavations at Elder Bush Cave, Wetton, Staffs, *N. Staffs. J. Field Studies* **4**, 46–60.

– – (1969). The Tenth Report on Fox Hole Cave, High Wheeldon, *Peakland Arch. Soc. Bull.* **25**, 8–10.

– – (1971). Excavations at Fox Hole Cave, High Wheeldon, 1961–1970, *Derbys. Arch. J.* **91**, 1–19.

— — (1973) *Archaeology in the Peak District: a Guide to the Region's Prehistory*, Buxton.

Breuil, H. (1922a). Observations on the Pre-Neolithic Industries of Scotland, *Proc. Soc. Ant. Scot.* **56**, 261–81.

— — (1922b). Remarques sur les divers niveaux archéologiques du gisement de Spy (Belgique), *Rev. Anth.* **22**, 126–29.

— — (1950). Glânes conchyliologiques en France (Nord et Sud-Ouest), *Congrès Préhist. France (Paris)* **13**, 191–240.

British Caenozoic Fossils (1963). *British Caenozoic Fossils (Tertiary and Quaternary): British Museum (Natural History)*, London.

Broecker, W.S., Thurber, D.C., Goddard, J., Ku, T.L., Mathews, R.K. and Mesolella, K.J. (1968). Milankovitch Hypothesis Supported by Precise Dating of Coral Reefs and Deep-Sea Sediments, *Science* **159**, 297–300.

Brothwell, D. (1972). Palaeodemography and Earlier British Populations, *World Arch.* **4**, 75–87.

Bryan, R.B. (1970). Parent Materials and Texture of Peak District Soils. *Zeitschrift f. Geomorph.* **14**, 262–74.

Buckland, W. (1823). *Reliquiae Diluvianae: or Observations on the Organic Remains Contained in Caves, Fissures, and Diluvial Gravel and on other Geological Phenomena Attesting the Action of an Universal Deluge*, London.

Buckley, J.D., Trautman, M.A. and Willis, E.H. (1968). Isotopes' Radiocarbon Measurements VI, *Radiocarbon* **10**, 246–94.

Buckley, J.D. and Willis, E.H. (1969). Isotopes' Radiocarbon Measurements VII, *Radiocarbon* **11**, 53–105.

— — (1970). Isotopes' Radiocarbon Measurements VIII, *Radiocarbon* **12**, 87–129.

— — (1972). Isotopes' Radiocarbon Measurements IX, *Radiocarbon* **14**, 114–39.

Burch, E.S. (1972). The Caribou/Wild Reindeer as a Human Resource, *Amer. Antiquity* **37**, 339–68.

Burkitt, M.C. (1938). Description of a Flint Implement from a Digging in the Gravel of the Eastington Pit, *Proc. Cotteswold Nat. Field Club* **26**, 296–97.

Burleigh, R., Switzur, V.R. and Renfrew, C. (1973). The Radiocarbon Calendar Recalibrated Too Soon? *Antiquity* **47**, 309–17.

Butzer, K.W. (1964). *Environment and Archaeology: an Introduction to Pleistocene Geography*, Chicago.

— — (1971). *Environment and Archeology: an Ecological Approach to Prehistory*, Chicago.

Callow, W.J., Baker, M.J. and Hassall, G.I. (1965). National Physical Laboratory Radiocarbon Measurements III, *Radiocarbon* **7**, 156–61.

— — (1966). National Physical Laboratory Radiocarbon Measurements IV, *Radiocarbon* **8**, 340–47.

Callow, W.J., Baker, M.J. and Pritchard, D.H. (1963). National Physical Laboratory Radiocarbon Measurements I, *Radiocarbon* **5**, 34–38.

— — (1964). National Physical Laboratory Radiocarbon Measurements II, *Radiocarbon* **6**, 25–30.

Callow, W.J. and Hassall, G.I. (1969). National Physical Laboratory Radiocarbon Measurements VI, *Radiocarbon* **11**, 130–36.

Campbell, J.B. (1969). Excavations at Creswell Crags: Preliminary Report, *Derbys. Arch. J.* **89**, 47–58.

— — (1970). The Upper Palaeolithic Period, *in* J. Campbell, D. Elkington, P. Fowler and L. Grinsell, *The Mendip Hills in Prehistoric and Roman Times*, Bristol, pp. 5–11.

— — (1971). The Upper Palaeolithic of Britain: a Study of British Upper Palaeolithic Cultural Material and its Relation to Environmental and Chronological Evidence, 2 vols. Oxford Univ. D. Phil. dissertation.

Campbell, J.B. and Pohl, M. (1971). Appendix. Pollen Analysis of the Loessic Deposits at La Cotte de St Brelade, Jersey, C.I., *in* C.B.M. McBurney and P. Callow, The Cambridge Excavations at La Cotte de St Brelade, Jersey – a Preliminary Report, *Proc. Prehist. Soc.* 37(2), 204–07.

Campbell, J.B. and Sampson, C.G. (1971). A New Analysis of Kent's Cavern, Devonshire, England, *Univ. Oregon Anth. Papers*, no. 3.

Campbell, J.M. (1968). Territoriality among Ancient Hunters: Interpretations from Ethnography and Nature, *in* B. Meggers (ed.), *Anthropological Archeology in the Americas*, Washington, pp. 1–21.

Campo, M. van (1969). Végétation würmienne en France. Données bibliographiques. Hypothèse, *Bull. Assoc. franç. Etude Quaternaire*, suppl., pp. 104–11.

Casteel, R.W. (1972). Some Archaeological Uses of Fish Remains, *Amer. Antiquity* 37, 404–19.

– – (1974). On the Number and Sizes of Animals in Archaeological Faunal Assemblages, *Archaeometry* **16**, 238–43.

Catt, J.A., Corbett, W.M., Hodge, C.A.H., Madgett, P.A., Tatler, W. and Weir, A.H. (1971). Loess in the Soils of North Norfolk, *J. Soil Sci.* 22, 444–52.

Charlesworth, J.K. (1929). The South Wales End-moraine, *Quart. J. Geol. Soc.* **85**, 335–58.

– – (1956). The Late-glacial History of the Highlands and Islands of Scotland, *Trans. Roy. Soc. Edinb.* 62, 769–928.

– – (1963). Some Observations on the Irish Pleistocene, *Proc. Roy. Irish Acad. (B)* **62**, 295–322.

Chauchat, C. and Thibault, C. (1968). La station de plein air du Basté à Saint-Pierre d'Irube (Basses-Pyrénées), *Bull. Soc. préhist. franç.* **65**, 295–318.

Chisholm, M. (1962). *Rural Settlement and Land Use: An Essay in Location*, London.

Chmielewski, W. (1961). *Civilisation de Jerzmanovice*, Inst. Hist. Kult. Mat. Polskiej Akad. Nauk, Warszawa-Kraków (Warsaw-Cracow).

– – (1972). The Continuity and Discontinuity of the Evolution of Archaeological Cultures in Central and Eastern Europe between the 55th and 25th Millenaries B.C., *in* F. Bordes (ed.), *The Origin of* Homo sapiens, *Ecology and Conservation* 3, Paris: Unesco, pp. 173–79.

Clapham, A.R., Tutin, T.G. and Warburg, E.F. (1962). *Flora of the British Isles*, Cambridge.

Clapham, W.B. (1973). *Natural Ecosystems*, New York.

Clark, A. McF. (1974). The Athapaskans: Strangers of the North, *in* N.J. Boudreau (ed.), *The Athapaskans: Strangers of the North* (*National Museum of Man*), Ottawa, pp. 17–42.

Clark, J.G.D. (1932). *The Mesolithic Age in Britain*, Cambridge.

– – (1938). Reindeer Hunters' Summer Camps in Britain? *Proc. Prehist. Soc.* **4**, 229.

– – (1948). The Development of Fishing in Prehistoric Europe, *Antiquaries J.* **28**, 45–85.

– – (1952). *Prehistoric Europe: The Economic Basis*, London.

– – (1954). *Excavations at Star Carr, an Early Mesolithic Site at Seamer near Scarborough, Yorkshire*, Cambridge.

– – (1972). Star Carr: a Case Study in Bioarchaeology, *Addison-Wesley Modular Publ.* 10, 1–42.

Clarke, D.L. (1968). *Analytical Archaeology*, London.

Clarke, W.G. (1914). Norfolk Implements of Palaeolithic "Cave" Types, *Proc. Prehist. Soc. E. Ang.* 1, 338–45.

Clay, R.C.C. (1927). Excavations at Chelm's Combe, Cheddar: Report on the Bone and Flint Implements, *Proc. Som. Arch. & Nat. Hist. Soc.* 72, 113–15.

Clifford, E.M., Garrod, D.A.E. and Gracie, H.S. (1954). Flint Implements from Gloucestershire, *Antiquaries J.* **34**, 178–83.

Coles, J.M. *et al.* (1971). The Early Settlement of Scotland: Excavations at Morton, Fife. *Proc. Prehist. Soc.* 37(2), 284–366.

Coles, J.M. and Higgs, E.S. (1969). *The Archaeology of Early Man*, London.

Collins, D. (1973). Early Man, *in* D. Collins, R. Whitehouse, M. Henig and D. Whitehouse, *Background to Archaeology: Britain in its European Setting*, Cambridge, pp. 1–25.

Collins, E.R. (1933). Upper Palaeolithic Sites in Nidderdale, *Proc. Prehist. Soc. E. Ang.* **7**, 185–87.

Conolly, A.P., Godwin, H. and Megaw, E.M. (1950). Studies in the Post-glacial History of British Vegetation XI: Late-glacial Deposits in Cornwall, *Phil. Trans. Roy. Soc. (B)* **234**, 397–469.

Coon, C.S. (1972). *The Hunting Peoples*, London.

Coope, G.R. (1959). A Late Pleistocene Insect Fauna from Chelford, Cheshire, *Proc. Roy. Soc. (B)* **151**, 70–86.

— — (1961). On the Study of Glacial and Interglacial Insect Faunas, *Proc. Linnean Soc. London* **172**, 62–65.

— — (1962). A Pleistocene Coleopterous Fauna with Arctic Affinities from Fladbury, Worcestershire, *Quart. J. Geol. Soc.* **118**, 103–23.

— — (1965a). Fossil Insect Faunas from Late Quaternary Deposits in Britain, *Advance. Sci.* **21**, 564–75.

— — (1965b). The Response of the British Insect Fauna to Late Quaternary Climatic Oscillations, *Proc. 12th Internat. Congr. Entomol. (London)* **173**, 444–45.

— — (1967). The Value of Quaternary Insect Faunas in the Interpretation of Ancient Ecology and Climate, *in* E.J. Cushing and H.E. Wright (eds.), *Quaternary Paleoecology*, New Haven.

— — (1968a). Coleoptera from the 'Arctic Bed' at Barnwell Station, Cambridge, *Geol. Mag.* **105**, 482–86.

— — (1968b). Insect Fauna from Middle Weichselian Deposit at Brandon, Warwickshire, *Phil. Trans. Roy. Soc. (B)* **254**, 425–56.

— — (1969a). The Contribution that the Coleoptera of Glacial Britain Could Have Made to the Subsequent Colonisation of Scandinavia, *Opuscula Entomol.* **34**, 95–108.

— — (1969b). Late Weichselian Coleoptera in Britain, *Résumés des Communications VIII Congrès INQUA (Paris), 1969*, p. 146.

— — (1969c). The Response of Coleoptera to Gross Thermal Changes during the Mid-Weichselian Interstadial, *Mitt. Int. Verein. Limnol.* **17**, 173–83.

— — (1970). Interpretations of Quaternary Insect Fossils, *Ann. Rev. Entomol.* **15**, 97–120.

— — (1975). Climatic Fluctuations of North West Europe since the Last Interglacial Indicated by Fossil Assemblages of Coleoptera, *Geol. J.*, special issue no. 6 (Ice Ages: Ancient and Modern), pp. 133–68.

Coope, G.R., Morgan, A. and Osborne, P.J. (1971). Fossil Coleoptera as Indicators of Climatic Fluctuations during the Last Glaciation in Britain, *Palaeogeogr., Palaeoclimatol., Palaeoecol.* **10**, 87–101.

Coope, G.R., and Sands, C.H. (1966). Insect Faunas of the Last Glaciation from the Tame Valley, Warwickshire, *Proc. Roy. Soc. (B)* **165**, 389–412.

Coope, G.R., Shotton, F.W. and Strachan, I. (1961). A Late Pleistocene Fauna and Flora from Upton Warren, Worcestershire, *Phil. Trans. Roy. Soc. (B)* **244**, 379–421.

Corbet, G.B. (1966). *The Terrestrial Mammals of Western Europe*, London.

Cordy, J.–M. (1974). La faune aurignacienne de la Grotte Princesse Pauline à Marche-les-Dames, *Bull. Soc. Roy. Belge Anth. et Préhist.* **85**, 243–52.

Cornwall, I.W. (1958). *Soils for the Archaeologist*, London.

Cowling, E.T. (1946). *Rombalds Way*, Otley.

Dale, M.B. (1970). Systems Analysis and Ecology, *Ecology* **51**, 2–16.

Damblon, F. (1974). Observations palynologiques dans la grotte de Remouchamps, *Bull. Soc. Roy. Belge Anth. et Préhist.* **85**, 131–55.

Dansgaard, W., Johnsen, S.J., Clausen, H.B. and Langway, C.C. (1971). Climatic Record Revealed by the Camp Century Ice Core, *in* K.K. Turekian (ed.), *The Late Cenozoic Glacial Ages*, New Haven (Conn.), pp. 37–56.

Danthine, H. (1961). Fouilles dans un gisement préhistorique du Domaine de Presle: Rapport préliminaire, *Documents et Rapports Soc. Roy. Arch. et Paléontol. Charleroi* **50** (premier fasc., 1955–60), 3–39.

Darwin, C. (1859). *On the Origin of Species, by Means of Natural Selection*, London.

Dauchot-Dehon, M. and Heylen, J. (1975). Institut Royal du Patrimoine Artistique Radiocarbon Dates V, *Radiocarbon* **17**, 1–3.

Davies, H.N. (1904). The Discovery of Human Remains under the Stalagmite Floor of Gough's Cavern, Cheddar, *Quart. J. Geol. Soc.* **60**, 335–48.

Davies, J.A. (1921). Aveline's Hole, Burrington Coombe, an Upper Palaeolithic Station, *Proc. Univ. Bristol Spel. Soc.* **1**, 61–82.

— — (1922). Second Report on Aveline's Hole, *Proc. Univ. Bristol Spel. Soc.* **1**, 113–25.

— — (1923). Third Report on Aveline's Hole, *Proc. Univ. Bristol Spel. Soc.* 2, 5–39.

— — (1925). Fourth Report on Aveline's Hole, *Proc. Univ. Bristol Spel. Soc.* 2, 104–21.

— — (1926). Notes on Upper Palaeolithic Implements from some Mendip Caves, *Proc. Univ. Bristol Spel. Soc.* 2, 261–73.

Davies, M. (1974). Some Further Cave Excavations, *William Pengelly Cave Studies Trust Newsletter*, **23**, 6–7.

Davis, M.B. (1968). Pollen Grains in Lake Sediments: Redeposition Caused by Seasonal Water Circulation, *Science* **162**, 796–98.

Dawkins, W.B. (1862). On a Hyaena Den at Wookey Hole, near Wells, *Quart. J. Geol. Soc.* **18**, 115–26.

— — (1863a). On a Hyaena Den at Wookey Hole, near Wells, *Quart. J. Geol. Soc.* **19**, 260–74.

— — (1863b). Wookey Hole Hyaena Den, *Proc. Soms. Arch. & Nat. Hist. Soc.* **11**, 196–219.

— — (1874). *Cave Hunting, Researches on the Evidence of Caves Respecting the Early Inhabitants of Europe*, London.

— — (1876). On the Mammalia and Traces of Man found in the Robin-Hood Cave, *Quart. J. Geol. Soc.* **32**, 245–58.

— — (1877). On the Mammal Fauna of the Caves of Creswell Crags, *Quart. J. Geol. Soc.* **33**, 589–612.

— — (1880). *Early Man in Britain and his Place in the Tertiary Period*, London.

Dawkins, W.B. and Mello, J.M. (1879). Further Discoveries in the Creswell Caves, *Quart. J. Geol. Soc.* **35**, 724–35.

Deacon, J. (1972). A Data Bank of Quaternary Plant Fossil Records, *New Phytol.* **71**, 1227–32.

Deevey, E.S., Flint, R.F. and Rouse, I. (eds.) (1967). *Radiocarbon Measurements: Comprehensive Index, 1950–1965*, New Haven (Conn.).

Delibrias, G., Guillier, M.T. and Labeyrie, J. (1971). Gif Natural Radiocarbon Measurements VI, *Radiocarbon* **13**, 213–54.

Delibrias, G. and Larsonneur, C. (1966). Datation absolue de dépôts organiques würmiens en Normandie, *Comptes rendus (Paris)* **263**, 1023.

Derville, H. and Firtion, F. (1951). Sur la palynologie du dépôt de comblement d'un abri-sous-roche de Haute-Auvergne, *Comptes rendus (Paris)* **233**, 423–24.

Dewez, M.C. (1970). Contribution à la technologie lithique du Paléolithique supérieur final, *Bull. Soc. Roy. Belge Anth. et Préhist.* **81**, 39–59.

— — (1974). New Hypotheses Concerning Two Engraved Bones from La Grotte de Remouchamps, Belgium, *World Arch.* **5**, 337–45.

— — (1975). Nouvelles recherches à la grotte du Coléoptère à Bomal-sur-Ourthe: Rapport provisoire de la premiere campagne de fouille, *Helinium* **15**, pp. 105–33.

Dewez, M.C., Brabant, H., Bouchud, J., Callut, M., Damblon, F., Degerbøl, M., Ek, C., Frère, H. and Gilot, E. (1974). Nouvelles recherches à la grotte de Remouchamps, *Bull. Soc. Roy. Belge Anth. et Préhist.* **85**, 5–161.

Dimbleby, G.W. (1967). *Plants and Archaeology*, London.

Donner, J.J. (1970). Land/Sea Level Changes in Scotland, *in* D. Walker and R.G. West (eds.), *Studies in the Vegetational History of the British Isles*, Cambridge, pp. 23–39.

Donner, J.J. and Kurtén, B. (1958). The Floral and Faunal Succession of Cueva del Toll, Spain, *Eiszeitalter u. Gegenw.* 9, 72–82.

Donovan, D.T. (1955). The Pleistocene Deposits at Gough's Cave, Cheddar, Including an Account of Recent Excavations, *Proc. Univ. Bristol Spel. Soc.* 7, 76–104.

— — (1962). Sea Levels of the Last Glaciation, *Geol. Soc. Amer. Bull.* 73, 1297–98.

— — (ed.) (1968). *Geology of Shelf Seas*, Edinburgh.

Dowie, H.G. (1925). The Excavation of a Cave at Torbryan, *Trans. & Proc. Torquay Nat. Hist. Soc.* 4, 261–68.

— — (1928). Note on Recent Excavations in Kent's Cavern, *Proc. Prehist. Soc. E. Ang.* 5, 306–07.

Dowie, H.G. and Ogilvie, A.H. (1927). Second Report on the Excavations in Kent's Cavern, Torquay: October 1926–June 1927, *Rep. Brit. Assoc. Advmt. Sci. (Leeds)*, p. 305.

Draper, J.C. (1962). Flints of Upper Palaeolithic Aspect from Long Island, Hants., *Proc. Hants. Field Club* 22, 105–06.

Dresser, P.Q. and McAulay, I.R. (1974). Dublin Radiocarbon Dates II, *Radiocarbon* **16**, 6–9.

Dudley, H. (1949). *Early Days in North-Western Lincolnshire*, Scunthorpe.

Dumait, P., Marceau, L., Devin, C. and Campo, M. van (1963). Nouvelle méthode de concentration des pollens dans les sédiments pauvres par microflottation, *Comptes rendus (Paris)* **256**, 231–33.

Dupont, E. (1865). Etude sur les Cavernes des bords de la Lesse et de la Meuse explorées jusqu'au mois d'octobre 1865, *Bull. Acad. Roy. Belgique* 20, 824–50.

— — (1867). *Notices préliminaires sur les Fouilles exécutées sous les auspices du Gouvernement belge dans les Cavernes de la Belgique*, vols. 1 and 2, Brussels.

— — (1873). *L'Homme pendant les Ages de la Pierre dans les Environs de Dinant-sur-Meuse*, Brussels.

Ellis, A.E. (1951). Census of the Distribution of British Non-marine Mollusca, *J. Conch.* **23**, 171–244.

Ellwood, J. (1818). Ground Plan of a Cave at Darren-y-kille in the Parish of Llangattock, near Crickhowell, Brecknockshire, Wales, August, 1818, unpubl. Oxford Univ. Museum.

Eloy, L. (1956). Le Proto Solutréen dans le bassin de la Meuse, en Belgique, *Bull. Soc. préhist. franç.* **53**, 532–39.

Embleton, C. and King, C.A.M. (1968). *Glacial and Periglacial Geomorphology*, London.

Emiliani, C. (1972). Quaternary Paleotemperatures and the Duration of the High-temperature Intervals, *Science* **178**, 398–401.

Engleheart, G.H. (1923). Surface Implements from Wiltshire, *Antiquaries J.* 3, 144–45.

Engstrand, L.G. (1967). Stockholm Natural Radiocarbon Measurements VII, *Radiocarbon* 9, 387–438.

Evans, J. (1872). *The Ancient Stone Implements, Weapons and Ornaments of Great Britain*, London; revised 1897.

Evans, J.G. (1972). *Land Snails in Archaeology, with Special Reference to the British Isles*, London.

Evans, J.G. and Jones, H. (1973). Subfossil and Modern Land-snail Faunas from Rock-rubble Habitats, *J. Conch.* 8, 103–29.

Faegri, K. and Iversen, J. (1964). *Textbook of Pollen Analysis*, Copenhagen.

Falconer, H. (1868). *Palaeontological Memoirs: Volume 2*, London.

Fisher, J. (1966). *The Shell List of British and Irish Birds*, London.

Forde, C.D. (1934). *Habitat, Economy and Society: a Geographical Introduction to Ethnology*, London.
Fox, W.S. (1910). Ravencliffe Cave, *J. Derbys. Arch. & Nat. Hist. Soc.* **32**, 141–46.
— — (1911). Derbyshire Cave-Men of the Roman Period, *J. Derbys. Arch. & Nat. Hist. Soc.* **33**, 115–26.
Franks, J.W. and Pennington, W. (1961). The Late-glacial and Post-glacial Deposits of the Esthwaite Basin, North Lancashire, *New Phytol.* **60**, 27–42.
Frenzel, B. (1964). Zur Pollenanalyse von Lössen, *Eiszeitalter u. Gegenw.* **15**, 5-39.
— — (1968a). *Grundzüge der Pleistozänen Vegetationsgeschichte Nord-Eurasiens*, Wiesbaden.
— — (1968b). The Pleistocene Vegetation of Northern Eurasia, *Science* **161**, 637–49.
Garrod, D.A.E. (1926). *The Upper Palaeolithic Age in Britain*, Oxford.
— — (1927). Excavations at Langwith Cave, Derbyshire, April 11–27, 1927, *Rep. Brit. Assoc. Advmt. Sci. (Leeds)*.
Geer, G. de (1912). A Geochronology of the Last 12,000 Years, *C.R. 11th Int. Geol. Congress (Stockholm)* **1**, 241–53.
— — (1934). Geology and Geochronology, *Geogr. Annaler* **16**, 1–52.
— — (1940). Geochronologia Suecica Principles, *Kungl. Svenska Vetensk. Handl.* **18**, no. 6.
Gilbertson, D.D. and Hawkins, A.B. (1974). Upper Pleistocene Deposits and Landforms at Holly Lane, Clevedon, Somerset (ST 419727), *Proc. Univ. Bristol. Spel. Soc.* **13**, 349–60.
Godwin, H. (1956). *The History of the British Flora: a Factual Basis for Phytogeography*, Cambridge.
— — (1959). Studies of the Post-glacial History of British Vegetation, XIV: Late-glacial Deposits at Moss Lake, Liverpool, *Phil. Trans. Roy. Soc. (B)* **242**, 127–49.
Godwin, H. and Switsur, V.R. (1966). Cambridge University Natural Radiocarbon Measurements VIII, *Radiocarbon* **8**, 390–400.
Godwin, H., Walker, D. and Willis, E.H. (1957). Radiocarbon Dating and Post-glacial Vegetational History: Scaleby Moss, *Proc. Roy. Soc. (B)* **147**, 352–66.
Godwin, H. and Willis, E.H. (1959). Cambridge University Natural Radiocarbon Measurements I, *Radiocarbon* **1**, 63–75.
— — (1960). Cambridge University Natural Radiocarbon Measurements II, *Radiocarbon* **2**, 62–72.
— — (1961). Cambridge University Natural Radiocarbon Measurements III, *Radiocarbon* **3**, 60–76.
— — (1962). Cambridge University Natural Radiocarbon Measurements V, *Radiocarbon* **4**, 57–70.
— — (1964). Cambridge University Natural Radiocarbon Measurements VI, *Radiocarbon* **6**, 116–37.
Godwin, H., Willis, E.H. and Switsur, V.R. (1965). Cambridge University Natural Radiocarbon Measurements VII, *Radiocarbon* **7**, 205–12.
Gould, R.A. (1969). *Yiwara: Foragers of the Australian Desert*, London.
— — (1971). The Archaeologist as Ethnographer: a Case from the Western Desert of Australia, *World Arch.* **3**, 143–77.
Graindor, M.–J. (1972). Upper Palaeolithic Rock Engravings at Gouy (France), *World Arch.* **3**, 243–51.
Grayson, D.K. (1973). On the Methodology of Faunal Analysis, *Amer. Antiquity* **38**, 432–39.
Grimes, W.F. (1933). Priory Farm Cave, Monkton, Pembrokeshire, *Arch. Cambrensis* **88**, 88–100.
Grimes, W.F. and Cowley, L.F. (1935). Coygan Cave, Llansadyrnin, *Arch. Cambrensis* **90**, 95–111.
Gross, H. (1940). Die Renntierjäger-Kulturen Ostpreussens, *Prähist. Zeitschrift* **30–31**, 39–67.
Gubser, N.J. (1965). *The Nunamiut Eskimos: Hunters of Caribou*, New Haven (Conn.).

Hahn, J. (1972). Aurignacian Signs, Pendants and Art Objects in Central and Eastern Europe, *World Arch.* **3**, 252–66.

— — (1973). Eine jungpaläolithische Feuerstelle aus Lommersum, Kreis Euskirchen, *Rheinische Ausgrabungen* **11**, 56–80.

— — (1974). Lommersum, Kr. Euskirchen, *in* G. Bosinski, K. Brunnacker, L. Fiedler, J. Hahn, H. Löhr, W. Schol, H. Thieme and G. Weiss, *Altsteinzeitliche Fundplätze des Rheinlandes*, Köln (Cologne), pp. 53–55.

Hahn, J., Müller-Beck, H. and Taute, W. (1973). *Eiszeithöhlen im Lonetal: Archäologie einer Landschaft auf der Schwäbischen Alb*, Stuttgart.

Hall, E.R. and Kelson, K.R. (1959). *The Mammals of North America*, vols. 1 and 2, New York.

Hallam, J.S., Edwards, B.J.N., Barnes, B. and Stuart, A.J. (1973). The Remains of a Late Glacial Elk Associated with Barbed Points from High Furlong, near Blackpool, Lancashire, *Proc. Prehist. Soc.* **39**, 100–28.

Halls, H.H. (1914). Implements from a Station at Cranwich, Norfolk, *Proc. Prehist. Soc. E. Ang.* **1**, 454–57.

Hammen, T. van der, Maarleveld, G.C., Vogel, J.C. and Zagwijn, W.H. (1967). Stratigraphy, Climatic Succession and Radiocarbon Dating of the Last Glacial in the Netherlands, *Geol. en Mijnb.* **46**, 79–95.

Harkness, D.D. and Burleigh, R. (1974). Possible Carbon–14 Enrichment in High Altitude Wood, *Archaeometry* **16**, 121–27.

Harkness, D.D. and Wilson, H.W. (1973). Scottish Universities Research and Reactor Centre Radiocarbon Measurements I, *Radiocarbon* **15**, 554–65.

— — (1974). Scottish Universities Research and Reactor Centre Radiocarbon Measurements II, *Radiocarbon* **16**, 238–51.

Harmer, F.W. (1914–19). *The Pliocene Mollusca of Great Britain*, London.

Harrison, S.G., Masefield, G.B. and Wallis, M. (1969). *The Oxford Book of Food Plants*, Oxford.

Hawkes, C.J., Tratman, E.K. and Powers, R. (1970). Decorated Piece of Rib Bone from the Palaeolithic Levels at Gough's Cave, Cheddar, Somerset, *Proc. Univ. Bristol Spel. Soc.* **12**, 137–42.

Heinzelin, J. de (1973). An Example of Upper Palaeolithic Geometry, *Antiquity* **47**, 297–98.

Helm, J. (1968). The Nature of Dogrib Socioterritorial Groups, *in* R.B. Lee and I. DeVore (eds.), *Man the Hunter*, Chicago, pp. 118–25.

Henderson, A.H. (1973). Flint and Chert Implements from Froggatt, Derbyshire, *Trans. Hunter Arch. Soc.* **10**, 138–42.

Hewer, T.F. (1926). King Arthur's Cave, *Proc. Univ. Bristol Spel. Soc.* **2**, 221–28.

Hickerson, H. (1965). The Virginia Deer and Intertribal Buffer Zones in the Upper Mississippi Valley, *in* A. Leeds and A.P. Vayda (eds.), *Man, Culture, and Animals: The Role of Animals in Human Ecological Adjustments*, Washington, pp. 43–65.

— — (1970). *The Chippewa and their Neighbors: a Study in Ethnohistory*, New York.

Hicks, H. (1886). Results of some Recent Researches in some Bone-caves in North Wales, *Quart. J. Geol. Soc.* **42**, 3–11.

— — (1888). On the Cae Gwyn Cave, North Wales, *Quart. J. Geol. Soc.* **44**, 562–77.

Hubbs, C.L., Bien, G.S. and Suess, H.E. (1965). La Jolla Natural Radiocarbon Measurements IV, *Radiocarbon* **7**, 66–117.

Hulshof, A.K., Jungerius, P.D. and Riezebos, P.A. (1968). A Late Glacial Volcanic Ash Deposit in South-east Belgium, *Geol. en Mijnb.* **47**, 106–11.

Isaac, G. (1971). The Diet of Early Man: Aspects of Archaeological Evidence from Lower and Middle Pleistocene Sites in Africa, *World Arch.* **2**, 300–20.

Isenbart, H.–H. (1969). *The Kingdom of the Horse*, London.

Jackson, J.W. (1945). A Lance-point of Upper Palaeolithic Type from Victoria Cave, Settle, Yorkshire, *Antiquaries J.* **25**, 147–48.

— — (1962). Archaeology and Palaeontology, *in* C.H.D. Cullingford (ed.), *British Caving: an Introduction to Speleology*, London, pp. 252–346.

— — (1967). The Creswell Caves, *Cave Science: J. Brit. Spel. Assoc.* **6**, 8–23.

Jackson, J.W. and Mattinson, W.K. (1932). A Cave on Giggleswick Scars, near Settle, Yorkshire, *The Naturalist*, pp. 5–9.

Jarman, M.R. (1972). A Territorial Model for Archaeology: a Behavioural and Geographical Approach, *in* D.L. Clarke (ed.), *Models in Archaeology*, London, pp. 705–33.

John, B.S. (1970). Pembrokeshire, *in* C.A. Lewis (ed.), *The Glaciations of Wales and Adjoining Regions*, London, pp. 229–65.

— — (1971a). Glaciation and the West Wales Landscape, *Nature in Wales* **12**, 138–55.

— — (1971b). The 'Red Lady' of Paviland: a Comment, *Antiquity* **45**, 141–44.

Jongsma, D. (1970). Eustatic Sea Level Changes in the Arafura Sea, *Nature* **228**, 150.

Kaszab, Z. (1966). New Sighting of Przewalski Horses, *Oryx: J. Fauna Preservation Soc.* **8**, 345–47.

Keeley, L. (1974). Technique and Methodology in Microwear Studies: a Critical Review, *World Arch.* **5**, 323–36.

Keith, A. (1925). *The Antiquity of Man*, London.

— — (1926). Report on a Human Skull Found near the North Entrance to Kent's Cavern, *Trans. Torquay Nat. Hist. Soc.* **4**, 289–94.

Kerney, M.P. (1963). Late-glacial Deposits on the Chalk of South-east England, *Phil. Trans. Roy. Soc. (B)* **246**, 203–54.

— — (1966). Snails and Man in Britain, *J. Conch.* **26**, 3–14.

Kitching, J.W. (1963). *Bone, Tooth and Horn Tools of Palaeolithic Man: an Account of the Osteodontokeratic Discoveries in Pin Hole Cave, Derbyshire*, Manchester.

Klein, R.G. (1969a). *Man and Culture in the Late Pleistocene: a Case Study*, San Francisco.

— — (1969b). The Mousterian of European Russia, *Proc. Prehist. Soc.* **35**, 77–111.

Klíma, B. (1955). Beitrag der neuen paläolithischen Station im Pavlov zur Problematik der ältesten landwirtschaftlichen Geräte, *Památky Arch.* **46**, 28–29.

— — (1963). *Dolní Věstonice: Výzkum tábořiště lovců mamutů v letech 1947–1952*, Praha (Prague).

Kurtén, B. (1968). *Pleistocene Mammals of Europe*, London.

Lacaille, A.D. (1954). *The Stone Age in Scotland*, London.

Lacaille, A.D. and Grimes, W.F. (1955). The Prehistory of Caldey, *Arch. Cambrensis* **104**, 85–165.

Lambert, C.A., Pearson, R.G. and Sparks, B.W. (1962). Flora and Fauna from Late Pleistocene Deposits at Sidgwick Avenue, Cambridge, *Proc. Linnean Soc. London* **174**, 13–29.

Large, N.F. and Sparks, B.W. (1961). The Non-marine Mollusca of the Cainscross Terrace near Stroud, Gloucestershire, *Geol. Mag.* **98**, 423–26.

Lartet, E. and Christy, H. (1865–75). *Reliquiae Aquitanicae; being Contributions to the Archaeology and Palaeontology of Périgord and the adjoining Provinces of Southern France*, London.

Layard, N.F. (1927). A Late Palaeolithic Settlement in the Colne Valley, Essex, *Antiquaries J.* **7**, 500–14.

Leach, A.L. (1916). Nanna's Cave, Isle of Caldey, *Arch. Cambrensis* **16**, 157–72.

— — (1918). *Some Prehistoric Remains in the Tenby Museum*, Tenby.

Leakey, M.D. (1971). *Olduvai Gorge; Volume 3: Excavations in Beds I and II, 1960–1963*, Cambridge.

Lee, R.B. (1968). What Hunters Do for a Living, or, How to Make Out on Scarce Resources, *in* R.B. Lee and I. DeVore (eds.), *Man the Hunter*, Chicago, pp. 30–48.

— — (1969). !Kung Bushman Subsistence: an Input-output Analysis, *in* A.P. Vayda (ed.), *Environment and Cultural Behavior: Ecological Studies in Cultural Anthropology*, New York, pp. 47–79.

— — (1972). Work Effort, Group Structure and Land-use in Contemporary Hunter-gatherers, *in* P.J. Ucko, R. Tringham and G.W. Dimbleby (eds.), *Man, Settlement and Urbanism*, London, pp. 177–85.

Lee, R.B. and DeVore, I. (eds.) (1968). *Man the Hunter*, Chicago.

Leeds, A. (1965). Reindeer Herding and Chukchi Social Institutions, *in* A. Leeds and A.P. Vayda (eds.), *Man, Culture, and Animals: The Role of Animals in Human Ecological Adjustments*, Washington. pp. 87–128.

Leroi-Gourhan, And. (1968). *The Art of Prehistoric Man in Western Europe*, London.

— — (1971). La spatule aux poissons de la grotte du Coucoulu à Calviac (Dordogne), *Gallia Préhist.* **14**, 253–59.

Leroi-Gourhan, And. and Brézillon, M. (1972). Fouilles de Pincevent: Essai d'analyse ethnographique d'un habitat magdalénien, *Gallia Préhist.*, suppl. 7.

Leroi-Gourhan, Arl. (1956). Notes sur l'analyse pollinique des sédiments quaternaires des grottes, *Comptes rendus Congrès préhist. franç. (Poitiers)*, pp. 671–75.

— — (1959). Résultats de l'analyse pollinique de la grotte d'Isturitz, *Bull. Soc. préhist. franç.* **56**, 619–24.

— — (1965). Les analyses polliniques sur les sédiments des grottes, *Bull. Assoc. franç. Etude Quaternaire* **2**, 145– 52.

— — (1966). La grotte de Prélétang (commune de Presles, Isère); II. Analyse pollinique des sédiments, *Gallia Préhist.* **9**, 85–92.

— — (1968). L'abri du Facteur à Tursac (Dordogne); III. Analyse pollinique, *Gallia Préhist.* **11**, 123–31.

Leroi-Gourhan, Arl. and And. (1964). Chronologie des Grottes d'Arcy-sur-Cure (Yonne), *Gallia Préhist.* 7, 1–64.

Letocart, L. (1970). Un gisement du Paléolithique final à Obourg "St-Macaire" (Hainaut), *in* K. Gripp, R. Schütrumpf and H. Schwabedissen (eds.), *Frühe Menschheit und Umwelt: Teil I (Fundamenta A2)*, Köln (Cologne), pp. 352–61.

Levin, B., Ives, P.C., Oman, C.L. and Rubin, M. (1965). U.S. Geological Survey Radiocarbon Dates VIII, *Radiocarbon* 7, 372–98.

Libby, W.F. (1952). Chicago Radiocarbon Dates III, *Science* **116**, 673–81.

Lindroth, C.H. (1970). Survival of Animals and Plants on Ice-free Refugia during the Pleistocene Glaciations, *Endeavour* **29**, 129–34.

Loë, A. de and Rahir, E. (1911). Nouvelles fouilles à Spy, grotte de la Betche-aux-Rotches, *Bull. Soc. Anth. Bruxelles*, pp. 40–58.

Longin, R. (1971). New Method of Collagen Extraction for Radiocarbon Dating, *Nature* **230**, 241–42.

Lösch, A. (1954). *The Economics of Location*, New Haven (Conn.).

Lowerison, B. (1914). An Aurignacian Floor at Heacham, Norfolk, *Proc. Prehist. Soc. E. Ang.* **1**, 475.

Lozek, V. (1971). Profily 'Pod Skálou' a jejich stratigrafický význam, *Československý Kras* **23**, 17–32.

Lubbock, J. (1865). *Pre-historic Times as Illustrated by Ancient Remains and the Manners and Customs of Modern Savages*, London.

Lumley, H. de *et al.* (1969). Une Cabane Acheuléene dans la Grotte du Lazaret (Nice), *Mém. Soc. préhist. franç.* 7.

— — (1972). La Grotte de l'Hortus (Valflaunès Hérault): Les chasseurs néandertaliens et leur milieu de vie, *Etudes Quaternaires Mém. Univ. Provence (Marseille)*, 1.

Lyell, C. (1863). *The Geological Evidences of the Antiquity of Man with Remarks on Theories of the Origin of Species by Variation*, London.

McAulay, I.R. and Watts, W.A. (1961). Dublin Radiocarbon Dates I, *Radiocarbon* 3, 26–38.

McBurney, C.B.M. (1959). Report on the First Season's Fieldwork on British Upper Palaeolithic Cave Deposits, *Proc. Prehist. Soc.* **25**, 260–69.

— — (1961). Two Soundings in the Badger Hole near Wookey Hole, in 1958, and their Bearings on the Palaeolithic Finds of the Late H.E. Balch, *Rep. Wells Nat. Hist. & Arch. Soc.*, pp. 19–27.

— — (1965). The Old Stone Age in Wales, *in* G. Daniel and I.L. Foster (eds.), *Prehistoric and Early Wales*, London, pp. 17–34.

— — (1967). *The Haua Fteah (Cyrenaica) and the Stone Age of the South-east Mediterranean*, Cambridge.

McBurney, C.B.M. and Callow, P. (1971). The Cambridge Excavations at La Cotte de St Brelade, Jersey – a Preliminary Report, *Proc. Prehist. Soc.* 37 (2), 167–207.

Mace, A. (1959). An Upper Palaeolithic Open-site at Hengistbury Head, Christchurch, Hants., *Proc. Prehist. Soc.* **25**, 233–59.

Manby, T.G. (1966). Creswellian Site at Brigham, East Yorkshire, *Antiquaries J.* **46**, 211–28.

Manker, E. (1954). *Les Lapons des Montagnes suèdoises*, Paris.

Manley, G. (1952). *Climate and the British Scene*, London.

Margalef, R. (1968). *Perspectives in Ecological Theory*, Chicago.

Marshack, A. (1971). *Notation dans les gravures du Palèolithique supèrieur: nouvelles mèthodes d'analyse*, Bordeaux.

— — (1972a). Cognitive Aspects of Upper Paleolithic Engraving, *Current Anth.* 13, 445–77.

— — (1972b). *The Roots of Civilization: The Cognitive Beginnings of Man's First Art, Symbol and Notation*, New York.

— — (1972c). Upper Paleolithic Notation and Symbol, *Science* **178**, 817–28.

— — (1975). Exploring the Mind of Ice Age Man, *National Geographic* **174**(1), 62–89.

Martin, P.S. and Guilday, J.E. (1967). A Bestiary for Pleistocene Biologists, *in* P.S. Martin and H.E. Wright (eds.), *Pleistocene Extinctions: The Search for a Cause*, New Haven (Conn.), pp. 1–62.

Martin, Y. and Graindor, M.–J. (1972). *L'Art palèolithique de Gouy*, St-Etienne-de-Rouvray (French private publ.).

Mathiassen, T. (1928). *Material Culture of the Iglulik Eskimos*, Copenhagen.

Matolcsi, J. (ed.) (1973). *Domestikationsforschung und Geschichte der Haustiere*, Budapest.

Meggers, B.J. (1971). *Amazonia: Man and Culture in a Counterfeit Paradise*, Chicago.

Mellars, P.A. (1969). Radiocarbon Dates for a New Creswellian Site, *Antiquity* **43**, 308–10.

— — (1970). An Antler Harpoon-head of 'Obanian' Affinities from Whitburn, County Durham, *Arch. Aeliana*, **48**, 337–46.

— — (1974). The Palaeolithic and Mesolithic, *in* C. Renfrew (ed.), *British Prehistory: a New Outline*, London, pp. 41–99.

Mello, J.M. (1875). On Some Bone-caves in Creswell Crags, *Quart. J. Geol. Soc.* **31**, 679–91.

— — (1876). The Bone-caves of Creswell Crags, *Quart. J. Geol. Soc.* **32**, 240–44.

— — (1877). The Bone-caves of Creswell Crags, *Quart. J. Geol. Soc.* **33**, 579–88.

Mercer, J. (1974). New C14 Dates from the Isle of Jura, Argyll, *Antiquity* **48**, 65–66.

Mercer, J.H. (1970). Antarctic Ice and Interglacial High sea Levels, *Science* **168**, 1605-06.

Metcalfe, C.R. and Chalk, L. (1950). *Anatomy of the Dicotyledons*, Oxford.

Michael, H.N. and Ralph, E.K. (1974). University of Pennsylvania Radiocarbon Dates XVI, *Radiocarbon* **16**, 198–218.

Milliman, J.D. and Emery, K.O. (1968). Sea Levels during the Past 35,000 Years, *Science* **162**, 1121–23.

Miskovsky, J.–C. (1969). Sèdimentologie des couches supèrieures de la grotte du Lazaret, *in* H. de Lumley *et al.*, Une Cabane Acheulèenne dans la Grotte du Lazaret (Nice), *Mem. Soc. prèhist. franç.* 7, 25–51.

Mitchell, G.F. (1958). A Late-glacial Deposit near Ballaugh, Isle of Man, *New Phytol.* 57, 256–63.

— — (1971). The Larnian Culture: a Minimal View, *Proc. Prehist. Soc.* 37(2), 274–83.

Mitchell, G.F., Penny, L.F., Shotton, F.W. and West, R.G. (1973). A Correlation of Quaternary Deposits in the British Isles, *Geol. Soc. London Special Rep.* **4**.

Mitchell, G.F. and Watts, W.A. (1970). The History of the Ericaceae in Ireland during the Quaternary Epoch, *in* D. Walker and R.G. West (eds.), *Studies in the Vegetational History of the British Isles*, Cambridge, pp. 13–21.

Mohr, E. (1971). *The Asiatic Wild Horse:* Equus przevalskii *Poliakoff, 1881*, London.

Moir, J.R. (1922). Four Suffolk Flint Implements, *Antiquaries J.* **2**, 114–17.

— — (1923). A Series of Solutré Blades from Suffolk and Cambridgeshire, *Proc. Prehist. Soc. E. Ang.* **4**, 71–81.

— — (1927). *The Antiquity of Man in East Anglia*, Cambridge.

— — (1929). Some hitherto Unpublished Implements, *Antiquaries J.* **9**, 8–12.

— — (1931). Ancient Man in the Gipping-Orwell Valley, Suffolk, *Proc. Prehist. Soc. E. Ang.* **6**, 182–221.

— — (1938). Four Flint Implements, *Antiquaries J.* **18**, 258–61.

Moir, J.R. and Burchell, J.P.T. (1930). Flint Implements of Upper Palaeolithic Facies from beneath the Uppermost Boulder Clay of Norfolk and Yorkshire, *Antiquaries J.* **10**, 359–83.

Moore, J.W. (1950). Mesolithic Sites in the Neighbourhood of Flixton, North-east Yorkshire, *Proc. Prehist. Soc.* **16**, 101–08.

— — (1954). Excavations at Flixton, Site 2, *in* J.G.D. Clark, *Excavations at Star Carr: an Early Mesolithic Site at Seamer near Scarborough, Yorkshire*, Cambridge, pp. 192–94.

Morgan, A. (1969). A Pleistocene Fauna and Flora from Great Billing, Northamptonshire, England, *Opuscula Entomol.* **34**, 109–29.

Morgan, A.V. (1973). The Pleistocene Geology of the Area North and West of Wolverhampton, Staffordshire, England, *Phil. Trans. Roy. Soc. (B)* **265**, 233–97.

Morgan, W.LL. (1913). Bacon Hole, Gower, *Arch. Cambrensis* **13**, 173–80.

Mörner, N.–A. (1969). Eustatic and Climatic Changes during the Last 15,000 Years, *Geol. en Mijnb.* **48**, 384–85.

Mörner, N.–A., Lanser, J.P. and Hospers, J. (1971). Late Weichselian Palaeomagnetic Reversal, *Nature* **234**, 173–74.

Movius, H.L. (1942). *The Irish Stone Age: Its Chronology, Development and Relationships*, Cambridge.

— — (1975). Abri Pataud, *Peabody Mus. Papers*, 6 vols., Cambridge (Mass.).

Movius, H.L. and Brooks, A.S. (1971). The Analysis of Certain Major Classes of Upper Palaeolithic Tools: Aurignacian Scrapers, *Proc. Prehist. Soc.* **37**(2), 253–73.

Movius, H.L., David, N.C., Bricker, H.M. and Clay, R.B. (1968). The Analysis of Certain Major Classes of Upper Palaeolithic Tools, *Bull. Amer. School Prehist. Research* **26**.

Müller-Karpe, H. (1966). *Handbuch der Vorgeschichte, Band I: Altsteinzeit*, München (Munich).

Mullins, E.H. (1913). The Ossiferous Cave at Langwith, *J. Derbys. Arch. & Nat. Hist. Soc.* **35**, 135–58.

Nédervelde, J. van, Davies, M. and John, B.J. (1973). Radiocarbon Dating from Ogof-yr-Ychen, a New Pleistocene Site, West Wales, *Nature* **245**, 453–55.

Newcomer, M.H. (1971). Some Quantitative Experiments in Handaxe Manufacture, *World Arch.* **3**, 85–94.

— — (1974). Study and Replication of Bone Tools from Ksar Akil (Lebanon), *World Arch.* **6**, 138–53.

Noten, F. van (1967). Le Tjongérien en Belgique, *Bull. Soc. Roy. Belge Anth. et Préhist.* **78**, 197–236.

— — (1975). Meer II: Verdere Opgravingen op de Tjongervindplaats, *Arch. belgica*, no. 172.

Oakley, K.P. (1939). Geology and Palaeolithic Studies, *in* K.P. Oakley, W.F. Rankine and A.W.G. Lowther, *A Survey of the Prehistory of the Farnham District (Surrey)*, Guildford, pp. 3–58.

— — (1958). The Antiquity of the Skulls Reputed to be from Flint Jack's Cave, Cheddar, Somerset, *Proc. Univ. Bristol Spel. Soc.* **8**, 77–82.

— — (1968). The Date of the 'Red Lady' of Paviland, *Antiquity* **42**, 306–07.

— — (1971). British Isles, *in* K.P. Oakley, B.G. Campbell and T. Molleson, *Catalogue of Fossil Hominids, Part II: Europe*, London.

Oakley, K.P. and Baden-Powell, D.F.W. (1963). Néogène et Pléistocène, *in* W.F. Whittard and S. Simpson (eds.), *Lexique Stratigraphique International, Volume I: Europe; Fascicule 3a: Angleterre, Pays de Galles et Ecosse*, Fasc. 3a XIII.

Odum, E.P. (1971). *Fundamentals of Ecology*, Philadelphia.

Olson, E.A. and Broecker, W.S. (1961). Lamont Natural Radiocarbon Measurements VII, *Radiocarbon* **3**, 141–75.

Olsson, I.U. (ed.) (1970). *Nobel Symposium 12: Radiocarbon Variations and Absolute Chronology*, Stockholm.

Otlet, R.L. and Slade, B.S. (1974). Harwell Radiocarbon Measurements I, *Radiocarbon* **16**, 178–91.

Ottaway, B. (1973). Dispersion Diagrams: a New Approach to the Display of Carbon-14 Dates, *Archaeometry* **15**, 5–12.

Ottaway, B. and J.H. (1972). The Suess Calibration Curve and Archaeological Dating, *Nature* **239**, 512–13.

Otte, M. (1969–70). L'Aurignacien de Spy, Mémoire de Licence, Univ. de Liège.

— — (1974a). L'industrie osseuse aurignacienne de la grotte de la Princesse à Marche-les-Dames, province de Namur, Belgique, *Bull Soc. Roy. Belge Anth. et Préhist.* **85**, 209–41.

— — (1974b). Les pointes à retouches plates du Paléolithique supérieur initial de Belgique, *Etudes & Recherches Arch. Univ. Liège*, sér. A, no. 2.

Paddayya, K. (1973). A *Federmesser* Site with Tanged Points at Norgervaart, Province of Drenthe (Netherlands), *Palaeohistoria* **15**, 167–213.

Paine, R. (1972). The Herd Management of Lapp Reindeer Pastoralists, *in* W. Irons and N. Dyson-Hudson (eds.), *Perspectives on Nomadism*, Leiden, pp. 76–87.

Palmer, L.S. and Hinton, M.A.C. (1928). Some Gravel Deposits at Walton near Clevedon, *Proc. Univ. Bristol Spel. Soc.* **3**, 154–61.

Palmer, S. (1967). Upper Palaeolithic Artifacts from Portland, *Proc. Dorset Nat. Hist. & Arch. Soc.* **89**, 117–19.

— — (1968). A Mesolithic Site at Portland Bill: Excavations 1966, *Proc. Dorset Nat. Hist. & Arch. Soc.* **90**, 183–206.

— — (1970). The Stone Age Industries of the Isle of Portland, Dorset, and the Utilisation of Portland Chert as Artifact Material in Southern England, *Proc. Prehist. Soc.* **36**, 82–115.

Paquereau, M.M. (1969). Etude palynologique du Würm I du Pech de l'Azé (Dordogne), *Quaternaria* **11**, 227–35.

Parry, R.F. (1928). Recent Excavations at the Cheddar Caves, *Rep. Wells Nat. Hist. & Arch. Soc.*, pp. 32–36.

— — (1929). Excavation at the Caves, Cheddar, *Proc. Soms. Arch. & Nat. Hist. Soc.* **74**, 102–21.

— — (1931). Excavations at Cheddar, *Proc. Soms. Arch. & Nat. Hist. Soc.* **76**, 46–62.

Passemard, E. (1922). La caverne d'Isturitz (Basses-Pyrénées), *Rev. arch.* **15**, 1–45.

Patten, B.C. (ed.) (1971). *Systems Analysis and Simulation in Ecology*, New York.

Pearson, A. (1964). *Animals and Plants of the Cenozoic Era: Some Aspects of the Faunal and Floral History of the Last Sixty Million Years*, London.

Pearson, R.G. (1962a). The Coleoptera from a Detritus Deposit of Full-glacial Age at Colney Heath, near St. Albans, *Proc. Linnean Soc. London* **173**, 37–55.

— — (1962b). Coleoptera from a Late-glacial Deposit at St. Bees, West Cumberland, *J. Animal Ecol.* **31**, 129–50.

— — (1963). Coleopteran Associations in the British Isles during the Late Quaternary Period, *Biol. Rev.* **38**, 334–63.

Pengelly, W. (1865–80). Kent's Cavern Exploration Journal, 5 vol. unpubl. Diary of excavations, Torquay Nat. Hist. Soc. Museum.

— — (1868). The Literature of Kent's Cavern, Part I, *Trans. Devons. Assoc.* **1**, 469–522.

— — (1869). The Literature of Kent's Cavern, Part II, *Trans. Devons. Assoc.* 3,191–482.

— — (1871a). The Ash Hole and Bench Bone-Caverns at Brixham, South Devon, *Rep. Devon. Assoc.* 4, 73.

— — (1871b). The Literature of Kent's Cavern, Part III, *Trans. Devons. Assoc.* 4, 467–90.

— — (1878). The Literature of Kent's Cavern, Part IV, *Trans. Devons. Assoc.* 10, 141–81.

— — (1884). The Literature of Kent's Cavern, Part V, *Trans. Devons. Assoc.* 14, 189–434.

Pennington, W. (1969). *The History of British Vegetation*, London.

Penny, L.F. (1964). A Review of the Last Glaciation in Great Britain, *Proc. Yorks. Geol. Soc.* 34, 387–411.

Penny, L.F., Coope, G.R. and Catt, J.A. (1969). Age and Insect Fauna of the Dimlington Silts, East Yorkshire, *Nature* 224, 65–67.

Pepin, C.E. (1967). *Hengistbury Head: an Environmental Study*, Bournemouth.

Peters, E. (1930). *Die altsteinzeitliche Kulturstätte Petersfels*, Augsburg.

Peters, E. and Toepfer, V. (1932). Der Abschluss der Grabungen am Petersfels bei Engen im badischen Hegau, *Prähist. Zeitschrift* 23, 155–99.

Peterson, N. (1975). Hunter-Gatherer Territoriality: The Perspective from Australia, *Amer. Anth.* 77(1), 53–68.

Polunin, N. (1959). *Circumpolar Arctic Flora*, Oxford.

Pradel, L. (1973). Traces d'usage sur les burins du Paléolithique supérieur, *Bull. Soc. préhist. franç. (C.R.S.M.)* 70, 90–96.

Puydt, M. de and Lohest, M. (1886). L'Homme contemporain du mammouth à Spy, *Ann. Féd. Arch. & Hist. Belgique (Congrès de Namur)*, pp. 207–40.

Radley, J. (1964). Late Upper Palaeolithic and Mesolithic Surface Sites in South Yorkshire, *Trans. Hunter Arch. Soc.* 9, 38–50.

— — (1967). Excavations at a Rock Shelter at Whaley, Derbyshire, *Derbys. Arch. J.* 87, 1–17.

— — (1968). A Mesolithic Structure at Sheldon, with a Note on Chert as a Raw Material on Mesolithic Sites in the Southern Pennines, *Derbys. Arch. J.* 88, 26–36.

— — (1969). A Note on Four Maglemosian Bone Points from Brandesburton, and a Flint Site at Brigham, Yorkshire, *Antiquaries J.* 49, 377–78.

Radley, J., Tallis, J.H. and Switsur, V.R. (1974). The Excavation of Three 'Narrow Blade' Mesolithic Sites in the Southern Pennines, England, *Proc. Prehist. Soc.* 40, 1–19.

Raftery, J. (1970). Prehistoric Coiled Basketry Bags, *J. Roy. Soc. Antiq. Ireland* 100, 167-68.

Renault-Miskovsky, J. (1972). La végétation pendant le Würmien II, aux environs de la grotte de l'Hortus (Valflaunès, Hérault) d'après l'étude des pollens, *in* H. de Lumley *et al.*, La grotte de l'Hortus (Valflaunès, Hérault), *Etudes Quaternaires (Marseille)* 1, 313–24.

Renfrew, C. (1970). The Tree-ring Calibration of Radiocarbon: an Archaeological Evaluation, *Proc. Prehist. Soc.* 36, 280–311.

— — (1971). Carbon 14 and the Prehistory of Europe, *Sci. Amer.* 225(4), 63–72.

— — (ed.) (1973). *The Explanation of Culture Change: Models in Prehistory*, London.

(ed.) (1974). *British Prehistory: a New Outline*, London.

Rich, D. (1974). The Mesolithic of Scotland: a Summary, unpubl. Edinburgh Univ. M.A. thesis.

Riek, G. (1959). Das federmesserführende Magdalenien der Burckardtshöhle bei Westerhein in Kreis Münsingen (Schwäbische Alb), *Fundberichte aus Schwaben* 15.

Riley, D.N. (1967). Flint Implements Found on Totley Moor, 1962, *Trans. Hunter Arch. Soc.* 9, 170–78.

Roe, D.A. (1968). A Gazetteer of British Lower and Middle Palaeolithic Sites, *Council Brit. Arch. Research Rep.* 8.

— — (1970). *Prehistory: an Introduction*, London.

Rogers, E.H. (1955). Stratification of the Cave Earth in Kent's Cavern, *Proc. Devon Arch. Explor. Soc.* **5**, 1–25

Rona, E. and Emiliani, C. (1969). Absolute Dating of Caribbean Cores, *Science* **163**, 66–68.

Rose, J., Turner, C. and Wymer, J.J. (1973). Sproughton, *in* J. Rose and C. Turner (eds.), Quaternary Research Assoc., Easter Field Meeting 1973, Clacton, unpubl. handbook.

Rosenfeld, A. (1964). Excavations in the Torbryan Caves, Devonshire, II: Three Holes Cave, *Trans. Devon Arch. Explor. Soc.* 22, 3–26.

— — (1969). Palaeolithic and Mesolithic, *in* F. Barlow (ed.), *Exeter and its Region*, Exeter, pp. 129–37.

— — (1971). The Examination of Use Marks on Some Magdalenian End Scrapers, *British Museum Quarterly* 35, 176–82.

Rowlands, B.M. (1971). Radiocarbon Evidence of the Age of an Irish Sea Glaciation in the Vale of Clwyd, *Nature: Physical Science* **230**, 9–11.

Rust, A. (1937). *Das altsteinzeitliche Rentierjägerlager Meiendorf*, Neumünster.

— — (1943). *Die alt- und mittelsteinzeitlichen Funde von Stellmoor*, Neumünster.

— — (1958). *Die jungpaläolithischen Zeltanlagen von Ahrensburg*, Neumünster.

— — (1962). *Vor 20000 Jahren: Rentierjäger der Eiszeit*, Neumünster.

Rutter, J. (1829). *Delineation of the North-Western Division of the County of Somerset and of its Antediluvian Bone-Caverns*, London.

Rutter, J.G. and Allen, E.E. (1948). *Gower Caves*, Swansea.

Ryder, M.L. (1968). *Animal Bones in Archaeology*, Oxford.

Sackett, J.R. (1966). Quantitative Analysis of Upper Paleolithic Stone Tools, *in* J.D. Clark and F.C. Howell (eds.), Recent Studies in Paleoanthropology, *Amer. Anth.* **68**(2), 356–93.

Sahlins, M. (1972). *Stone Age Economics*, Chicago.

Saint-Périer, R. de (1936). La Grotte d'Isturitz, II, Le Magdalénien de la Grande Salle, *Archives Inst. Paléontol. Humaine*, Mém. 17.

— — (1952). La Grotte d'Isturitz, III, Les Solutréens, les Aurignaciens et les Moustériens, *Archives Inst. Paléontol. Humaine*, Mém. 25.

Sampson, C.G. (1968). *The Middle Stone Age Industries of the Orange River Scheme Area*, Bloemfontein.

Saunders, G.E. (1968). Glaciation of Possible Scottish Re-advance Age in North-west Wales, *Nature* **218**, 76–78.

Sauramo, M. (1923). Studies on the Quaternary Varve Sediments in Southern Finland, *Comm. géol. Finlande Bull.* **60**.

Savage, R.J.G. (1966). Irish Pleistocene Mammals, *Irish Nat. J.* 15, 117–30.

Savory, H.N. (1973). Excavations at the Hoyle, Tenby, in 1968, *Arch. Cambrensis* 122, 18–34.

Schmerling, P.C. (1833–34). *Recherches sur les Ossements fossiles découverts dans les Cavernes de la Province de Liège*, vols. 1 and 2, Liège.

Schmider, B. (1971). Les industries lithiques du Paléolithique supérieur en Ile de France, *Gallia Préhist.*, suppl. 6.

Schütrumpf, R. (1951). Die pollenanalytische Datierung der altsteinzeitlichen Funde, *in* A. Bohmers, Die Höhlen von Mauern, *Palaeohistoria* **1**, 10–20.

Schwabedissen, H. (1954). *Die Federmesser-Gruppen des nordwesteuropäischen Flachlandes. Zur Ausbreitung des Spät-Magdalénien*, Neumünster.

Seddon, B. (1957). Late-glacial Cwm Glaciers in Wales, *J. Glaciol.* 3, 94–99.

— — (1962). Late-glacial Deposits at Llyn Dwythwch and Nant Ffrancon, Caernarvonshire, *Phil. Trans. Roy. Soc. (B)* **244**, 459–81.

Segota, T. (1967). Paleotemperature Changes in the Upper and Middle Pleistocene, *Eiszeitalter u. Gegenw.* **18**, 127–41.

Seligman, C.G. and Parsons, F.G. (1914). The Cheddar Man: a Skeleton of Late Palaeolithic Date, *J. Roy. Anth. Inst.* **44**, 241–63.

Semenov, S.A. (1964). *Prehistoric Technology: an Experimental Study of the Oldest Tools and Artefacts from Traces of Manufacture and Wear*, London (transl. fr. Russian orig., 1957).

Service, E.R. (1966). *The Hunters*, Englewood Cliffs (New Jersey).

Shackley, M.L. (1972). The Use of Textural Parameters in the Analysis of Cave Sediments, *Archaeometry* **14**, 133–45.

Shotton, F.W. (1965). Movements of Insect Populations in the British Pleistocene, *Geol. Soc. Amer.*, Special Paper No. 84, 17–33.

— — (1967). Age of Irish Sea Glaciation of Midlands, *Nature* **215**, 1366.

Shotton, F.W., Blundell, D.J. and Williams, R.E.G. (1967). Birmingham University Radiocarbon Dates I, *Radiocarbon* **9**, 35–37.

— — (1968). Birmingham University Radiocarbon Dates II, *Radiocarbon* **10**, 200–06.

— — (1969). Birmingham University Radiocarbon Dates III, *Radiocarbon* **11**, 263–70.

— — (1970). Birmingham University Radiocarbon Dates IV, *Radiocarbon*, **12**, 385–99.

Shotton, F.W. and Williams, R.E.G. (1971). Birmingham University Radiocarbon Dates V, *Radiocarbon* **13**, 141–56.

— — (1973). Birmingham University Radiocarbon Dates VII, *Radiocarbon* **15**, 451–68.

Shotton, F.W., Williams, R.E.G. and Johnson, A.S. (1974). Birmingham University Radiocarbon Dates VIII, *Radiocarbon* **16**, 285–303.

Sieveking, G. de G. (1971). The Kendrick's Cave Mandible, *British Museum Quarterly* **35**, 230–50.

Sieveking, G. de G., Craddock, P.T., Hughes, M.J., Bush, P. and Ferguson, J. (1970). Characterization of Prehistoric Flint Mine Products, *Nature* **229**, 251–54.

Simpson, I.M. and West, R.G. (1958). On the Stratigraphy and Palaeobotany of a Late-Pleistocene Organic Deposit at Chelford, Cheshire, *New Phytol.* **57**, 239–50.

Singer, R., Wymer, J., Gladfelter, B.G. and Wolff, R.G. (1973). Excavation of the Clactonian Industry at the Golf Course, Clacton-on-Sea, Essex, *Proc. Prehist. Soc.* **39**, 6–74.

Sissons, J.B. (1967). Glacial Stages and Radiocarbon Dates in Scotland, *Scot. J. Geol.* **3**, 375–81.

Smith, A.G., Pearson, G.W. and Pilcher, J.R. (1971a). Belfast Radiocarbon Dates III, *Radiocarbon* **13**, 103–25.

— — (1971b). Belfast Radiocarbon Dates IV, *Radiocarbon* **13**, 450–67.

— — (1974). Belfast Radiocarbon Dates VII, *Radiocarbon* **16**, 269–76.

Smith, P.E.L. (1966). *Le Solutrèen en France*, Bordeaux.

Smith, W.G. (1894). *Man the Primeval Savage: His Haunts and Relics from the Hill-tops of Bedfordshire to Blackwall*, London.

Solecki, R.S. (1971). *Shanidar: The Humanity of Neanderthal Man*, London.

Sollas, W.J. (1911). *Ancient Hunters and their Modern Representatives*, London; revised 1915.

— — (1913). Paviland Cave: an Aurignacian Station in Wales, *J. Roy. Anth. Inst.* **43**, 325–74.

Sonneville-Bordes, D. de (1960). *Le Paléolithique supérieur en Pèrigord*, vols. 1 and 2, Bordeaux.

— — (1961). Le Paléolithique supérieur en Belgique, *L'Anthropologie* **65**, 421–43.

Sparks, B.W. and West, R.G. (1972). *The Ice Age in Britain*, London.

Spencer, R.F. (1959). *The North Alaskan Eskimo: a Study in Ecology and Society*, Washington (Smithsonian Inst. Bur. Amer. Ethnol. Bull. **171**).

Spooner, B. (1973). The Cultural Ecology of Pastoral Nomads, *Addison-Wesley Modular Publ.* **45**, 1–53.

Stamp, L.D. (1946). *Britain's Structure and Scenery*, London.

Stanner, W.E.H. (1965). Aboriginal Territorial Organization: Estate, Range, Domain and Regime, *Oceania* **36**, 1–26.

Stearns, C.E. and Thurber, D.L. (1965). Th^{230}/U^{234} Dates of Late Pleistocene Marine Fossils from the Mediterranean and Moroccan Littorals, *Quaternaria* **7**, 29–42.

Stephens, N. and Synge, F.M. (1966). Pleistocene Shorelines, *in* G.H. Dury (ed.), *Essays in Geomorphology*, London.

Steward, J.H. (1933). The Owen's Valley Paiute, *Univ. California Publ. Amer. Arch. & Ethnol.* **33**.

— — (1955). *Theory of Culture Change: the Methodology of Multilinear Evolution*, Urbana (Illinois).

Stringer, C.B. (1974). Population Relationships of Later Pleistocene Hominids: a Multivariate Study of Available Crania, *J. Arch. Sci.* **1**, 317–42.

Stuiver, M. (1971). Evidence for the Variation of Atmospheric C^{14} Content in the Late Quaternary, *in* K.K. Turekian (ed.), *The Late Cenozoic Glacial Ages*, New Haven (Conn.), pp. 57–70.

Sturge, W.A. (1914). Implements of the Later Palaeolithic 'Cave' Periods in East Anglia, *Proc. Prehist. Soc. E. Ang.* **1**, 210–32.

Suess, H.E. (1967). Bristlecone Pine Calibration of the Radiocarbon Time Scale from 4100 B.C. to 1500 B.C., *in Radioactive Dating and Methods of Low Level Counting*, Vienna (Internat. Atomic Energy Comm.).

Suggate, R.P. and West, R.G. (1959). On the Extent of the Last Glaciation in Eastern England, *Proc. Roy. Soc. (B)* **150**, 263–83.

Sutcliffe, A.J. and Zeuner, F.E. (1962). Excavations in the Torbryan Caves, Devonshire, I: Tornewton Cave, *Proc. Devon Arch. Explor. Soc.* **5**, 127–45.

Suttles, W. (1968). Coping with Abundance: Subsistence on the Northwest Coast, *in* R.B. Lee and I. DeVore (eds.), *Man the Hunter*, Chicago, pp. 56–68.

Swanton, E.W. (1914). *Bygone Haslemere*, London.

Switzur, V.R., Hall, M.A. and West, R.G. (1970). University of Cambridge Natural Radiocarbon Measurements IX, *Radiocarbon* **12**, 590–98.

Switsur, V.R. and West, R.G. (1972). University of Cambridge Natural Radiocarbon Measurements X, *Radiocarbon* **14**, 239–46.

— — (1973). University of Cambridge Natural Radiocarbon Measurements XI, *Radiocarbon* **15**, 156–64.

— — (1975). University of Cambridge Natural Radiocarbon Measurements XIII, *Radiocarbon* **17**, 35–51.

Symonds, W.S. (1871). On the Contents of a Hyaena's Den on the Great Doward, Whitchurch, Ross, *Geol. Mag.* **8**, 433.

Synge, F.M. (1956). The Glaciation of North-east Scotland, *Scot. Geogr. Mag.* 72, 129–43.

Synge, F.M. and Stephens, N. (1960). The Quaternary Period in Ireland – an Assessment, 1960, *Irish Geogr.* **4**, 121–30.

Tallis, J.H. (1964). The Pre-peat Vegetation of the Southern Pennines, *New Phytol.* **63**, 363–73.

Taute, W. (1968). *Die Stielspitzen-Gruppen im nördlichen Mitteleuropa*, Köln (Cologne).

Taylor, H. (1926). Fifth Report on Rowberrow Cavern, *Proc. Univ. Bristol. Spel. Soc.* **2**, 206–09.

— — (1928). King Arthur's Cave, near Whitchurch, Ross-on-Wye, *Proc. Univ. Bristol Spel. Soc.* **3**, 59–83.

Thomas, A.C. (1958). The Palaeolithic and Mesolithic Periods in Cornwall, *Proc. W. Cornwall Field Club* **2**, 5–12.

Thomas, D.H. (1974). *Predicting the Past: an Introduction to Anthropological Archaeology*, New York.

Thünen, J.H. von (1826). *Der Isolierte Staat in Beziehung auf Landwirtschaft und Nationalökonomie*, Rostock.

Tixier, J. (1972). Obtention de lames par débitage "sous le pied", *Bull. Soc. préhist. franç. (C.R.S.M.)* **69**, 134–39.

Toepfer, V. (1970). Stratigraphie und Ökologie des Paläolithikums, *in* H. Richter, G. Haase, I. Leiberoth and R. Ruske (eds.), *Periglazial – Löss – Paläolithikum im Jungpleistozän der Deutschen Demokratischen Republik*, Leipzig.

Tratman, E.K. (1955). Second Report on the Excavations at Sun Hole, Cheddar, *Proc. Univ. Bristol Spel. Soc.* **7**, 61–75.

— — (1963). Sun Hole Cave, Cheddar, Somerset: Pleistocene Fauna, *Proc. Univ. Bristol Spel. Soc.* **10**, 16–17.

— — (1964). Picken's Hole, Crook Peak, Somerset; a Pleistocene Site: Preliminary Note, *Proc. Univ. Bristol Spel. Soc.* **10**, 112–15.

Tratman, E.K., Donovan, D.T. and Campbell, J.B. (1971). The Hyaena Den (Wookey Hole), Mendip Hills, Somerset, *Proc. Univ. Bristol Spel. Soc.* **12**, 245–79.

Tratman, E.K. and Henderson, G.T.D. (1928). First Report on the Excavations at Sun Hole, Cheddar, *Proc. Univ. Bristol Spel. Soc.* **3**, 84–97.

Turekian, K.K. (ed.) (1971). *The Late Cenozoic Glacial Ages*, New Haven (Conn.).

Tyldesley, J.B. (1973). Long Range Transmission of Tree Pollen to Shetland: I, Sampling and Trajectories; II, Calculation of Pollen Deposition, *New Phytol.* **72**, 175–90.

Vachell, E.T. (1953). Kent's Cavern: Its Origin and History, *Proc. Torquay Nat. Hist. Soc.* **11**, 51–73.

VanStone, J.W. (1974). *Athapaskan Adaptations: Hunters and Fishermen of the Subarctic Forests*, Chicago.

Veenstra, H.J. (1970). Quaternary North Sea Coasts, *Quaternaria* **12**, 169–84.

Vértes, L. (1959). *Untersuchungen an Höhlensedimenten: Methode und Ergebnisse*, Budapest.

Vita-Finzi, C. and Higgs, E.S. (1970). Prehistoric Economy in the Mount Carmel Area of Palestine: Site Catchment Analysis, *Proc. Prehist. Soc.* **36**, 1–37.

Vivian, E. (1859). *Cavern Researches: Edited from the Original Manuscript Notes of the Late Revd. J. MacEnery*, London.

Vogel, J.C. and Waterbolk, H.T. (1963). Groningen Radiocarbon Dates IV, *Radiocarbon* **5**, 163–202.

— — (1967). Groningen Radiocarbon Dates VII, *Radiocarbon* **9**, 107–55.

— — (1972). Groningen Radiocarbon Dates X, *Radiocarbon* **14**, 6–110.

Vogel, J.C. and Zagwijn, W.H. (1967). Groningen Radiocarbon Dates VI, *Radiocarbon* **9**, 63–106.

Voous, K.H. (1960). *Atlas of European Birds*, London.

Walker, D. (1955). Late-glacial Deposits at Lunds, Yorkshire, *New Phytol.* **54**, 343–49.

Walker, D. and Godwin, H. (1954). Lake-stratigraphy, Pollen-analysis and Vegetational History, *in* J.G.D. Clark, *Excavations at Star Carr: an Early Mesolithic Site at Seamer near Scarborough, Yorkshire*, Cambridge, pp. 25–69.

Walker, D. and West, R.G. (eds.) (1970). *Studies in the Vegetational History of the British Isles*, Cambridge.

Walker, H.H. and Sutcliffe, A.J. (1967). James Lyon Widger, 1823–1892, and the Torbryan Caves, *Trans. Devons. Assoc.* **99**, 49–110.

Ward, G.K. (1974). A Systematic Approach to the Definition of Sources of Raw Material, *Archaeometry* **16**, 41–53.

Warren, S.H. (1938). The Correlation of the Lea Valley, Arctic Beds, *Proc. Prehist. Soc.* **4**, 328–29.

Watanabe, H. (1968). Subsistence and Ecology of Northern Food Gatherers with Special Reference to the Ainu, *in* R.B. Lee and I. DeVore (eds.), *Man the Hunter*, Chicago, pp. 69–77.

Watson, E. and S. (1967). The Periglacial Origin of the Drifts at Morfa-Bychan, near Aberystwyth, *Geol. J.* **5**, 419–40.

Weertman, J. (1964). Rate of Growth or Shrinkage of Nonequilibrium Ice Sheets, *J. Glaciol.* **5**, 145–58.

Weir, A.H., Catt, J.A. and Madgett, P.A. (1971). Postglacial Soil Formation in the Loess of Pegwell Bay, Kent (England), *Geoderma* **5**, 131–49.

Weiss, G. (1974). Magdalenahöhle, Gem. Gerolstein, Kr. Daun, *in* G. Bosinski, K. Brunnacker, L. Fiedler, J. Hahn, H. Löhr, W. Schol, H. Thieme and G. Weiss, *Altsteinzeitliche Fundplätze des Rheinlandes*, Köln (Cologne), pp. 63–65.

Welin, E., Engstrand, L. and Vaczy, S. (1971). Institute of Geological Sciences Radiocarbon Dates I, *Radiocarbon* **13**, 26–28.

— — (1972). Institute of Geological Sciences Radiocarbon Dates II, *Radiocarbon* **14**, 140–44.

— — (1973). Institute of Geological Sciences Radiocarbon Dates IV, *Radiocarbon* **15**, 299–302.

— — (1974). Institute of Geological Sciences Radiocarbon Dates V, *Radiocarbon* **16**, 95–104.

Wells, L.H. (1958). Human Remains from Flint Jack's Cave, Cheddar, Somerset, *Proc. Univ. Bristol Spel. Soc.* **8**, 83–88.

Welten, M. (1944). Pollenanalytische und stratigraphische Untersuchungen in der prähistorische Höhle des 'Chilchli' im Simmental, *Ber. geobot. Inst. Rübel Zürich*, p. 90.

West, R.G. (1968). *Pleistocene Geology and Biology, with Especial Reference to the British Isles*, London.

— — (1970). Pleistocene History of the British Flora, *in* D. Walker and R.G. West (eds.), *Studies in the Vegetational History of the British Isles*, Cambridge, pp. 1–11.

Wheat, J.B. (1972). The Olsen-Chubbuck Site: a Paleo-Indian Bison Kill, *Mem. Soc. Amer. Arch.* **26** (=*Amer. Antiquity* 37(1), pt. 2).

White, G.F. (1970). *Excavation of the Dead Man's Cave, North Anston*, Worksop.

White, J.P. and Thomas, D.H. (1972). What Mean These Stones? Ethno-taxonomic Models and Archaeological Interpretations in the New Guinea Highlands, *in* D.L. Clarke (ed.), *Models in Archaeology*, London, pp. 275–308.

Williams, B.J. (1974). A Model of Band Society, *Mem. Soc. Amer. Arch.* **29** (=*Amer. Antiquity* 39(4), pt. 2).

Willis, E.H., Tauber, H. and Münnich, K.O. (1960). Variations in the Atmospheric Radiocarbon Concentration over the Past 1300 Years, *Radiocarbon* **2**, 1–4.

Wilson, G.H. (1934). *Cave Hunting Holidays in Peakland*, Chesterfield.

— — (1937). Cave Work in the Manifold Valley, *Caves & Caving* **1**, 61–69.

Winbolt, S.E. (1929). A Late-Pleistocene Flint Point, *Antiquaries J.* **9**, 152–53.

Woldstedt, P. (1960). Die letzte Eiszeit in Nordamerika und Europa, *Eiszeitalter u. Gegenw.* **11**, 148–65.

— — (1967). The Quaternary of Germany, *in* K. Rankama (ed.), *The Geologic Systems: The Quaternary*, vol. 2, London, pp. 239–300.

Wood, R.H., Ashmead, P. and Mellars, P.A. (1970). First Report on the Archaeological Excavations at Kirkhead Cavern, *North West Speleology* **1**, 19–24.

Woodburn, J. (1968). An Introduction to Hadza Ecology, *in* R.B. Lee and I. DeVore (eds.), *Man the Hunter*, Chicago, pp. 49–55.

— — (1972). Ecology, Nomadic Movement and the Composition of the Local Group among Hunters and Gatherers: an East African Example and Its Implications, *in* P.J. Ucko, R. Tringham and G.W. Dimbleby (eds.), *Man, Settlement and Urbanism*, London, pp. 193–206.

Wymer, J.J. (1962). Excavations at the Maglemosian Sites at Thatcham, Berkshire, England, *Proc. Prehist. Soc.* **28**, 329–61.

— — (1968). *Lower Palaeolithic Archaeology in Britain as Represented by the Thames Valley*, London.

— — (1971). A Possible Late Upper Palaeolithic Site at Cranwich, Norfolk. *Norfolk Arch.* **35**, 259–63.

Wymer, J.J., Jacobi, R.M. and Rose, J. (1975). Late Devensian and Early Flandrian Barbed Points from Sproughton, Suffolk, *Proc. Prehist. Soc.* **41**, 235–41.

Yamasaki, F., Hamada, C. and Hamada, T. (1972). Riken Natural Radiocarbon Measurements VII, *Radiocarbon* **14**, 223–38.

— — (1974). Riken Natural Radiocarbon Measurements VIII, *Radiocarbon* **16**, 331–57.

Yamasaki, F., Hamada, T. and Hamada, C. (1969). Riken Natural Radiocarbon Measurements V, *Radiocarbon* **11**, 451–62.

Zamyatnin, S.N. (1934). Gagarino, *Bull. Acad. Hist. Culture Matérielle*, fasc. 88.

Zeuner, F. (1955). *11th Annual Rep., Univ. London Inst. Arch.*, no. Pl.–8.

— — (1963). *A History of Domesticated Animals*, London.

INDEX